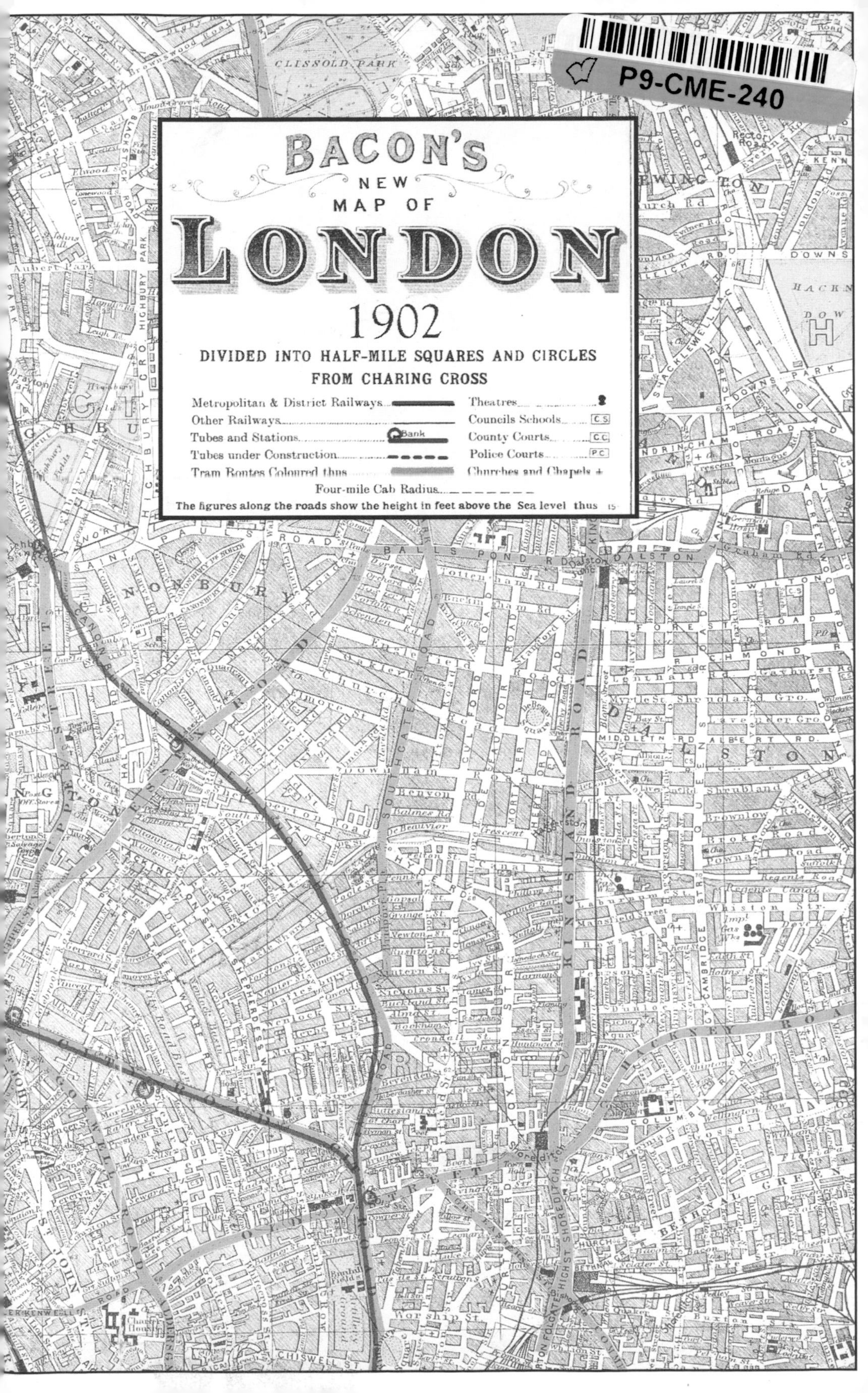
BACON'S
NEW
MAP OF
LONDON
1902
DIVIDED INTO HALF-MILE SQUARES AND CIRCLES
FROM CHARING CROSS
Metropolitan & District Railways
Other Railways
Tubes and Stations
Bank
Tubes under Construction
Tram Routes Coloured thus
Theatres
Councils Schools
C.S
County Courts
C.C
Police Courts
P.C
Churches and Chapels +
Four-mile Cab Radius
The figures along the roads show the height in feet above the Sea level thus 15
CLISSOLD PARK
CANONBURY
DALSTON
BALLS POND R.
KINGSLAND ROAD
HACKNEY ROAD
BETHNAL GREEN
SHOREDITCH
CHISWELL ST.

Thunderstruck

Also by Erik Larson

The Devil in the White City

Isaac's Storm

Lethal Passage

The Naked Consumer

Thunderstruck

Erik Larson

CROWN PUBLISHERS / NEW YORK

Published in the United States by Crown Publishers, an imprint of the Crown Publishing Group, a division of Random House, Inc., New York.

www.crownpublishing.com

CROWN is a trademark and the Crown colophon is a registered trademark of Random House, Inc.

Library of Congress Cataloging-in-Publication Data

Larson, Erik

Thunderstruck / Erik Larson.

Includes biographical references and index.

1. Crippen, Hawley Harvey, 1862–1910.

2. Murderers—England—London—Biography.

3. Murder—England—London—Case studies.

4. Murder—Investigation—Great Britain—Case Studies.

5. Telegraph, Wireless—Marconi system—History. I. Title.

HV6248.C75L37 2006

364.152'309421—dc22

2006011908

ISBN-13: 978-1-4000-8066-3

ISBN-10: 1-4000-8066-5

Printed in the United States of America

Design by Leonard Henderson

10 9 8 7 6 5 4 3 2 1

First Edition

For my wife and daughters,
and in memory of my mother,
who first told me about Crippen

Contents

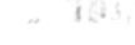

A Note to Readers

There is murder in this book, the second most famous in England, but what I intend here is more than a saga of violence. P. D. James in *The Murder Room* has one of her characters observe, "Murder, the unique crime, is a paradigm of its age." By chronicling the converging stories of a killer and an inventor, I hope to present a fresh portrait of the period 1900 to 1910, when Edward VII ruled the British Empire with a slightly pudgy cigar-stained hand, assuring his subjects that duty was important but so too was fun. "It doesn't matter what you do," he said, "so long as you don't frighten the horses."

The murder fascinated Raymond Chandler and so captivated Alfred Hitchcock that he worked elements into some of his movies, most notably *Rear Window.* Followed by millions of newspaper readers around the world, the great chase that ensued helped advance the evolution of a technology we today take utterly for granted. "It was hot news indeed," wrote playwright and essayist J. B. Priestley, himself a scion of the Edwardian age, "something was happening for the first time in world history." There was a poignancy as well, for the story unfolded during what many, looking back, would consider the last sunny time before World War I, or as Priestley put it, "before the real wars came, before the fatal telegrams arrived at every great house."

This is a work of nonfiction. Anything appearing between quotation marks comes from a letter, memoir, or other written document. I relied heavily on investigative reports from Scotland Yard, many of which as best I can tell have not previously been published. I ask readers to forgive my passion for digression. If, for example, you learn more than you need to know about a certain piece of flesh, I apologize in advance, though I confess I make that apology only halfheartedly.

Erik Larson
Seattle
2006

A safe but sometimes chilly way of recalling the past is to force open a crammed drawer. If you are searching for anything in particular you don't find it, but something falls out at the back that is often more interesting.

J. M. Barrie
"Dedication"
Peter Pan
1904

The Mysterious Passengers

On Wednesday, July 20, 1910, as a light fog drifted along the River Scheldt, Capt. Henry George Kendall prepared his ship, the SS *Montrose,* for what should have been the most routine of voyages, from Antwerp direct to Quebec City, Canada. At eight-thirty in the morning the passengers began streaming aboard. He called them "souls." The ship's manifest showed 266 in all.

Captain Kendall had a strong jaw and a wide mouth that bent easily into a smile, a trait that made him popular among all passengers but especially women. He told good stories and laughed easily. He did not drink. By standards prevailing at the time, he was young to have command of his own ship, only thirty-five years old, but he was by no means untested. Already he had lived a life as eventful as any imagined by Joseph Conrad, whose novels always were popular among passengers once the *Montrose* entered the vast indigo plain of the Atlantic, though thrillers and detective tales and the latest books warning of a German invasion were in demand as well.

As an apprentice seaman, Kendall had served aboard a brutal vessel with a charming name, *Iolanthe,* where he witnessed the murder of a shipmate by an unstable crew member, who then began stalking Kendall to silence him. Kendall fled the ship and tried his hand at mining gold in Australia, a pursuit that left him penniless and hungry. He stowed away on another ship, but the captain caught him and marooned him on Thursday Island, in the Torres Strait off Queensland. After a brief stint harvesting pearls, Kendall joined a small Norwegian barkentine—a three-masted sailing ship—carrying seagull excrement bound for farms in Europe, but storms tore away portions of its masts and turned the voyage into an epic of starvation and stench that lasted 195 days. His love of ships and the sea endured, however. He joined the crew of the *Lake Champlain,* a small steam-powered cargo ship owned by the Beaver Line of Canada but subsequently acquired by the Canadian Pacific Railway. He was its second officer in May 1901, when it became the first merchant vessel to be equipped with wireless. He caught the attention of his superiors and soon found himself first officer on the railway's flagship liner, the *Empress of Ireland.* In 1907 he gained command of the *Montrose.*

It was not the most glamorous of ships, especially when compared to the *Empress,* which was new and nearly three times as large and infinitely more luxurious. The *Montrose* was launched in 1897 and in succeeding years ferried troops to the Boer War and cattle to England. It had one funnel, painted Canadian Pacific's trademark colors—buff with a black top—and flew the line's red-and-white-checkered "house" flag. It carried only two classes, second and third, the latter known more commonly as "steerage," a term that originally denoted the belowdecks portion of a ship devoted to steering. A Canadian Pacific timetable from the era described the second-class quarters. "The Cabin accommodation on the MONTROSE is situated amidship, where least motion is felt. The staterooms are large, light and airy. There is a comfortable ladies' room and a smoking room, also a spacious promenade deck. An excellent table is provided. Surgeons and stewardesses are carried on all steamers." The line's motto was, "A little better than the best."

The manifest for the upcoming voyage listed only 20 passengers in second class but 246 in third, nearly all immigrants. In addition the *Montrose* carried a crew of 107, among them a wireless operator, Llewellyn Jones. Canadian Pacific had been aggressive about installing wireless on its transoceanic vessels, and the *Montrose,* despite its age and modest decor, carried the latest apparatus.

To be successful, Kendall knew, a captain needed more than skill at navigation and ship-handling. He had to dress well, be charming, and possess a knack for conversation, while also owning the mental wherewithal to monitor a thousand operational details, including whether the lifeboats were adequately secured, whether the correct foods and wines had come aboard, and—a new responsibility—whether the ship's Marconi set and aerial were in good repair and ready to receive the inevitable flurry of trivial messages that engulfed a liner upon departure. Although the jokes, bon voyages, and riddles were utterly predictable, they nonetheless reflected the wonder with which people still treated this new and almost supernatural means of communication. First-time passengers often seemed mesmerized by the blue spark fired with each touch of the key and the crack of miniature thunder that followed, though shipping lines had learned from experience that wonderment faded quickly for passengers whose cabins were too close to the wireless room. They learned too that it was prudent to locate Marconi sets a good distance from the wheelhouse so as not to distort the magnetic field registered by the ship's compass.

Before each voyage Kendall tried to read as many newspapers as he could to keep himself up to date on current events and thereby arm himself to meet his nightly obligation to host guests at his table. Amazing things were happening in the world, so there was a lot to talk about. A year earlier Louis Blériot had flown his airplane across the English Channel, from Calais to Dover. While on display at Selfridge's department store, the craft drew 120,000 admirers. Science seemed foremost on people's minds; talk of X-rays, radiation, vaccines, and so forth infused dinner conversation. If such talk ever lagged, there was always the compelling subject of Germany, which by the day seemed to grow more pompous and bellicose. Another foolproof way to inject life, if not violence, into a moribund conversation was to comment upon the apparent decline of morality, as made evident most shockingly in Bernard Shaw's recent play, *Misalliance,* which Beatrice Webb, the social reformer, called "brilliant but disgusting," with "everyone wishing to have sexual intercourse with everyone else." If all the above failed to ignite a good conversation, one could always talk about ghosts. The whole country seemed engaged in the hunt for proof of an afterlife, with the exploits of the venerable Society for

Psychical Research often in the news. And if by chance a conversation became too heated, too lively, one could recall anew how one felt upon the death of King Edward and remark upon how eerie it was that Halley's comet should appear at nearly the same time.

Shortly before his passengers were due to board, Kendall bought a copy of the continental edition of London's *Daily Mail,* an English-language newspaper distributed in Europe. The edition was full of fresh detail about the North London Cellar Murder and the escalating search for two suspects, a doctor and his lover. Back in London, the ship had been visited by two officers from Scotland Yard's Thames Division, patrolling the wharves in hopes of thwarting the couple's escape.

Everyone loved a mystery. Kendall knew at once that *this* would be the mainstay of conversation throughout the voyage—not aircraft or dead kings or haunted country houses, but murder at its most loathsome.

The question at the fore: Where were the fugitive lovers now?

The journey began in typical fashion, with Kendall greeting his second-class passengers as they came aboard. Passengers always seemed at their best at the start of a voyage. They dressed well, and their faces bore an appealing flush of excitement and apprehension. They stepped from the boarding ramp carrying few belongings, but this did not mean they were traveling light. The bulk of their baggage—typically multiple trunks and valises—was stowed belowdecks or delivered to their staterooms. Many chose to keep with them a small carrying case containing their most important belongings, such as personal papers, jewels, and keepsakes. Nothing about the passengers struck Kendall as unusual.

The *Montrose* eased from the wharf amid the usual squall of white handkerchiefs and began making its way down the River Scheldt toward the North Sea. Stewards helped passengers find the ship's library and its dining room and lounges, known as "saloons." Despite the modest proportions of the *Montrose,* its second-class travelers felt as pampered as they would have felt on the *Lusitania.* The stewards—and stewardesses—

brought blankets and books to passengers, and took orders for tea, Belgian cocoa, and scotch, and carried pads and envelopes upon which passengers could write messages for transmission via the Marconi room. Kendall made it a point to stroll the deck several times a day looking for untidy uniforms, tarnished fittings, and other problems, and trying always to greet passengers by name, a good memory being another attribute necessary for the captain of a liner.

Three hours into the voyage Kendall saw two of his passengers lingering by a lifeboat. He knew them to be the Robinsons, father and son, returning to America. Kendall walked toward them, then stopped.

They were holding hands, he saw, but not in the manner one might expect of father and son, if indeed one could ever expect a boy on the verge of manhood to hold hands with his father. The boy squeezed the man's hand with an intensity that suggested a deeper intimacy. It struck Kendall as "strange and unnatural."

He paused a moment, then continued walking until he came abreast of the two. He stopped and wished them a pleasant morning. As he did so, he took careful note of their appearance. He smiled, wished them a fine voyage, and moved on.

He said nothing about the passengers to his officers or crew but as a precaution ordered the stewards to gather up every newspaper on the ship and lock them away. He kept a revolver in his cabin for the worst kinds of emergencies; now he placed it in his pocket.

"I did not do anything further that day or take any steps because I wanted before raising an alarm to make sure I was making no mistake."

WITHIN TWENTY-FOUR HOURS CAPTAIN Kendall would discover that his ship had become the most famous vessel afloat and that he himself had become the subject of breakfast conversation from Broadway in New York to Piccadilly in London. He had stepped into the intersection of two wildly disparate stories, whose collision on his ship in this time, the end of the Edwardian era, would exert influence on the world for the century to come.

Part I

Ghosts and Gunfire

Guglielmo Marconi.

Hawley Harvey Crippen.

Distraction

In the ardently held view of one camp, the story had its rightful beginning on the night of June 4, 1894, at 21 Albemarle Street, London, the address of the Royal Institution. Though one of Britain's most august scientific bodies, it occupied a building of modest proportion, only three floors. The false columns affixed to its facade were an afterthought, meant to impart a little grandeur. It housed a lecture hall, a laboratory, living quarters, and a bar where members could gather to discuss the latest scientific advances.

Inside the hall, a physicist of great renown readied himself to deliver the evening's presentation. He hoped to startle his audience, certainly, but otherwise he had no inkling that this lecture would prove the most important of his life and a source of conflict for decades to come. His name was Oliver Lodge, and really the outcome was his own fault—another manifestation of what even he acknowledged to be a fundamental flaw in how he approached his work. In the moments remaining before his talk, he made one last check of an array of electrical apparatus positioned on a demonstration table, some of it familiar, most unlike anything seen before in this hall.

Outside on Albemarle Street the police confronted their usual traffic problem. Scores of carriages crowded the street and gave it the look of a great black seam of coal. While the air in the surrounding neighborhood of Mayfair was scented with lime and the rich cloying sweetness of hothouse flowers, here the street stank of urine and manure, despite the efforts of the young, red-shirted "street orderlies" who moved among the horses collecting ill-timed deposits. Officers of the Metropolitan Police directed drivers to be quick about exiting the street once their passengers had departed. The men wore black, the women gowns.

Established in 1799 for the "diffusion of knowledge, and facilitating the general introduction of useful mechanical improvements," the Royal Institution had been the scene of great discoveries. Within its laboratories Humphry Davy had found sodium and potassium and devised the miner's safety lamp, and Michael Faraday discovered electromagnetic induction, the phenomenon whereby electricity running through one circuit induces a current in another. The institution's lectures, the "Friday Evening Discourses," became so popular, the traffic outside so chaotic, that London officials were forced to turn Albemarle into London's first one-way street.

Lodge was a professor of physics at the new University College of Liverpool, where his laboratory was housed in a space that once had been the padded cell of a lunatic asylum. At first glance he seemed the embodiment of established British science. He wore a heavy beard misted with gray, and his head—"the great head," as a friend put it—was eggshell bald to a point just above his ears, where his hair swept back into a tangle of curls. He stood six feet three inches tall and weighed about 210 pounds. A young woman once reported that the experience of dancing with Lodge had been akin to dancing with the dome of St. Paul's Cathedral.

Though considered a kind man, in his youth Lodge had exhibited a cruel vein that, as he grew older, caused him regret and astonishment. While a student at a small school, Combs Rectory, he had formed a club, the Combs Rectory Birds' Nest Destroying Society, whose members hunted nests and ransacked them, smashing eggs and killing fledglings, then firing at the parent birds with slingshots. Lodge recalled once beating a dog with a toy whip but dismissed this incident as an artifact of childhood cruelty. "Whatever faults I may have," he wrote in his memoir, "cruelty is not one of them; it is the one thing that is utterly repugnant."

Lodge had come of age during a time when scientists began to coax from the mists a host of previously invisible phenomena, particularly in the realm of electricity and magnetism. He recalled how lectures at the

Royal Institution would set his imagination alight. "I have walked back through the streets of London, or across Fitzroy Square, with a sense of unreality in everything around, an opening up of deep things in the universe, which put all ordinary objects of sense into the shade, so that the square and its railings, the houses, the carts, and the people, seemed like shadowy unrealities, phantasmal appearances, partly screening, but partly permeated by, the mental and spiritual reality behind."

The Royal Institution became for Lodge "a sort of sacred place," he wrote, "where pure science was enthroned to be worshipped for its own sake." He believed the finest science was theoretical science, and he scorned what he and other like-minded scientists called "practicians," the new heathen, inventors and engineers and tinkerers who eschewed theoretical research for blind experimentation and whose motive was commercial gain. Lodge once described the patent process as "inappropriate and repulsive."

As his career advanced, he too was asked to deliver Friday Evening Discourses, and he reveled in the opportunity to put nature's secrets on display. When a scientific breakthrough occurred, he tried to be first to bring it to public notice, a pattern he had begun as early as 1877, when he acquired one of the first phonographs and brought it to England for a public demonstration, but his infatuation with the new had a corollary effect: a vulnerability to distraction. He exhibited a lofty dilettantism that late in life he acknowledged had been a fatal flaw. "As it is," he wrote, "I have taken an interest in many subjects, and spread myself over a considerable range—a procedure which, I suppose, has been good for my education, though not so prolific of results." Whenever his scientific research threatened to lead to a breakthrough, he wrote, "I became afflicted with a kind of excitement which caused me to pause and not pursue that path to the luminous end. . . . It is an odd feeling, and has been the cause of my not clinching many subjects, not following up the path on which I had set my feet."

To the dismay of peers, one of his greatest distractions was the world of the supernatural. He was a member of the Society for Psychical Research, established in 1882 by a group of level-headed souls, mostly scientists and philosophers, to bring scientific scrutiny to ghosts, séances, telepathy, and other paranormal events, or as the society stated in each issue of its *Journal,* "to examine without prejudice or prepossession and in a scientific spirit, those faculties of man, real or supposed, which appear to be inexplicable on any generally recognized hypothesis." The society's constitution stated that membership did not imply belief in "physical forces other than those recognized by Physical Science." That the SPR had a Committee on Haunted Houses deterred no one. Its membership expanded quickly to include sixty university dons and some of the brightest lights of the era, among them John Ruskin, H. G. Wells, William E. Gladstone, Samuel Clemens (better known as Mark Twain), and the Rev. C. L. Dodgson (with the equally prominent pen name Lewis Carroll). The roster also listed Arthur Balfour, a future prime minister of England, and William James, a pioneer in psychology, who by the summer of 1894 had been named the society's president.

It was Lodge's inquisitiveness, not a belief in ghosts, that first drove him to become a member of the SPR. The occult was for him just one more invisible realm worthy of exploration, the outermost province of the emerging science of psychology. The unveiling during Lodge's life of so many hitherto unimagined physical phenomena, among them Heinrich Hertz's discovery of electromagnetic waves, suggested to him that the world of the mind must harbor secrets of its own. The fact that waves could travel through the ether seemed to confirm the existence of another plane of reality. If one could send electromagnetic waves through the ether, was it such an outrageous next step to suppose that the spiritual essence of human beings, an electromagnetic soul, might also exist within the ether and thus explain the hauntings and spirit rappings that had become such a fixture of common legend? Reports of ghosts inhabiting country houses, poltergeists rattling abbeys, spirits knocking on tables during séances—all these in the eyes of Lodge and fellow members of the society seemed as worthy of dispassionate analysis as the invisible travels of an electromagnetic wave.

Within a few years of his joining the SPR, however, events challenged Lodge's ability to maintain his scientific remove. In Boston William

James began hearing from his own family about a certain "Mrs. Piper"—Lenore Piper—a medium who was gaining notoriety for possessing strange powers. Intending to expose her as a fraud, James arranged a sitting and found himself enthralled. He suggested that the society invite Mrs. Piper to England for a series of experiments. She and her two daughters sailed to Liverpool in November 1889 and then traveled to Cambridge, where a sequence of sittings took place under the close observation of SPR members. Lodge arranged a sitting of his own and suddenly found himself listening to his dead aunt Anne, a beloved woman of lively intellect who had abetted his drive to become a scientist against the wishes of his father. She once had told Lodge that after her death she would come back to visit if she could, and now, in a voice he remembered, she reminded him of that promise. "This," he wrote, "was an unusual thing to happen."

To Lodge, the encounter seemed proof that some part of the human mind persisted even after death. It left him, he wrote, "thoroughly convinced not only of human survival, but of the power to communicate, under certain conditions, with those left behind on the earth."

Partly because of his diverse interests and his delight in new discoveries, by June 1894 he had become one of the Royal Institution's most popular speakers.

THE EVENING'S LECTURE WAS ENTITLED "The Work of Hertz." Heinrich Hertz had died earlier in the year, and the institution invited Lodge to talk about his experiments, a task to which Lodge readily assented. Lodge had a deep respect for Hertz; he also believed that if not for his own fatal propensity for distraction, he might have beaten Hertz to the history books. In his memoir, Lodge stopped just short of claiming that he himself, not Hertz, was first to prove the existence of electromagnetic waves. And indeed Lodge had come close, but instead of pursuing certain tantalizing findings, he had dropped the work and buried his results in a quotidian paper on lightning conductors.

Every seat in the lecture hall was filled. Lodge spoke for a few moments, then began his demonstration. He set off a spark. The gunshot crack jolted the audience to full attention. Still more startling was the fact that this spark caused a reaction—a flash of light—in a distant, unattached electrical apparatus. The central component of this apparatus was a device Lodge had designed, which he called a "coherer," a tube filled with minute metal filings, and which he had inserted into a conventional electric circuit. Initially the filings had no power to conduct electricity, but when Lodge generated the spark and thus launched electromagnetic waves into the hall, the filings suddenly became conductors—they "cohered"—and allowed current to flow. By tapping the tube with his finger, Lodge returned the filings to their nonconductive state, and the circuit went dead.

Though seemingly a simple thing, in fact the audience had never seen anything like it: Lodge had harnessed invisible energy, Hertz's waves, to cause a reaction in a remote device, without intervening wires. The applause came like thunder.

Afterward Lord Rayleigh, a distinguished mathematician and physicist and secretary of the Royal Society, came up to Lodge to congratulate him. He knew of Lodge's tendency toward distraction. What Lodge had just demonstrated seemed a path that even he might find worthy of focus. "Well, now you can go ahead," Rayleigh told Lodge. "There is your life work!"

But Lodge did not take Lord Rayleigh's advice. Instead, once again exhibiting his inability to pursue one theme of research to conclusion, he left for a vacation in Europe that included a scientific foray into a very different realm. He traveled to the Ile Roubaud, a small island in the Mediterranean Sea off the coast of France, where soon very strange things began to happen and he found himself distracted anew, at what would prove to be a critical moment in his career and in the history of science.

For even as Lodge conducted his new explorations on the Ile Roubaud, far to the south someone else was hard at work—ingeniously, energetically, compulsively—exploring the powers of the invisible world, with the same tools Lodge had used for his demonstration at the Royal Institution, much to Lodge's eventual consternation and regret.

The Great Hush

It was not precisely a vision, like some sighting of the Madonna in a tree trunk, but rather a certainty, a declarative sentence that entered his brain. Unlike other lightning-strike ideas, this one did not fade and blur but retained its surety and concrete quality. Later Marconi would say there was a divine aspect to it, as though he had been chosen over all others to receive the idea. At first it perplexed him—the question, why him, why not Oliver Lodge, or for that matter Thomas Edison?

The idea arrived in the most prosaic of ways. In that summer of 1894, when he was twenty years old, his parents resolved to escape the extraordinary heat that had settled over Europe by moving to higher and cooler ground. They fled Bologna for the town of Biella in the Italian Alps, just below the Santuario di Oropa, a complex of sacred buildings devoted to the legend of the Black Madonna. During the family's stay, he happened to acquire a copy of a journal called *Il Nuovo Cimento,* in which he read an obituary of Heinrich Hertz written by Augusto Righi, a neighbor and a physics professor at the University of Bologna. Something in the article produced the intellectual equivalent of a spark and in that moment caused his thoughts to realign, like the filings in a Lodge coherer.

"My chief trouble was that the idea was so elementary, so simple in logic that it seemed difficult to believe no one else had thought of putting it into practice," he said later. "In fact Oliver Lodge had, but he had missed the correct answer by a fraction. The idea was so real to me that I did not realize that to others the theory might appear quite fantastic."

What he hoped to do—*expected* to do—was to send messages over long distances through the air using Hertz's invisible waves. Nothing in the laws of physics as then understood even hinted that such a feat might be possible. Quite the opposite. To the rest of the scientific world what he now proposed was the stuff of magic shows and séances, a kind of electric telepathy.

His great advantage, as it happens, was his ignorance—and his mother's aversion to priests.

WHAT MOST STRUCK PEOPLE on first meeting Guglielmo Marconi was that no matter what his true age happened to be at a given moment, he looked much older. He was of average height and had dark hair, but unlike many of his compatriots, his complexion was pale and his eyes were blue, an inheritance from his Irish mother. His expression was sober and serious, the sobriety amplified by his dark, level eyebrows and by the architecture of his lips and mouth, which at rest conveyed a mixture of distaste and impatience. When he smiled, all this changed, according to those who knew him. One has to accept this on faith, however. A search of a hundred photographs of him is likely to yield at best a single half-smile, his least appealing expression, imparting what appeared to be disdain.

His father, Giuseppe Marconi, was a prosperous farmer and businessman, somewhat dour, who had wanted his son to continue along his path. His mother, Anne Jameson, a daughter of the famous Irish whiskey empire, had a more impulsive and exploratory nature. Guglielmo was their second child, born on April 25, 1874. Family lore held that soon after his birth an elderly gardener exclaimed at the size of his ears—"*Che orecchi grandi ha!*"—essentially, "What big ears he has!"—and indeed his ears were larger than one might have expected, and remained one of his salient physical features. Annie took offense. She countered, "He will be able to hear the still, small voice of the air." Family lore also held that along with her complexion and blue eyes, her willful nature was transferred to the boy and established within him a turbulence of warring traits. Years later his own daughter, Degna, would describe him as "an aggregate of opposites: patience and uncontrollable anger, courtesy and harshness, shyness and pleasure in adulation, devotion to purpose and"—this last for her a point of acute pain—"thoughtlessness toward many who loved him."

Marconi grew up on the family's estate, Villa Griffone, in Pontecchio, on the Reno River a dozen or so miles south of Bologna, where the land begins to rise to form the Apennines. Like many villas in Italy, this one was a large stone box of three stories fronted with stucco painted the color of autumn wheat. Twenty windows in three rows punctuated its front wall, each framed by heavy green shutters. Tubs planted with lemon trees stood on the terrace before the main door. A loggia was laced with paulownia that bloomed with clusters of mauve blossoms. To the south, at midday, the Apennines blued the horizon. As dusk arrived, they turned pink from the falling sun.

Electricity became a fascination for Marconi early in his childhood. In that time anyone of a scientific bent found the subject compelling, and nowhere was this more the case than in Bologna, long associated with advances in electrical research. Here a century earlier Luigi Galvani had done awful things to dead frogs, such as inserting brass hooks into their spinal cords and hanging them from an iron railing to observe how they twitched, in order to test his belief that their muscles contained an electrical fluid, "animal electricity." It was in Bologna also that Galvani's peer and adversary, Count Alessandro Volta, constructed his famous "pile" in which he stacked layers of silver, brine-soaked cloth, and zinc and thereby produced the first battery capable of producing a steady flow of current.

As a child, Marconi was possessive about electricity. He called it "*my* electricity." His experiments became more and more involved and consumed increasing amounts of time. The talent he exhibited toward tinkering did not extend to academics, however, though one reason may have been his mother's attitude toward education. "One of the enduring mysteries surrounding Marconi is his almost complete lack of any kind of formal schooling," wrote his grandson, Francesco Paresce, a physicist in twenty-first-century Munich. "In my mind this had certainly something to do with Annie's profound distaste for the Catholic Church ingrained in her by her Protestant Irish upbringing and probably confirmed by her association with the late nineteenth-century society of Bologna." At the time the city was closely bonded to the Vatican. In a letter to her husband Annie sought assurances that Marconi would be allowed to learn "the good principles of my religion and that he not come into contact with the great superstition that is commonly taught to small children in Italy." The city's best schools were operated by Jesuits, and this from Annie's point of view made them inappropriate for Marconi. She made her husband swear that he would not let his son "be educated by the Priests."

She tutored Marconi or hired tutors for him and allowed him to concentrate on physics and electricity, at the expense of grammar, literature, history, and mathematics. She also taught him piano. He came to love Chopin, Beethoven, and Schubert and discovered he had a gift not just for reading music on sight but also for mentally transposing from one key to another. She taught him English and made sure he spoke it without flaw.

What schooling Marconi did have was episodic, occurring wherever the family happened to choose to spend its time, perhaps Florence or Livorno, an important Italian seaport known to the British as Leghorn. His first formal schooling began when he was twelve years old, when his parents sent him to Florence to the Istituto Cavallero, where his solitary upbringing now proved a liability. He was shy and had never learned the kind of tactics necessary for making and engaging friends that other children acquired in their first years in school. His daughter, Degna, wrote, "The expression on Guglielmo's face, construed by his classmates as arising from a sense of superiority, was actually a cover for shyness and worry."

At the *istituto* he discovered that while he had been busy learning English, his ability to speak Italian had degraded. One day the principal told him, "Your Italian is atrocious." To underscore the point, or merely to humiliate the boy, he then ordered Marconi to recite a poem studied in class earlier that day. "And speak up!" the principal said.

Marconi made it through one line, when the class erupted in laughter. As Degna put it, "His classmates began baying like hounds on a fresh scent. They howled, slapped their thighs, and embarked on elaborate pantomimes."

Years later one teacher would tell a reporter, "He always was a model of good behavior, but as to his brain—well, the least said, the soonest mended. I am afraid he got many severe smackings, but he took them like an angel. At that time he never could learn anything by heart. It was impossible, I used to think. I had never seen a child with so defective a memory." His teachers referred to Marconi as "the little Englishman."

Other schools and tutors followed, as did private lessons on electricity by one of Livorno's leading professors. Here Marconi was introduced to a retired telegrapher, Nello Marchetti, who was losing his eyesight. The two got along well, and soon Marconi began reading to the older man. In turn, Marchetti taught him Morse code and techniques for sending messages by telegraph.

Many years later scientists would share Marconi's wonder at why it was that he of all people should come to see something that the most august minds of his day had missed. Over the next century, of course, his idea would seem elementary and routine, but at the time it was startling, so much so that the sheer surprise of it would cause some to brand him a fraud and charlatan—worse, a *foreign* charlatan—and make his future path immeasurably more difficult.

To fully appreciate the novelty, one has to step back into that great swath of history that Degna later would call "The Great Hush."

IN THE BEGINNING, IN THE INVISIBLE realm where electromagnetic energy traveled, there was emptiness. Such energy did exist, of course, and traveled in the form of waves launched from the sun or by lightning or any random spark, but these emanations rocketed past without meaning or purpose, at the speed of light. When men first encountered sparks, as when a lightning bolt incinerated their neighbors, they had no idea of their nature or cause, only that they arrived with a violence unlike anything else in the world. Historians often place humankind's initial awareness of the distinct character of electrical phenomena in ancient Greece, with a gentleman named Thales, who discovered that by rubbing amber he could attract to it small bits of things, like beard hair and lint. The Greek word for amber was *elektron.*

As men developed a scientific outlook, they created devices that allowed them to generate their own sparks. These were electrostatic machines that involved the rubbing of one substance against another, either manually or through the use of a turning mechanism, until enough electrostatic charge—static electricity—built up within the machine to produce a healthy spark or, in the jargon of electrical engineers, a disruptive discharge. Initially scientists were pleased just to be able to launch a spark, as when Isaac Newton did it in 1643, but the technology quickly improved and, in 1730, enabled one Stephen Gray to devise an experiment that for sheer inventive panache outstripped anything that had come before. He clothed a boy in heavy garments until his body was thoroughly insulated but left the boy's hands, head, and feet naked. Using nonconducting silk strings, he hung the boy in the air, then touched an electrified glass tube to his naked foot, thus causing a spark to rocket from his nose.

The study of electricity got a big boost in 1745 with the invention of the Leyden jar, the first device capable of storing and amplifying static electricity. It was invented nearly simultaneously in Germany and in Leyden, the Netherlands, by two men whose names did not readily trip from the tongue: Ewald Jürgen von Kleist and Pieter van Musschenbroek. A French scientist, the Abbé Nollet, simplified things by dubbing the invention the Leyden phial, although for a time a few proprietary Germans persisted in calling it a von Kleist bottle. In its best-known iteration, the Leyden jar consisted of a glass container with coatings of foil on the inside and outside. A friction machine was used to charge, or fill, the jar with electricity. When a wire was used to link both coatings, the jar released its energy in the form of a powerful spark. In the interests of science Abbé Nollet went on to deploy the jar to make large groups of people do strange things, as when he invited two hundred monks to hold hands and then discharged a Leyden jar into the first man, causing an abrupt and furious flapping of robes.

Naturally a competition got under way to see who could launch the longest and most powerful spark. One researcher, Georg Richman, a Swede living in Russia, took a disastrous lead in 1753 when, in the midst of an attempt to harness lightning to charge an electrostatic device, a huge spark leaped from the apparatus to his head, making him the first scientist to die by electrocution. In 1850 Heinrich D. Ruhmkorff perfected a means of wrapping wire around an iron core and then rewrapping the assembly with more wire to produce an "induction coil" that made the creation of powerful sparks simple and reliable—and incidentally set mankind on the path toward producing the first automotive ignition coil. A few years later researchers in England fashioned a powerful Ruhmkorff coil that they then used to fire off a spark forty-two inches long. In 1880 John Trowbridge of Harvard launched a seven-footer.

Along the way scientists began to suspect that the sudden brilliance of sparks might mask deeper secrets. In 1842 Joseph Henry, a Princeton professor who later became the first director of the Smithsonian Institution, speculated that a spark might not be a onetime burst of energy but in fact a rapid series of discharges, or oscillations. Other scientists came to the same conclusion and in 1859 one of them, Berend Fedderson, proved it beyond doubt by capturing the phenomenon in photographs.

But it was James Clerk Maxwell who really shook things up. In 1873 in his *A Treatise on Electricity and Magnetism* he proposed that such oscillations produced invisible electromagnetic waves, whose properties he described in a series of famous equations. He also argued that these waves were much like light and traveled through the same medium, the mysterious invisible realm known to physicists of the day as ether. No one yet had managed to capture a sample of ether, but this did not stop Maxwell from calculating its relative density. He came up with the handy estimate that it had 936/1,000,000,000,000,000,000,000ths the density of water. In 1886 Heinrich Hertz proved the existence of such waves through laboratory experiments and found also that they traveled at the speed of light.

Meanwhile other scientists had discovered an odd phenomenon in which a spark appeared to alter the conducting properties of metal filings. One of them, Edouard Branly of France, inserted filings into glass tubes to better demonstrate the effect and discovered that simply by tapping the tubes he could return the filings to their nonconducting state. He published his findings in 1891 but made no mention of using his invention to detect electromagnetic waves, though his choice of name for his device was prophetic. He called it a radio-conductor. At first his work was ignored, until Oliver Lodge and his peers began to speculate that maybe Hertz's waves were what caused the filings to become conductive. Lodge devised an improved version of the Branly tube, his "coherer," the instrument he unveiled at the Royal Institution.

Lodge's own statements about his lecture reveal that he did not think of Hertzian waves as being useful; certainly the idea of harnessing them for communication never occurred to him. He believed them incapable of traveling far—he declared half a mile as the likely limit. It remained the case that as of the summer of 1894 no means existed for communicating without wires over distances beyond the reach of sight. This made for lonely times in the many places where wires did not reach, but nowhere was this absence felt more acutely than on the open sea, a fact of life that is hard to appreciate for later generations accustomed to πthe immediate world-grasp afforded by shortwave radio and cellular telephone.

The completeness of this estrangement from the affairs of land came home keenly to Winston Churchill in 1899 on the eve of the Boer War, when as a young war correspondent he sailed for Cape Town with the commander of Britain's forces aboard the warship *Dunottar Castle*. He wrote, "Whilst the issues of peace and war seemed to hang in their last flickering balance, and before a single irrevocable shot had been fired, we steamed off into July storms. There was, of course, no wireless at sea in those days, and, therefore, at this most exciting moment the Commander-in-Chief of the British forces dropped completely out of the world. After four days at sea, the ship called at Madeira where there was no news. Twelve days passed in silence and only when the ship was two days from Cape Town was another ship sighted coming from the 'land of knowledge' and bearing vital news. Signals"—visual signals—"were made to the steamer, a tramp, asking for news, upon which she altered course to pass within a hundred yards of the *Dunottar Castle,* and held up a blackboard bearing the words, 'Three battles. Penn Symonds killed.' Then she steamed on her way, and the Commander-in-Chief, whose troops had been in action without his knowledge, was left to meditate upon this very cryptic message."

BACK FROM THE ALPS, Marconi immediately set to work devising equipment to transform his idea into reality, with nothing to guide him but an inner conviction that his vision could be achieved. His mother recognized that something had changed. Marconi's tinkering had attained focus. She saw too that now he needed a formal space dedicated to his experiments, though she had only a vague sense of what it was that he hoped to achieve. She persuaded her husband to allow Marconi to turn a portion of the villa's third-floor attic into a laboratory. Where once Marconi's ancestors had raised silkworms, now he wound coils of wire and fashioned Leyden jars that snapped blue with electrical energy.

On hot days the attic turned into a Sahara of stillness. Marconi grew thin, his complexion paler than usual. His mother became concerned. She left trays of food on the landing outside the attic door. Marconi's father, Giuseppe, grew increasingly unhappy about Marconi's obsession and its jarring effect on family routine. He sought to reassert control by crimping his already scant financial support for his son's experiments. "Giuseppe was punishing Guglielmo in every way he knew," wrote Degna. "Characteristically he considered money a powerful weapon." At one point Marconi sold a pair of shoes to raise money to buy wire and batteries, but this clearly was a symbolic act meant to garner sympathy from his mother, for he had plenty of shoes to spare.

In his attic laboratory Marconi found himself at war with the physical world. It simply was not behaving as he believed it should. From his reading, Marconi knew the basic character of the apparatus he would need to build. A Leyden jar or Ruhmkorff coil could generate the required spark. For a receiver, Marconi built a coherer of the kind Branly had devised and that Lodge had improved, and he connected it to a galvanometer, a device that registered the presence of an electrical current.

But Marconi found himself stymied. He could generate the spark easily but could not cause a response in his coherer. He tinkered. He tried a shorter tube than that deployed by Lodge, and he experimented with different sizes and combinations of filings. At last he got a response, but the process proved fickle. The coherer "would act at thirty feet from the transmitter," Marconi wrote, but "at other times it would not act even when brought as close as three or four feet."

It was maddening. He grew thinner, paler, but kept at it. "I did not lose courage," he wrote. But according to Degna, "he did lose his youth" and took on a taciturnity that, by her account, would forever color his demeanor.

He wanted distance. He knew that if his telegraphy without wires was ever to become a viable means of communication, he would need to be able to send signals hundreds of miles. Yet here in his attic laboratory he sometimes could not detect waves even an arm's length from the spark. Moreover, established theory held that transmitting over truly long distances, over the horizon, simply was not possible. The true scholar-physicists, like Lodge, had concluded that waves must travel in the same manner as light, meaning that even if signals could be propelled for hundreds of miles, they would continue in a straight line at the speed of light and abandon the curving surface of the earth.

Another man might have decided the physicists were right—that long-range communication was impossible. But Marconi saw no limits. He fell back on trial and error, at a level of intensity that verged on obsession. It set a pattern for how he would pursue his quest over the next decade. Theoreticians devised equations to explain phenomena; Marconi cut wire, coiled it, snaked it, built apparatus, and flushed it with power to see what would happen, a seemingly mindless process but one governed by the certainty that he was correct. He became convinced, for example, that the composition of the metal filings in the coherer was crucial to its performance. He bought or scavenged metals of all kinds and used a chisel to scrape loose filings of differing sizes, then picked through the filings to achieve uniformity. He tried nickel, copper, silver, iron, brass, and zinc, in different amounts and combinations. He inserted each new mixture into a fragile glass tube, added a plug of silver at each end, then sealed the apparatus and placed it within his receiving circuit.

He tested each mixture repeatedly. No instrument existed to monitor the strength or character of the signals he launched into space. Instead, he gauged performance by instinct and accident. He did this for days and weeks on end. He tried as many as four hundred variations before settling on what he believed to be the best possible combination for his coherer: a fine dust that was 95 percent nickel and 5 percent silver, with a trace of mercury.

At first he tried to use his transmitter to ring a bell at the far side of his laboratory. Sometimes it worked, sometimes not. He blamed the Branly-style coherer, calling it "far too erratic and unreliable" to be practical. Between each use he had to tap it with his finger to return the filings to their nonconducting state. He tried shrinking the size of the tube. He emptied thermometers, heated the glass, and shaped it. He moved the silver plugs within the tube closer and closer together to reduce the expanse of filings through which current would have to flow, until the entire coherer was about an inch and a half long and the width of a tenpenny nail. He once stated that it took him a thousand hours to build a single coherer. As a future colleague would put it, he possessed "the power of continuous work."

Marconi's obsession with distance deepened. He moved the bell to the next room and discovered how readily the waves passed through obstacles. As he worked, a fear grew within him, almost a terror, that one day he would awaken to discover that someone else had achieved his goal first. He understood that as research into electromagnetic waves advanced, some other scientist or inventor or engineer might suddenly envision what he had envisioned.

And in fact he was right to be concerned. Scientists around the world were conducting experiments with electromagnetic waves, though they still focused on their optical qualities. Lodge had come closest, but inexplicably had not continued his research.

The Scar

The young woman who now presented herself at the Brooklyn, New York, office of Dr. Hawley Harvey Crippen, and who was destined to cause such tumult in his life, was named Cora Turner. At least that was her name for the present. She was seventeen years old, Crippen thirty and already a widower, but the distance between them was not as great as chronology alone suggested, for Miss Turner had the demeanor and physical presence of a woman much older. Her figure was full and inevitably drew forth the adjective *voluptuous*. Her eyes were alight with a knowledge not of books but of how hardship made morality more fungible than the clerics of Brooklyn's churches might have wanted parishioners to believe. She was a patient of the physician who owned the practice, a Dr. Jeffrey, and she had come in for a problem described with Victorian reticence as "female."

Crippen was lonely, and genetic fate had conspired to keep him that way. He was not handsome, and his short stature and small bones conveyed neither strength nor virility. Even his scalp had betrayed him, his hair having begun a brisk retreat years before. He did have a few assets, however. Though he was nearsighted, his eyes were large and conveyed warmth and sympathy—provided he was wearing his glasses. Lately he had grown a beard in a narrow V, which imparted a whiff of continental sophistication. He dressed well, and the sharp collars and crisp-cut suits that tailors of the day favored gave him definition against the passing landscape, the way a line of India ink edged a drawing. Also, he was a doctor. Medicine in this era was becoming a more scientific profession, one that conveyed intellect and character, and, increasingly, prosperity.

Crippen fell for Cora Turner immediately. He saw her youth as no obstacle and began courting her, taking her out to lunch and dinner and for walks. Gradually he learned her story. Her father, a Russian Pole, had died when she was a toddler; her German mother had remarried, but now she too was dead. Cora was fluent both in German and in English. Her stepfather, Fritz Mersinger, lived on Forrest Avenue in Brooklyn. Crippen learned that for one of her birthdays Mersinger had taken her into Manhattan to hear an opera, and the experience had ignited an ambition to become one of the world's great divas.

As Crippen got to know her, he learned too that her passion had become obsession, which in turn had led her down a path that diverged from the savory. She lived alone in an apartment paid for by a man named C. C. Lincoln, a stove maker who was married and lived elsewhere. He paid for her food and clothing and voice lessons. In return he received sex and the companionship of a woman who was young, vivacious, and physically striking. But a complication arose: She became pregnant. The problem that brought her to the office of Crippen's employer, Dr. Jeffrey, was not some routine female complaint. "I believe she had had a miscarriage, or something of that kind," Crippen said. But this may have been code for a circumstance even more wrenching.

Nonetheless Crippen was entranced, and Cora knew it. With each new encounter, she came increasingly to see him as a tool to help her break from Lincoln and achieve her dream of operatic stardom. She knew how to get his attention. During one of their outings she told him that Lincoln had just asked her to run away with him. Whether true or not, the news had the desired effect.

"I told her I could not stand that," Crippen said.

A few days later, on September 1, 1892, the two exchanged vows in a private ceremony at the home of a Catholic priest in Jersey City, New Jersey. Presumably the priest knew nothing of the past pregnancy.

Soon after the wedding Cora gave Crippen his first glimpse of a trait in her character that would become more salient as the years passed: She liked secrets. She told him her real name was not Cora Turner—though the name she now gave seemed hardly real, more like something concocted by a music hall comedian. Her true name, she said, was Kunigunde Mackamotzki.

She planned, however, to keep calling herself Cora. It had been her nickname since childhood, but more importantly Kunigunde Mackamotzki was hardly a name to foster success in the world of Grand Opera.

Almost immediately the newlyweds found themselves battered by failed decisions and forces beyond their control.

HAWLEY HARVEY CRIPPEN was born in Coldwater, Michigan, in 1862, in the midst of two wars, the distant Civil War and closer to home the war against Satan, an enemy deemed by most people in the town to be as real, if not as tangible, as the gray-uniformed men of the South.

The Crippen clan came to Coldwater early and in force, their arrival described in a nineteenth-century history of Branch County, Michigan, as "the coming of a colony of methodists." They spent generously toward the construction of a Methodist church in Coldwater, though at least one prominent member of the family was a Spiritualist. In this he had company, for Coldwater was known as a hotbed not just of Protestant but also of Spiritualist belief, the latter apparently a product of migration. Like so many of their neighbors, the Crippens had moved to Michigan from western New York, a region eventually nicknamed the "burnt-over district" for its willingness to succumb to new and passionate religions.

Crippen's grandfather, Philo, arrived in 1835 and after courting with alacrity married a Miss Sophia Smith later the same year. He founded a dry goods store, which expanded to become one of the most important businesses in town and a significant presence on Chicago Street, the main commercial corridor, where the Chicago Turnpike sliced through. Soon Crippens seemed to be taking over. One operated a flouring mill on Pearl Street; another opened a store that sold produce as well as general merchandise. A Crippen named Hattie played the organ at the Methodist church, and still another, Mae, became a principal in the city's schools. There was a Crippen Building and a Crippen Street.

The town of Coldwater grew rapidly, thanks to its location both on the turnpike and on the main line of the Lake Shore Michigan Southern Railroad, and Chicago Street became the center of commerce in southern Michigan. A man with money strolling the street could buy nearly anything from an array of specialized shops that sold boots, guns, hats, watches, jewelry, and locally made cigars and carriages, for which the town was becoming increasingly famous. The most prestigious industry was horse breeding. One farm cultivated racing horses that achieved fame nationwide, among them Vermont Hero, Hambletonian Wilkes, and the most famous, Green Mountain Black Hawk.

Coldwater was wealthy, and its residents built houses to reflect the fact, studding the town's core with graceful homes of wood, brick, and stone, many destined to survive into the twenty-first century and transform Coldwater into a mecca for students of Victorian architecture. One early edition of the Coldwater city directory observed, "The pleasant drives and parks shaded perfectly by magnanimous maples, the many miles of well kept walks, the brilliantly lighted streets, the neat and substantial residences, the urban appearing business places all in part go to make up this city so fortunately inhabited with well educated, intelligent and thoughtful people whose actions both publicly and privately are devotedly American."

Philo Crippen and his wife soon had a son, Myron, who in turn married a woman named Andresse Skinner and eventually took over the family's dry goods empire, while also serving as tax assessor for the city's first ward and an agent for a sewing machine company. In 1862 Myron and Andresse had their first child, Hawley Harvey, who arrived in the midst of national tumult. Every day the Coldwater newspapers reported the ebb and flow of shed blood, as the country's breeders shipped off horses to support the Union effort, in all three thousand horses by the war's end. The *Coldwater Union Sentinel* of Friday, April 29, 1864, when Hawley Harvey was two years old, cast gloom over the town. "The spring campaign seems to have a bad send off," the newspaper reported: "—disaster, defeat or retreat has attended thus far every effort of our armies." Branch County's young men came back with foreshortened limbs and grotesque scars and told stories of heroic maneuvers and bounding cannonballs. In these times conversation at Philo's dry goods store did not lack for exuberance and gore.

DESPITE THE WAR HAWLEY enjoyed a childhood of privilege. He grew up in a house at 66 North Monroe, one block north of Chicago Street, at the edge of an avenue columned with straight-trunked trees having canopies as dense and green as broccoli. In summer sunlight filtered to the ground and left a paisley of blue shadow that cooled the mind as well as the air.

Sundays were days of quiet. There was no Sunday newspaper. Townspeople crossed paths as they headed for their favored churches. In the heat of summer cicadas clicked off a rhythm of somnolence and piety. Hawley's grandfather, Philo, was a man of austere mien, which was not unusual in Coldwater among the town's older set. A photograph taken sometime after 1870 shows a gathering of about twenty of Coldwater's earliest residents, including Philo himself. It is true that in this time people set their faces hard for photographs, partly from custom, partly because of deficits in photographic technology, but this crowd might not have smiled for the better part of a century. The women seem suspended in a state somewhere between melancholy and fury and are surrounded by old men in strange beards that look as if someone had dabbed glue at random points on their faces, then hurled buckets of white hair in their direction. The day on which this photograph was taken must have been breezy because the longest and strangest beard, stuck on the oldest living citizen, Allen Tibbits, is a white blur resembling a cataract. Grandfather Philo stands in the back row, tall and bald at his summit, with tufts of white hair along the sides of his head and the ridge of his jaw. He has prominent ears, which fate and blood passed downward to Hawley.

A man of strong opinion and Methodist belief, Grandfather Philo exerted on the Crippen clan a force like gravity, suppressing passion and whim whenever he entered the room. He hunted evil in every corner.

Once he asked the Methodist council to "prevent the ringing of bells for auctions." In those days members of the congregation paid the pastor for their seats, with the highest-priced pews in front selling for forty dollars a year. It was an impressive sum—over $400 today—but Philo paid it and made Hawley and other members of the family join him every Sunday. In matters involving the Lord, no expense was too great.

Worship did not end with the pastor's amen. At home Grandfather Philo read the Bible aloud, with special emphasis on the gloomy fates that awaited sinners, in particular female sinners. Years later Hawley would tell an associate, "The devil took up residence in our household when I was a child, and never left it."

Somewhere in these three generations a weakening occurred. Grandfather Philo and his brother, Lorenzo, had faces that expressed strength and endurance and over time came to resemble dynamited rock more than human flesh. Two generations later Hawley emerged, pale, small, and myopic, harried now and then by bullies but himself gentle in manner and unscarred by hard work. His childhood progressed at a saunter, his days routine save for widely spaced moments of community excitement, such as the installation in 1866 of a toboggan slide and the fire in 1881 that destroyed Coldwater's Armory Hall, the town's only theater. The disaster prompted one of the town's leading cigar-makers, Barton S. Tibbits, to build a striking new opera house, and soon Coldwater began drawing the likes of James Whitcomb Riley, who read his poetry from the stage, and an array of less high-brow entertainers, including the Haverly's Minstrels with their "$10,000 Acting Dogs," various traveling companies hell-bent on performing *Uncle Tom's Cabin,* innumerable mind readers and mediums, and most memorably Duncan Clark's Lady Minstrels and New Arabian Nights, described by the *Coldwater Republican* as "eight females, scantily dressed"; the *Courier* called it "the vilest show that ever appeared in Coldwater."

Crippen enrolled in the University of Michigan's School of Homeopathy in 1882, when homeopathy was a mode of medicine that enjoyed great popularity among doctors and the public. The founder of homeopathy was a German physician named Samuel Hahnemann, whose name subsequently became applied to many hospitals around the United States. His treatise, *Organon of Rational Therapeutics,* first published in 1810, became the bible of homeopathy, positing that a doctor could cure a patient's ills by using various medicines and techniques to conjure the same symptoms as those evoked by whatever disease or condition had assailed the patient. He distilled his beliefs to three words, *similia similibus currentur:* Like cures like.

Crippen left the school in 1883 without graduating and sailed for London in hopes of continuing his medical education there. The English medical establishment greeted him with skepticism and disdain but did allow him to attend lectures and work as a kind of apprentice at certain hospitals, among them the Hospital of St. Mary of Bethlehem. An asylum for the insane, its name had shrunk through popular usage to Bedlam, which eventually entered dictionaries as a lowercase word used to describe scenes of chaos and confusion. It was here that Crippen felt most welcome, for the treatment of the insane was a realm in which few doctors cared to practice. Nothing cured madness. The most doctors could do was sedate resident lunatics to keep them from hurting themselves and others. In an environment where nothing worked, anything new that offered hope had to be considered.

Crippen brought with him an array of skills and a knowledge of compounds that asylum officials saw as useful. As a homeopath, he knew the powers not just of ordinary opiates but also of poisons such as aconite, from the root of the plant monkshood; atropine, from belladonna (or deadly nightshade); and rhus toxin from poison ivy. In large doses each could prove fatal, but when administered in tiny amounts, typically in combination with other agents, such compounds could produce a useful palette of physical reactions that mimicked the symptoms of known diseases.

At Bethlehem Hospital Crippen added a new drug to his basket, hydrobromide of hyoscine, derived from an herb of the nightshade family, *Hyoscyamus niger,* known more commonly as henbane. He used it there for the first time, though he long had known of the drug from his studies back home in America, where it was employed in asylums as a sedative to quell patients suffering delirium and mania, and to treat alcoholics suffering delirium tremens. Doctors injected the drug in tiny amounts of one-hundredth of a grain or less (a grain being a unit of measure based historically on the average weight of a single grain of wheat but subsequently set more precisely at 0.0648 grams or 0.002285 ounces). Crippen also knew that henbane was used commonly in ophthalmic treatments because of its power to dilate the pupils of the eye both in humans and animals, including cats, a property that would prove particularly important to Crippen's future. Any miscalculation would have been dangerous. Just a quarter-grain—that is, 0.0162 grams or 0.0005712 ounces—was likely to prove fatal.

Crippen did not stay long in London. Overall he had found his reception to be as chilly as the city's climate. He returned to the United States and enrolled in medical school at Cleveland Homeopathic Hospital. He studied surgery but said, later, that his training was purely theoretical—that he had never really operated on patients, alive or dead. Later he had occasion to insist, "I have never performed a postmortem examination in my life."

The city of Coldwater expected much from Crippen. He was not a man's man, like his uncles Lorenzo and General Fisk, but rather the cerebral sort, and medicine seemed a good career for him to pursue. The local papers tracked his travels; on March 21, 1884, the *Coldwater Courier* noted "Hawley Crippen, son of Myron Crippen, is in the city." He had returned for the funeral of his grandmother, Mrs. Philo Crippen, who had died a few days earlier. Supposedly, if improbably, her last words had been, "Blessed hope of a glorious immortality." An item in the next day's paper noted that Hawley Crippen "graduates at the Medical College of Cleveland next week."

After graduation Crippen opened a homeopathic practice in Detroit, but two years later he moved to New York to study ocular medicine at the New York Ophthalmic Hospital, a homeopathic institution at Third Avenue and Twenty-third Street. A few decades earlier the hospital had undergone a traumatic shift from allopathic medicine—where doctors sought to cure disease by conjuring symptoms *opposite* to those suffered by patients—to homeopathy, in the process jettisoning all its surgeons by giving them "permanent leave of absence." Under the new protocol and guided by a new cadre of physicians, "the success of the institution was as remarkable as its previous failure had been," according to *History of Homeopathy,* published in 1905 by a devotee, Dr. William Harvey King. One of the most important of the school's new leaders had the unfortunate surname Deady. School records show that Crippen graduated in 1887, one of the few students to do so each year. Wrote King, "The real worth of the hospital is measured by the good accomplished in the relief of suffering humanity rather than by the number of graduates who receive its coveted diploma."

Now in his mid-twenties, Crippen began an internship at the Hahnemann Hospital in New York, and there he met a student nurse named Charlotte Jane Bell, who had come to America from Dublin. Soon the *Coldwater Courier* had some exciting news: Shortly before Christmas 1887, Hawley Harvey Crippen had gotten married.

He and Charlotte left New York for San Diego, where Crippen opened an office. The two reveled in the absence of winter and in the blue clarity of the coast. Crippen's parents, Myron and Andresse, by now had moved from Coldwater to Los Angeles, a day's train ride north. Charlotte became pregnant and on August 19, 1889, gave birth to a son, Otto. Crippen and family moved again, this time to Salt Lake City, where Charlotte again became pregnant. In January 1892, shortly before this baby's expected arrival, Charlotte died suddenly, the cause attributed to apoplexy. Crippen sent Otto, now a toddler, to Los Angeles to live with his grandfather and grandmother, then himself fled back to New York. It was then that he joined the practice of Dr. Jeffrey, and took lodging in the doctor's house, and met the woman who was to alter the course and character of his life.

THEIR MARRIAGE BARELY UNDER WAY, Crippen and Cora moved from New York to St. Louis, where Crippen became an eye doctor in the office of an optician. They did not stay long in St. Louis. The city lacked the boisterous glory of New York and had little to offer a woman intent on a life in the world's embrace. No doubt at Cora's urging, the couple moved back to New York.

Cora's "female complaint" now worsened. There was pain and bleeding. She saw a doctor, who told her the problem lay in her ovaries. He recommended removal by surgery: an ovariectomy. Crippen had misgivings. He had seen enough surgeries and their results to know that while surgical skills had advanced greatly since the barbaric practices of the Civil War, an operation was not something to be done on a whim. Though progress with disinfectants had reduced the incidence of catastrophic infection and though improvements in anesthetics had made the whole process endurable, surgery remained a dangerous undertaking. But Cora's discomfort was too great. She agreed to have the operation.

Soon afterward Cora paid a visit to her sister, Mrs. Teresa Hunn, at her home on Long Island, and showed her the scar. It was still "fresh," a seam of angry red skin. During a subsequent visit Mrs. Hunn saw the scar again. It was still sufficiently impressive to score itself into Mrs. Hunn's memory, so that even years later she was able to detail its appearance: "it was healed much better than it was the first time I saw it. It would be about 4 or 5 inches long and about 1 inch wide, but I could not quite exactly say. It was more a cream colour than the rest of her skin, and paler looking. The outside, near the flesh, was paler than the centre of the scar." One detail Mrs. Hunn failed to note was whether the operation had required the removal of Cora's navel, a procedure that commonly accompanied such surgery.

The operation meant that Cora would never bear children, which became for her a source of grief. A close friend, Mrs. Adeline Harrison, later would say, "There was only one little shadow in their lives of which I was aware. They both were passionately fond of children, and she was childless." Whether Crippen truly shared his wife's longing is open to debate, however, given that he had sent his own son to live in Los Angeles and did not now bring him back.

One of Cora's half-sisters, Mrs. Louise Mills, said that her sister "craved motherhood" and that the lack of children cast a shadow over their marriage. "When I visited them four years ago they appeared to be perfectly happy," she reported, except that Cora "would bemoan the fact that she had no child. I fear that in the latter part of her married life she became more and more lonesome."

In a letter to another of her half-sisters, Cora wrote, "I love babies. I am certain that a baby makes a great deal of difference in a family. In fact, it is not complete without a baby. So I envy you. Oh, I tell you, it makes a great deal of difference when it is your own."

PRESSURES ACCUMULATED. AFTER RECOVERING from her surgery, Cora threw herself into her singing lessons, which Crippen was glad to pay for. He liked seeing his young wife happy. In May 1893 the nation slid into a deep depression—the Panic of '93—and the demand for Crippen's medical expertise plummeted. He paid for Cora's music lessons as long as he could but soon was compelled to tell her the lessons had to cease, at least for a while. They moved to less expensive rooms. As their income dwindled, they moved again, and again, until at last they found themselves forced to make a decision that Cora at the time of her marriage to the young and prosperous-seeming Dr. Crippen never imagined she would have to face. She should have been on stage by now, living well in an apartment in Manhattan or for that matter in London, Paris, or Rome. Instead she and the doctor found themselves not just back in Brooklyn, a discouraging condition in itself, but having to surrender to an even more humiliating state of affairs. They moved in with Fritz Mersinger, Cora's stepfather.

For Cora this was a turning point. First there had been St. Louis, little more than a smoke-grimed outpost. Then came a steady rung-by-rung decline as the panic deepened and people lost jobs, and parents struggled to provide their families with food and heat.

Cora pushed Crippen to find work that would yield a better standard of life and get them out of Mersinger's house and out of Brooklyn, closer to the world she had glimpsed in that first opera of her life, the men in their black suits and capes and tall hats, the women whose diamonds gleamed from the opera house boxes like constellations in a winter sky. Legitimate medicine—and homeopathy still was considered legitimate, though its appeal was fading—had failed to generate the required level of income.

It is likely, given Crippen's temperament, that he would have preferred simply to wait for better times, when once again a visit to the doctor would be perceived as necessary and affordable, not as a luxury to be done without.

Cora, however, could not wait. To do so, to be patient and accept what fate had to offer, would have been out of character. Filson Young—his full name was Alexander Bell Filson Young—a prominent journalist and author at the start of the twentieth century, described Cora as "robust and animal. Her vitality was of that loud, aggressive, and physical kind that seems to exhaust the atmosphere round it, and is undoubtedly exhausting to live with."

When Crippen first met her in Dr. Jeffrey's office, what immediately had drawn his attention besides her beauty and lush proportions was her impulsive, buoyant nature, her energy, and her determination not to let herself be crushed by the exigencies of late-nineteenth-century urban life. But increasingly what had seemed impulsive and charming began to appear volatile and wearing, even alarming.

Years later, referring to Cora in this first phase of their marriage, Crippen said that "she was always rather hasty in her temper." He knew, however, that others rarely saw this aspect of her personality: "to the outside world," he said, "she was extremely amiable and pleasant."

The tension in their marriage increased.

Strange Doings

The Ile Roubaud stood among a group of small islands in the Mediterranean, off the coast of France, and had only two houses, one occupied by a lighthouse keeper, the other, at the opposite side of the island, by a scientist named Charles Richet, a physiologist who a decade later would win the Nobel Prize for his discovery of anaphylaxis, the extreme reaction ignited in some people by bee stings, peanuts, and other triggering agents. The house served him primarily as an escape from the heat of the mainland, but now even the island was hot. The month, August 1894, would be remembered long afterward for the exceptional temperatures it brought throughout Europe. Those gathered at Richet's house, however, quickly found themselves distracted by a series of events that would have sent any ordinary mortal rowing for the mainland.

The evening was clear, the air hot and still and scented with brine. Oliver Lodge, Richet, and two others—a man and a woman—collected in the dining room of the house, while a fifth member of the party sat outside in the yard just below a window, notebook in hand, to record observations called to him from inside. Curtains of a sheer, ethereal material framed the window but did not move, testimony to the heat and the lack of breeze.

The woman was an Italian named Eusapia Palladino, and she now took a seat at the table in the center of the room. The men placed a device under her feet that would sound an alarm if either foot lost contact. To prevent her from simply using one foot "to do the duty of two," as Lodge put it, the men installed a screen around each. They extinguished lamps until the room became darker than the night outside, the windows rectangles of soft blue light.

Richet took a seat on one side of Palladino. Another man, a friend of Lodge's, sat at Palladino's other side. He was Frederic W. H. Myers, a poet and school inspector and one of the founders of the Society for Psychical Research. Myers had coauthored for the society a catalog of reports of ghostly and telepathic doings called *Phantasms of the Living,* published in 1886 in two large volumes containing what the authors believed to be dispassionate analysis of seven hundred incidents. This had led Myers and several fellow members to produce a "Census of Hallucinations," for which 410 people around the world distributed a survey which began: "Have you ever, when believing yourself to be completely awake, had a vivid impression of seeing or being touched by a living or inanimate object, or of hearing a voice; which impression, so far as you could discover, was not due to any external physical cause?" Twelve percent of the women surveyed and 7.8 percent of the men answered yes. The authors concluded, "Between deaths and the apparitions of dying persons a connection exists which is not due to chance alone. This we hold as a proved fact."

Now, in the darkened dining room of Richet's summer house, Lodge walked to the table and stood behind Eusapia Palladino. Richet took her right hand, Myers her left. Lodge placed his hands on opposite sides of her head and held fast.

PALLADINO WAS FORTY YEARS OLD. By most accounts, she was illiterate, or nearly so. She claimed that she had been an orphan for much of her childhood—that her mother had died giving her birth and that when she was twelve her father had been murdered by bandits. She then went to live with a family in Naples and earned her keep doing laundry. The family had a Spiritualist bent and in the evening often convened séances, inviting Eusapia to participate. At one such gathering the family learned in a vivid way that there was more to Eusapia than met the eye. As the séance progressed, furniture began to move.

Word of Palladino's alleged gift spread quickly, and soon she found herself in demand. In the lexicon of paranormal research, she was a "physical" medium as opposed to a trance medium. Trance mediums served merely as a kind of telephone to the beyond. Physical mediums also entered trances but then busied themselves conjuring forces that squeezed hands, touched faces, and moved furniture. During sittings by both types a psychical entity known as a "control" was said to guide communication with those beyond the veil.

Palladino had the right powers at the right time. Spiritualism was gaining adherents around the world, and reports of ghosts and poltergeists and premonitions-come-true became commonplace. Families acquired Ouija boards and scared themselves silly. Legendary mediums emerged, including two of the most famous, Madame Helena Blavatsky, eventually exposed as a fraud, and D. D. Home, whose talents convinced even skeptics.

By 1894 Eusapia Palladino too had achieved global fame. Lodge, Myers, and Richet now planned to put her powers to the test.

THE MEN HELD TIGHT. The room was dark and hot and very still. As Palladino entered her trance, a spirit entity named "John King," her control, took over the séance. "I am not presuming to judge what John King really was," Lodge wrote, "but the phenomena were certainly *as if* she were controlled by a big powerful man."

With each new manifestation, the men called out to each other, and to the secretary outside the window, to describe what had happened and to confirm that Palladino's hands and head remained under their grasp. They reported in French, "constantly ejaculating for the benefit of the others, whenever anything occurred," as Lodge put it.

Myers shouted, *"J'ai la main gauche."*

I have the left hand.

Richet: *"J'ai la main droite."*

In the darkness Lodge felt the sensation of having his hands squeezed, even though Palladino's hands were restrained.

"On me touche!" he said.

Something is touching me.

Lodge wrote, "It was as if there was something or someone in the room, which could go about and seize people's arms or the back of their necks, and give a grip; just as anybody might who was free to move about. These grips were very frequent, and everyone at the table felt them sooner or later." At one point Lodge felt "a long hairy beard" brush the top of his head. "It was said to be John King's beard, and the feeling was certainly eerie on my head, which even then was incipiently bald."

A writing desk stood against one wall. In the darkness, with the men still holding Palladino's hands, she gestured toward it. "Every time she did this, the piece of furniture tilted back against the wall, just as if she had had a stick in her hand and was pushing it." The tilting occurred three times. To Lodge this was perplexing though apparently not terrifying. "There must be some mechanical connexion to make matter move: mental activity could never do it," he wrote. The tiltings suggested the existence of "some structure unknown to science, which could transmit force to a distance."

As the séance progressed, Lodge wrote, "there appeared to emanate from her side, through her clothes, a sort of supernumerary arm." It was a ghostly extension, pale, barely visible, yet to Lodge unmistakably present and fluid, not the static shuddering appearance that might be expected from some device hidden underneath Palladino's clothes. Lodge—renowned physicist, professor at University College of Liverpool, member of the Royal Institution, revered lecturer, soon-to-be principal of Birmingham University, and destined for knighthood—wrote: "I saw this protuberance gradually stretching out in the dim light, until ultimately it reached Myers, who was wearing a white jacket. I saw it approach, recede, hesitate, and finally touch him."

Myers said, "*On me touche*" and calmly reported the sensation of a hand gripping his ribs. History is silent on why Myers did not leap from his chair and run screaming into the night.

LODGE STRUGGLED TO HAUL THESE occurrences back from the world of ghosts and into the realm of mechanical law. "As far as the physics of the movements were concerned," he wrote, "they were all produced, I believe, in accordance with the ordinary laws of matter." The emanations from Palladino's body prompted Richet to invent a new word to describe such phenomena: *ectoplasm.* Lodge wrote, "The ectoplasmic formation which operated was not normal; but its abnormality belongs to physiology or anatomy—it is something which biologists ought to study." He acknowledged that this was dicey territory and cautioned that care had to be taken to distinguish between real manifestations and those easily faked. "Let it be noted that ectoplasm proper is more than a secretion or extrusion of material: if genuine, it has powers of operating, it can exert force, and exhibit forms. A mere secretion from the mouth, which hangs down and does nothing, is of no interest."

The events on the island persuaded Lodge that some element of the human mind was able to exist after the body had died. In his formal report to the society he wrote, "Any person without invincible prejudice who had had the same experience would come to the same broad conclusion, viz., that things hitherto held impossible do actually occur."

Lodge became more and more committed to the exploration of the ether, where he believed the convergence of physical law and psychical phenomena might be found. "Whether there is any physical medium for telepathic communication, whether the ether of space serves for this also, and whether our continued existence is associated with that substance instead of with matter, we do not yet know for certain," he wrote. "The departed seem to think it is so, and as far as my knowledge goes they may be right."

Eusapia Palladino's apparent powers had evoked once again Lodge's lifelong vulnerability to distraction. Up to this point this flaw in his character had caused him no great harm.

Gunfire

Each moment mattered. Marconi shrank the size of his coherer until the space containing the filings was little more than a slit between the two silver plugs. He tried heating the glass tube just before he sealed it so that once the air within cooled and contracted to room temperature it would create a partial vacuum. This by itself caused a marked improvement in the coherer's sensitivity.

A persistent irritant was the need to tap the coherer to return it to a state where it once again could respond to passing waves. No telegraph system could survive such a laborious and imprecise procedure.

Marconi devised a clapper like that from a bell and inserted it into the receiving circuit. "Every time I sent a train of electric waves," Marconi wrote, "the clapper touched the tube and so restored the detector at once to its pristine state of sensibility."

He took his experiments outdoors. He managed to send the Morse letter S, three dots, to a receiver on the lawn in front of the villa. With additional tinkering and tweaking to improve the efficiency of his circuits, he increased his range to several hundred yards. He continued trying new adjustments but could send no farther.

One day, by chance or intuition, Marconi elevated one of the wires of his transmitter on a tall pole, thus creating an antenna longer than anything he previously had constructed. No theory existed that even hinted such a move might be useful. It was simply something he had not yet done and that was therefore worth trying. As it happens, he had stumbled on a means of dramatically increasing the wavelength of the signals he was sending, thus boosting their ability to travel long distances and sweep around obstacles.

"That was when I first saw a great new way open before me," Marconi said later. "Not a triumph. Triumph was far distant. But I understood in that moment that I was on a good road. My invention had taken life. I had made an important discovery."

It was a "practician's" discovery. He had so little grasp of the underlying physics that later he would contend that the waves he now harnessed were not Hertzian waves at all but something different and previously unidentified.

Enlisting the help of his older brother, Alfonso, and some of the estate's workers, he experimented now with different heights for his antennas and different configurations. He grounded each by embedding a copper plate in the earth. At the top he attached a cube or cylinder of tin. He put Alfonso in charge of the receiver and had him carry it into the fields in front of the house.

He began to see a pattern. Each increase in the height of his antenna seemed to bring with it an increase in distance that was proportionately far greater. A six-foot antenna allowed him to send a signal sixty feet. With a twelve-foot antenna, he sent it three hundred feet. This relationship seemed to have the force of physical law, though at this point even he could not have imagined the extremes to which he would go to test it.

Eventually Marconi sent Alfonso so far out, he had to equip him with a tall pole topped with a handkerchief, which Alfonso waved upon receipt of a signal.

The gain in distance was encouraging. "But," Marconi said, "I knew my invention would have no importance unless it could make communication possible across natural obstacles like hills and mountains."

Now it was September 1895, and the moment had come for the most important test thus far.

HE SAT AT THE WINDOW of his attic laboratory and watched as his brother and two workers, a farmer named Mignani and a carpenter named Vornelli, set off across the sun-blasted field in front of the house. The carpenter and the farmer carried a receiver and a tall antenna. Alfonso carried a shotgun.

The plan called for the men to climb a distant hill, the Celestine Hill, and continue down the opposite flank until completely out of sight of the house, at which point Marconi was to transmit a signal. The distance was greater than anything he had yet attempted—about fifteen hundred yards—but far more important was the fact that it would be his first try at sending a signal to a receiver out of sight and thus beyond the reach of any existing optical means of communication. If Alfonso received the signal, he was to fire his shotgun.

The attic was hot, as always. Bees snapped past at high velocity and confettied the banks of flowers below. In a nearby grove silver-gray trees stood stippled with olives.

Slowly the figures in the field shrank with distance and began climbing the Celestine Hill. They continued walking and eventually disappeared over its brow, into a haze of gold.

The house was silent, the air hot and still. Marconi pressed the key on his transmitter.

An instant later a gunshot echoed through the sun-blazed air.

At that moment the world changed, though a good deal of time and turmoil would have to pass before anyone was able to appreciate the true meaning of what just had occurred.

Easing the Sore Parts

Despite the Panic of '93, one branch of medicine expanded: the patent medicine industry. The depression may even have driven the industry's growth, as people who felt they could not afford to pay a doctor decided instead to try healing themselves through the use of home remedies that could be ordered through the mail or bought at a local pharmacy. That the industry was indeed booming was hard to miss. All Crippen had to do was open a newspaper to see dozens of advertisements for elixirs, tonics, tablets, and salves that were said to possess astonishing properties. "Does your head feel as though someone was hammering it; as though a million sparks were flying out of the eyes?" one company asked. "Have you horrible sickness of the stomach? Burdock Blood Bitters will cure you."

One of the most prominent patent medicine advertisers was the Munyon Homeopathic Home Remedy Co., headquartered in Philadelphia. Photographs and sketched portraits of its founder and owner, Prof. J. M. Munyon, appeared in many of the company's advertisements and made his face one of the most familiar in America and, increasingly, throughout the world. The Munyon glaring out from company advertisements was about forty, with a scalp thickly forested with dark unruly hair and a forehead so high and broad that the rest of his features seemed pooled by gravity at the bottom of his face. His firm-set mouth anchored an expression of sobriety and determination, as if he had sworn to wipe out illness the world over. "I will guarantee that my Rheumatism Care will relieve lumbago, sciatica and all rheumatic pains in two or three hours, and cure in a few days." A vial of the stuff could be found, he promised, at "all druggists" for twenty-five cents, and indeed small wooden cabinets produced by his company stood in almost every pharmacy, packed with cures for all manner of ailments but highlighting his most famous product, a hemorrhoid salve called Munyon's Pile Ointment, "For Piles, blind or bleeding, protruding or internal. Stops Itching almost immediately, allays inflammation and gives ease to sore parts. We recommend it for Fissure, Ulcerations, Cracks and such anal troubles."

In other advertisements Professor Munyon allied his remedies with the Good Lord himself. Wearing the same stern expression, he thrust his arm toward the heavens and urged readers not simply to buy his products but also to "Heed the Sign of the Cross." Later, during the Spanish-American War, he would publish sheet music for "Munyon's Liberty Song," with photographs of Pres. William McKinley, Adml. George Dewey, and other important officials on the front cover, but a single large photograph of himself on the back, implicitly tying his name to the great men of the age.

In 1894 Crippen applied for a job at Munyon's New York office, on East Fourteenth Street off Sixth Avenue, at that time one of New York's wealthier neighborhoods. Something about Crippen appealed to Munyon—his homeopathic credentials, perhaps, or his experience in London treating patients at the world's best-known lunatic asylum—for he offered him a position and, further, invited Crippen and his wife to live in rooms upstairs from the office.

Crippen accepted. He proved adept at preparing Munyon's existing line of treatments and at devising formulations for new products. Munyon was impressed. He called Crippen "one of the most intelligent men I ever knew, so proficient I gave him a position readily, nor have I ever regretted it."

What also impressed him was the gentleness in Crippen's character. Munyon described him as being "as docile as a kitten." But Cora was another story. She was, Munyon said, "a giddy woman who worried her husband a great deal."

He detected in Crippen signs of a deepening unhappiness and attributed it to the behavior of his wife. She engaged other men in conversations of candor and energy, flexing the power of her personality and physical presence. She conveyed appetite. Crippen was growing jealous, and Munyon believed any man would have felt likewise. Munyon's son, Duke, also noticed. He said, "She liked men other than her husband, which worried the doctor greatly."

Two for London

Marconi understood that the time had come to bring his invention into the world. His first thought, or so legend holds, was to offer it to the Italian government, specifically Italy's post and telegraph authority, only to have his offer rebuffed. But in a brief memoir, his grandson, Francesco Paresce, challenged this account. "No matter how much one might enjoy this idea or how plausible it might seem in a place like Italy even today, there is actually no hard proof whatsoever that he ever did so." The legend was too tidy and did not give enough weight to the fact that at twenty-one Marconi possessed the shrewd demeanor of a businessman twice his age. Always tuned to the importance of being first to stake claim to a new idea, Marconi may have had his next step in mind all along.

He resolved to bring his invention to London. It was the center of the world, yes, but also the locus of a patent system that granted broad rights to whomever was first to apply for them, not necessarily the inventor or discoverer of the underlying technology.

Marconi's mother endorsed Marconi's plans and persuaded her husband that the journey was necessary. In February 1896 mother and son left for London, Marconi carrying a locked box containing his apparatus. He wore a deerstalker cap of the kind that later would be identified with Sherlock Holmes.

On his arrival in England the alert agents of the customs house immediately confiscated his equipment, fearing it was a bomb or other device capable of placing the queen at risk.

In the course of their inspection they destroyed the apparatus.

AT MUNYON'S, CRIPPEN PROSPERED. His career advanced quickly. After a few months in New York he was transferred to Philadelphia, where he and Cora lived for about a year. Next Professor Munyon sent him to Toronto, to manage the company's office there. He and Cora stayed for six months, then returned to Philadelphia.

All this was good for Crippen's career, but Cora grew restive. She had spent the better part of a decade married to Crippen and still was no closer to her dream of becoming a diva. She told Crippen she wanted to renew her studies of opera. She wanted only the best and insisted on going to New York.

Ever indulgent, Crippen agreed to pay for an apartment and all her expenses. By this time Professor Munyon was paying him well. Patent medicine was a lucrative field, and money flowed into the company in a torrent. Crippen could afford the cost of Cora's lessons and her life in New York. What made him uneasy was the prospect of her living alone, without his presence to keep her from associating with other men. Subsequent events suggest that for Cora such freedom may in fact have been as important as the caliber of her song masters.

In 1897 Munyon assigned Crippen his biggest responsibility yet, to take over management of the company's London office. Munyon proposed to pay him $10,000 a year, an astonishing sum equivalent to about $220,000 in twenty-first-century dollars, and paid in an era when federal income tax did not exist. Crippen told Cora the news, expecting that she would find a move to London an irresistible prospect.

He was wrong. She complained that she could not give up her lessons and told him he would have to go alone, that she would join him later. On the question of when, she was disconcertingly vague.

As always he assented, though now an ocean would separate them and her freedom would be complete.

Full of sorrow and unease, the little doctor sailed to England. It was April 1897. Intent on establishing permanent residence, he brought all his belongings, including supplies of his favored poisons.

He arrived without challenge.

IN LONDON, CRIPPEN AND MARCONI both entered a realm of unaccustomed anxiety. Outwardly the stones of empire remained in place, snug and solid and suitably grimed, but there existed in certain quarters a perception that the world was growing unruly and that Britain and her increasingly frail queen had seen their best days.

London was still the biggest, most powerful city in the world. Its four and a half million people inhabited 8,000 streets, many only a block long, and did their drinking in 7,500 public houses and rode in 11,000 cabs pulled by 20,000 horses, the cab population divided among four-wheeled "growlers" and two-wheeled hansoms, named not for their appearance, which was indeed handsome, but rather for their inventor, Joseph Hansom. Crippen joined an estimated 15,000 of his countrymen already living in London, which number by coincidence equaled the total of known lunatics residing within the city's five asylums. Most important, London formed the seat of an empire that controlled one-quarter of the world's population and one-quarter of its land.

Cab whistles screamed for attention, one blast to summon a growler, two for a hansom. Horse-drawn omnibuses clotted the streets. The buses had two levels, with open-air "garden seating" on top, reached by a stairwell spiraling in a manner that allowed ladies to ascend without concern. Motorcars, or simply "motors," added a fresh layer of noise and stench and danger. In 1896 their increasing use forced repeal of a law that had limited speed to a maximum of two miles per hour and required a footman to walk ahead carrying a red flag. The new Locomotives (on Highways) Act raised the speed limit to fourteen and, wisely, did away with the footmen. Underneath the city there was hell in motion. Passengers descending to the subterranean railways encountered a seismic roar produced by too much smoke, too much steam, and too much noise packed into too small an enclosure, the Tube, into which the trains fit as snugly as pistons in a cylinder.

There was fog, yes, often days of it on end, and with a depth so profound as to classify it as a species distinct from the fogs of elsewhere. Londoners nicknamed these fits of opacity "London particulars." The fog came thick and sulfurous and squeezed the flare of gas lamps to cats' eyes of amber that left the streets so dark and sinister that children of the poor hired themselves out as torchbearers to light the way for men and women bent on walking the city's darker passages. The light formed around the walkers a shifting wall of gauze, through which other pedestrians appeared with the suddenness of ghosts. On some nights the eeriness was particularly acute, especially after an evening of engaging in what had become a common pastime, the holding of séances. The walk home afterward could be a long one, suffused with sorrow and grief and marked by the occasional glance behind.

This turn toward the veil was largely Darwin's fault. By reducing the rise of man to a process that had more to do with accident than with God, his theories had caused a shock to the faith of late Victorian England. The yawning void of this new "Darwinian darkness," as one writer put it, caused some to embrace science as their new religion but turned many others into the arms of Spiritualism and set them seeking concrete proof of an afterlife in the shifting planchettes of Ouija boards. In the mid-1890s Britain had 150 Spiritualist societies; by 1908 there would be nearly 400. Queen Victoria herself was rumored to have consulted often with a medium who claimed to be in touch with her dear dead husband, Albert, the prince consort.

There were other signs that the confidence and contentment that had suffused Britain under the queen were beginning ever so slightly to erode. Britain's birthrate was falling rapidly. The Panic of '93 had rattled the princes of industry. Britain and France seemed on the verge of war, though in fact events under way in Germany, as yet largely unnoticed by the public, soon would refocus the nation's attention and bring an end to its long-standing policy of "splendid isolation," rooted in the perception that because of its military and economic power the empire needed no alliances.

Unsettling, too, was the rising clamor from suffragists seeking the vote for women. The hostility that greeted the movement masked a deeper fear of an upwelling of sexual passion and power. It was kept quiet, but illicit sex occurred everywhere, at every level of society. It was on people's minds and in their hearts; it took place in back alleys and in elegant canopied beds at country homes. The new scientists of the mind studied sex, and in keeping with the revolution ushered in by Darwin, they sought to reduce it to sequences of stimuli and adaptive needs. Starting in 1897, Henry Havelock Ellis devoted six volumes to it: his pioneering *Studies in the Psychology of Sex,* sprinkled with case studies of unexpected explicitness and perversity. One memorable phrase from volume four, *Sexual Selection in Man:* "the contact of a dog's tongue with her mouth alone afterward sufficed to evoke sexual pleasure."

There existed, too, a rising consciousness of poverty and of the widening difference between how the rich and the poor lived. The Duke and Duchess of Devonshire owned an estate, Chatsworth, so large it could house more than four hundred weekend guests and the squads of servants that accompanied them. The wealthy served meals of extravagance, recalled J. B. Priestley, "probably including, if the chef were up to it, one of those quasi-Roman idiocies, in which birds of varying sizes were cooked one inside the other like nests of Oriental boxes." Barbara Tuchman, in *The Proud Tower,* recounted how at one luncheon at the Savoy Hotel for the diva Nellie Melba, guests enjoyed a dessert of fresh peaches, then "made a game of throwing them at passers-by beneath the windows."

With this new awareness of the great rift between rich and poor came the fear that extremists would seek to exploit class divisions and set Britain tumbling toward revolution. Anarchism had flamed into violence throughout Europe, often with an Italian holding the match. In late 1892 Scotland Yard arrested two Italians who confessed to planning to blow up the Royal Stock Exchange. The aptly named Errico Malatesta—literally, "evil headed"—preached revolution throughout Europe and found a willing audience. On June 24, 1894, a young Italian baker, Sante Caserio, attacked the president of France, Sadi Carnot, with a newly bought dagger and stabbed him to death. A bomb exploded in posh

Mayfair but hurt no one. Many in England feared that worse was yet to come and blamed the unrest on policies that allowed too many foreigners to seek refuge within the nation's borders. There were so many French anarchists in London that one, Charles Malato, published a guide with information about how to avoid the police, including a brief dictionary of useful phrases, among them *"Je vous tirerai le nez,"* meaning "I will pull your nose."

But these fears and pressures existed as a background tremolo, audible mainly to the writers, journalists, and reformers who made it their business to listen. Otherwise Britons had much to be pleased with. Though the murder rate was up, overall crime was on the wane. The Metropolitan Police, known more commonly as Scotland Yard, had grown larger and moved to new headquarters at Whitehall on the Victoria Embankment, on the north bank of the Thames. The building and the department became known as *New* Scotland Yard. The new location at first proved a bit problematic, albeit appropriately so. While excavating its foundation, in the midst of the terror raised by Jack the Ripper, workers had found the torso of a woman, without head, arms, or legs, triggering fears that this too had been Jack's work. The story got grislier still. A police surgeon tried fitting the torso to a severed arm, with armpit, that had been retrieved from the Thames several weeks earlier. It fit. Next, a reporter used a dog to search the excavation and turned up a left leg. This too fit. Soon afterward a second leg was fished from the Thames.

It did not fit.

Upon examination it proved to be another left leg, causing speculation that a medical student had tossed it into the river as a prank. The case became known as the Whitehall Mystery and was never solved. When the police moved into their new headquarters, one of the departments they left behind at their previous address in Great Scotland Yard was their lost and found division, with 14,212 orphaned umbrellas.

Overall there was a lightening of the British spirit. If any one individual symbolized this change, it was the Prince of Wales, Albert Edward, heir to the throne. In the spring of 1897 he was fifty-six years old and notorious for having an empire-sized appetite for fun, food, and women, the latter despite his thirty-four-year marriage to his wife, Alexandra. That the prince had had sexual dalliances with other women was considered a fact but not a topic for public conversation. Nor was his weight. He drank modestly but adored food. He loved pigeon pie and turtle soup and deer pudding and grouse, partridge, woodcock, and quail, and when the season allowed he consumed mounds of grilled oysters. No one called him fat to his face, for it hurt his feelings, but in private his friends referred to him with affection as "Tum Tum." When not eating, he was smoking. Before breakfast the prince allowed himself a single small cigar and two cigarettes. Through the remainder of a typical day he smoked twenty cigarettes and a dozen more cigars the diameter of gun barrels.

The prince hated being alone and loved parties and clubs and, especially, going out with friends to the music halls of London. Here he had much company. By the late 1890s music halls with their variety acts had become the most popular form of entertainment in Britain and were fast shedding the seedy image they had acquired earlier in the Victorian era. The number of variety theaters within London multiplied rapidly until the city had five hundred, including such familiar names as Tivoli, Empire, Pavilion, Alhambra, and Gaiety. On any given night a typical variety bill would feature dozens of short acts, called "turns," including comedy, acrobatics, ventriloquism, mind reading, and acts in which men pretended to be women, and women to be men.

Overseeing this changing empire was Queen Victoria. In 1896 she celebrated her seventy-seventh birthday. She had reigned for nearly sixty years, during which the empire had grown to be the biggest and most powerful ever known. Meanwhile she herself had grown frail. For over three decades she had lived in a state of perpetual mourning over the death in 1861 of her husband. Ever since then, the Widow of Windsor had worn only black satin. She kept a cast of his hand by her bedside, so that she could hold it when she needed comfort. Now her eyesight was failing, and she was plagued by periods of profound sleepiness. She had ruled so long and in such a benign, maternal way that it was hard to think forward to a future in which she did not exist. A man born in the year of her accession, 1837, would by 1897 be on the verge of old age. Yet queen or not, the laws of nature applied. Victoria would die and, given her health and age, probably soon.

As the end of the century approached, a question lay in the hearts of Britons throughout the empire's eleven million square miles: Without Victoria, what would the world be like?

What would happen then?

Part II

Betrayal

Belle Elmore.

The Secret Box

It is tempting to imagine the arrival of Marconi and his mother in London as something from a Dickens novel—the two entering a cold and alien realm, overpowered by the immensity and smoke and noise of the city—but in fact they stepped directly into the warm embrace of the Jameson family and into the center of a skein of blood and business connections that touched a good portion of the British Empire. They were met at Victoria Station by one of Marconi's cousins, Henry Jameson Davis, and were drawn immediately into the silk and flannel world of London's upper class, with its high teas, derby days, and Sunday carriage rides through Hyde Park. This inventor had not yet starved, except by choice and obsession, and would not starve now.

The delay caused by the destruction of his equipment amplified his ever-present fear that some other inventor with an apparatus as good as his own or better might suddenly appear. With Jameson Davis's help, Marconi acquired materials for his apparatus and set to work on reconstruction. He demonstrated the finished product to his cousin and to others in the Jameson diaspora. The effect was as startling as if a dead relative's voice had just emerged from the mouth of a medium. Here was a means of communicating not just across space but through walls.

They talked of what to do next. A patent was necessary, of course. And a sponsor would help—perhaps the British Post Office, which controlled all telegraphy in Britain.

Here the Jameson network proved invaluable. Through an intermediary, Jameson Davis arranged to have Marconi meet with William Preece, chief electrician of the British Post Office. By dint of his position,

Preece, two years from the post office's retirement age of sixty-five, was the most prominent man in British telegraphy and one of the empire's best-known lecturers. He was well liked by fellow engineers and employees but was loathed by Oliver Lodge and his allies, who together comprised a cadre of theoretical physicists known as "Maxwellians" for their reverence for Clerk Maxwell and his use of mathematics to posit the existence of electromagnetic waves. To the Maxwellians, Preece was the king of "practicians." He and Lodge had more than once come to metaphoric blows over whether theory or everyday experience had more power to uncover scientific truth.

Marconi knew of Preece and knew that he had attempted with some success to signal across short distances using induction, the phenomenon whereby one circuit can generate a sympathetic current in another. Preece had never heard of Marconi but with characteristic generosity agreed to see him.

Soon afterward Marconi arrived at post office headquarters, three large buildings on St. Martin's le Grand, just north of St. Paul's Cathedral. One building, named General Post Office East, occupied the east side of the street and managed the processing and delivery of 2,186,800,000 letters a year throughout the United Kingdom, 54.3 letters per resident, with deliveries in London up to a dozen times a day. Across the street stood General Post Office West, which housed the Telegraph Department, Preece's bailiwick, where anyone with a proper introduction from "a banker or other well-known citizen" could visit the Telegraph Instrument Galleries and see the heart of Britain's telegraphic empire. Here in a room measuring 27,000 square feet stood five hundred telegraphic instruments and their operators, the largest telegraph station in the world. Four large steam engines powered pneumatic tubes that allowed the immediate dispatch of telegrams from the galleries direct to offices throughout London's financial center, the City, and its neighboring district, the Strand, named for the boulevard that fronted the Thames.

Marconi carried two large bags of equipment. He set out his induction coil, spark generator, coherer, and other equipment, but apparently he had not brought with him a telegraph key. One of Preece's assistants, P. R. Mullis, found one and together he and Marconi set up sending and receiving circuits on two tables. At this point Preece pulled out his watch and said quietly, "It has gone twelve now. Take this young man over to the refreshment bar and see that he gets a good dinner on my account, and come back here again by two o'clock."

Mullis and Marconi had lunch and sipped tea, then strolled along Farringdon Road, where Marconi took particular interest in the wheelbarrows of street traders "with their loads of junk, books, and fruit." By Mullis's description, this lunch was one of relaxation and ease. Marconi would have described it differently. Ever anxious that someone would beat him to his goal, he now found himself dining and walking for two hours as his apparatus lay in Preece's office open to inspection by anyone.

At two they returned and rejoined Preece. Marconi was young, thin, and of modest height, but his manner was compelling. He spoke perfect English and dressed well, in a good suit with razor creases. His explanations of the various components of his apparatus were lucid. He did not smile. Anyone happening to glance at him would have gotten the impression that he was much older, though on closer inspection would have noted the smooth skin and clear blue eyes.

Marconi adjusted his circuits. He pressed the telegraph key. A bell rang on the opposite table. He tapped the coherer with his finger and pressed the key again. Again the bell rang.

Mullis looked at his boss. "I knew by the Chief's quiet manner and smile that something unusual had been effected."

PREECE LIKED MARCONI. He recognized that Marconi's coherer was a modification of devices already demonstrated by others, including Lodge, but he saw too that Marconi had put them together in an elegant way, and if the man—this boy—were to be believed, he had succeeded at something that Lodge and the Maxwellians considered impossible, the sending of legible signals not just over long distances but to a point out of optical range.

Preece and Marconi were kindred spirits. Both understood the power of work and everyday practice to reveal truths—useful, practical truths—about the forces that drove the world. In the battle of practice versus theory, Marconi held the promise of becoming Preece's secret weapon. Marconi was an inventor, an amateur, hardly even an adult, yet he had bested some of the great scientific minds of the age. Lodge had said that half a mile was probably the farthest that electromagnetic waves could travel, yet Marconi claimed to have sent signals more than twice as far and now, in Preece's office, forecast transmissions to much greater distances with a confidence that Preece found convincing.

Preece recognized that his own efforts to use induction to produce a crude form of wireless communication had reached their practical limits. Most recently he had attempted to establish communication with a lightship guarding the notoriously deadly Goodwin Sands off the English coast. He had strung wire around the hull of the ship and laid a spiral of wire on the sea floor large enough that no matter where the wind, tide, and waves moved the ship, it always was positioned over part of the spiral. By interrupting the current in the spiral, he hoped to induce matching interruptions in the coil on the ship, and in so doing send Morse messages back and forth. The experiment failed. Later Preece would state that Marconi "came to me at a very fortunate time for myself, for I was just then smarting under the disappointment of having made a failure in communicating with the East Goodwin lightship."

Two years from retirement, Preece understood that his discovery of Marconi might be the last shining thing that history would remember about his long tenure at the British Post Office. Far better to exit as the man who helped introduce the world to a revolution in communication than as an engineer whose own attempts at telegraphy without wires had failed.

The day came to an end when Preece's coachman appeared and Preece set out in his brougham for his home in Wimbledon, the beat of hooves keeping time in the cool spring air.

In a letter to his father Marconi wrote about the meeting and disclosed a bit of news that must have amazed the elder Marconi, who only a year earlier had been so skeptical of his son's electrical adventures. "He promised me that, if I wanted to perform experiments, then he would allow me the use of any necessary building belonging to the telegraphic administration in any city or town in the whole of the United Kingdom, as well as ensuring the help (at no cost, of course) of any personnel employed by the administration mentioned above that I might need. He added that he has ships on which I could install and try my equipment in case I wanted to perform an experiment between vessels at sea."

Preece assigned engineers from his staff to assist Marconi and recruited instrument-builders in the post office mechanics' shop to modify Marconi's equipment to make it more robust. Immediately Preece began arranging demonstrations for other government officials.

Soon Marconi found himself on the roof of the post office, sending signals from one rooftop to another, the spark of his transmitter snapping so loudly as to be audible on the street below. In July 1896 he achieved a distance of three hundred yards, well short of what he had done at the Villa Griffone but still impressive to Preece and his engineers.

Preece arranged the most important demonstration yet, one for observers from the army and navy, to take place on the military proving ground Salisbury Plain, near Stonehenge. By day's end he managed to transmit legible signals to a distance of one mile and three quarters.

The success of the demonstration raised Marconi onto another plane. The War Office wanted more demonstrations; Preece, nearly as delighted as Marconi, reiterated his pledge to provide as much assistance and equipment as Marconi needed. Until this point the intensity with which Marconi pursued his idea had been stoked only from within; now, suddenly, there were expectations from the outside.

"*La calma della mia vita ebbe allora fine,*" he said—The calm of my life ended then.

MARCONI REALIZED IT WAS NOW crucial to file for a patent for his apparatus. The number of people who had seen his invention was multiplying, and his fear that another inventor might come forward increased in step. He filed a "provisional specification" that established the date of filing and asserted that he was first to achieve the things he claimed. He would have to submit a more complete filing later.

Convinced now that Marconi truly had accomplished something remarkable, Preece decided to announce Marconi's breakthrough to the world.

In quick succession he gave a series of important lectures, during which he introduced Marconi as the inventor of a wholly new means of communication. He gave the first in September 1896 before a meeting of the British Association for the Advancement of Science, known best simply as the British Association, during which he revealed that "an Italian had come up with a box giving a quite new system of space telegraphy." He gave a brief description and disclosed that it had proven a great success at Salisbury Plain. In the audience were many of Britain's most famous scientists and, of course, Oliver Lodge and some of his Maxwellian allies, including a prominent physicist named George FitzGerald. Even in the best of circumstances Lodge and FitzGerald found the experience of listening to Preece to be the intellectual equivalent of hearing a fingernail scrape across a blackboard, but now they heard him describe Marconi as if he were the first man ever to experiment with Hertzian waves, and they were outraged. Both believed Lodge had done as much in his June 1894 lecture on Hertz at the Royal Institution.

FitzGerald wrote to a friend, "On the last day but one Preece surprised us all by saying that he had taken up an Italian adventurer who had done no more than Lodge & others had done in observing Hertzian radiations at a distance. Many of us were very indignant at this overlooking of British work for an Italian manufacturer. Science 'made in Germany' we are accustomed to but 'made in Italy' by an unknown firm was too bad." Lodge wrote to Preece and complained, "There is nothing new in what Marconi attempts to do."

The news may have been stale to Lodge and his friends, but it was not to the world at large. Word spread rapidly about this Italian who had *invented* wireless telegraphy. Newspapers referred to "Marconi waves," which to Lodge and his allies represented a cruel slight to the memory of Hertz. This Italian had invented nothing, they argued. If anyone could claim to be the inventor, it was Lodge.

Preece knew he had angered the Maxwellians, and he likely reveled in the fact, for he did not back off. Far from it—he resolved to devote his next big lecture entirely to Marconi and his wireless. It was scheduled for December 12, 1896, at London's Toynbee Hall, a settlement house dedicated to social reform based in London's impoverished East End, in Jack the Ripper's old hunting ground. Here, Preece knew, his lecture would draw not just scientists but a broad swath of the city's intellectual community and representatives of the daily press. The British Association had been mere preamble.

As the date approached, Preece roughed out his lecture. He sought maximum effect. Physicists had become increasingly knowledgeable about Hertzian waves, but not the public. A demonstration of telegraphy without wires was likely to strike the Toynbee audience as so magical as to verge on the supernatural.

Marconi agreed to a demonstration but expressed concern about revealing the secrets of his apparatus. His outlook was more that of a magician protecting his tricks than a scientist unveiling a new discovery to peers. He wrote, "I think it desirable just now that no explanation be given as to the means which I employ for obtaining the effects, as I fear it may give rise to discussions which I would rather avoid until my whole study can be laid before some scientific society."

Marconi satisfied his need for secrecy by concealing his apparatus. He constructed two boxes and painted them black. In one he installed his transmitter, in the other his receiver, with a bell attached. At the start of the lecture one box was at the podium, the other at the far side of the room.

Preece began the lecture with a brief summary of his own efforts to harness induction to signal across bodies of water. But tonight, he said, he would reveal a remarkable discovery made by a young Italian inventor, Guglielmo Marconi. And then, in the finest tradition of late-nineteenth-century scientific lectures, the demonstration began.

First Preece pressed the key at the box that housed the transmitter.

The audience heard the loud crack of a spark. At the same instant, the bell in the receiver box rang.

Nearly everyone in the audience had seen magic acts, and many doubtless had attended at least one of the famed shows at the Egyptian Hall in Piccadilly, "England's Home of Mystery," directed by a magician named Nevil Maskelyne. Compared to women sawn in half or men levitated to the ceiling, this at first glance was nothing special. It took a moment or two for the audience to absorb the fact that what they were witnessing was not a magic trick but a scientific effect conjured by the great William Preece of the British Post Office, who stood before them exuding as always absolute credibility, his eyes large behind thick glasses, his great beard marking each motion of his head with a tangled whoosh of gray-white whiskers. Still, many in the audience reacted the way Marconi's father once had, wondering by what clever means Marconi had hidden the wire connecting the two boxes.

Now Preece and Marconi launched a second phase of the demonstration, meant to quash any lingering skepticism. On a cue from Preece, Marconi picked up the black box that housed his receiver and walked with it through the lecture hall. The spark cracked, the bell rang, over and over, but now the audience could see, clearly, that no wire trailed Marconi as he moved. The audience also saw that Marconi was barely an adult, which only increased the wonder of the moment. No matter where Marconi walked, the bell rang.

Fame came suddenly. The lay press sought a name for Marconi's technology and called it space telegraphy or aetheric telegraphy or simply telegraphy without wires. *The Strand Magazine* sent a writer, H.J.W. Dam, to interview Marconi at his home.

Dam wrote, "He is a tall, slender young man, who looks at least thirty and has a calm, serious manner and a grave precision of speech which further gives the idea of many more years than are his."

Marconi told Dam that it was possible that he had discovered a kind of wave different from what Hertz had found. Asked to explain the difference, Marconi said, "I don't know. I am not a professional scientist, but I doubt if any scientist can tell you."

He declined to talk about the components of his apparatus, but he did tell Dam that his waves could "penetrate everything," including the hull of an ironclad battleship. This caught the interviewer's attention. "Could you not from this room explode a box of gunpowder placed across the street in that house yonder?"

"Yes," Marconi said, as matter of fact as always. He explained, however, that first he would need to insert two wires or metal plates into the powder to produce the spark necessary for detonation.

Reports of Marconi's feats now circulated abroad. Military representatives from Austria-Hungary asked for, and received, a demonstration. In Germany Kaiser Wilhelm also took notice and, as soon would be apparent, resolved that this technology needed further, deeper investigation. The Italian ambassador to England invited Marconi to dinner, after which the ambassador and Marconi traveled in an embassy coach to the post office for a demonstration. In a letter to his father, Marconi reported that the ambassador "even apologized a little for not having dedicated his attention to the matter sooner."

Lodge and his allies of course were enraged, but more broadly, within Britain's higher social tiers and within the scientific establishment as a whole, there were many others who looked upon Marconi with suspicion, even distaste. He was a troubling character, and not just because he had laid claim to apparatus that Lodge and other scientists had used first. He was something new in the landscape. As he himself had admitted, he was not a scientist. His grasp of physical theory was minimal, his command of advanced mathematics nonexistent. He was an entrepreneur of a kind that would become familiar to the world only a century or so later, with the advent of the so-called "start-up" company. In his time the closest models for this kind of behavior were unsavory—for example, the men who made fortunes selling quack medicines, immortalized in H. G. Wells's novel *Tono-Bungay.*

His obsession with secrecy rankled. Here he was, this young Italian, staking claim to a new and novel technology yet at the same time violating all that British science held dear by refusing to reveal details of how his apparatus worked. Marconi had succeeded in doing something believed to be impossible, but *how* had he done it? Why was he, a mere boy, able to do what no one else could? And why was he so unwilling to publish openly his work, as any other scientist would do as a matter of course? Lodge wrote, with oblique malice, that "the public has been educated by a secret box more than it would have been by many volumes of *Philosophical Transactions* and Physical Society Proceedings."

To add insult to injury, Marconi was *a foreigner* at a time when Britons were growing concerned about the increasing number of anarchists, immigrants, and refugees on British soil.

In the face of all this, Marconi remained confident. His early letters to his father were full of cool calculation. Somehow he had developed a belief in his vision that nothing could shake. His chief worry was whether he could develop his wireless quickly enough to outstrip the other inventors who, now that the news of his success was circling the globe, surely would intensify their own work on electromagnetic waves.

In this race he saw no room for loyalty, not to Preece, not to anyone.

Anarchists and Semen

Crippen found quarters in St. John's Wood, near Regent's Park. His Munyon's office was a distance away on Shaftesbury Avenue, which ran a soft serpentine between Bloomsbury and Piccadilly Circus among shops, offices, and restaurants and past side streets inhabited by actors, musicians, French and German émigrés, and other "foreigners," as well as a few prostitutes. The avenue was home also to three of London's best-known theaters, the Palace, the Shaftesbury, and the Lyric. The Munyon's office stood opposite the Palace.

Crippen made sure his wife had all the money she needed to live well in New York and to pursue her opera lessons. But Cora was growing disenchanted with opera, acknowledging at last what her teachers had recognized long before, that she had neither the voice nor the stage presence to succeed in so lofty a pursuit. She wrote to Crippen that she now planned to try making a career doing "music hall sketches." In America it was known as Vaudeville; the British called it Variety.

This troubled Crippen. Vaudeville seemed tawdry compared to opera, and even compared to variety, which as Crippen knew was popular in London and becoming increasingly respectable. Even the Prince of Wales was said to enjoy a good night of variety turns. Though some music halls still served as points of commerce for prostitutes and pickpockets, most had become clean and safe. Sarah Bernhardt, Marie Lloyd, and Vesta Tilley did turns, and within a decade so too would Anna Pavlova and the Russian ballet, first introduced to Britain at the Palace.

Crippen wrote to Cora and urged her to reconsider. He recommended that she come right away to London. Here at least she could perform variety without taint.

She agreed to join him, though probably neither love nor Crippen's plea had much to do with her decision. More likely her career doing musical sketches in New York had also been a failure, and now she wanted to try her hand in London, where she could sing before a sophisticated audience more appreciative of her true talent. Her impending arrival meant that Crippen had to find new lodgings that were large enough and luxurious enough to accommodate a wife with so swollen a sense of self-regard and need. He chose an apartment in Bloomsbury on a pretty street in the shape of a half-circle, one of London's many "crescents." This was South Crescent, off Tottenham Court Road, one block from the British Museum and an easy walk to the Munyon's office on Shaftesbury.

Cora arrived in August, and at once Crippen sensed a difference. "I may say that when she came to England from America her manner towards me was entirely changed, and she had cultivated a most ungovernable temper, and seemed to think I was not good enough for her, and boasted of the men of good position traveling on the boat who had made a fuss of her, and, indeed, some of these visited her at South Crescent, but I do not know their names."

IN BLOOMSBURY CRIPPEN HAD CHOSEN a neighborhood in which an array of forces then driving deep change in Britain were fully at play. Just east lay Bloomsbury Square and Bloomsbury Road, where within a few years Virginia and Vanessa Stephen, critic Roger Fry, John Maynard Keynes, and other members of their cadre of writers, poets, and gleaming personalities would become legendary as the Bloomsbury Group. Virginia would marry and take her husband's name, Woolf. A few blocks to the west, across Tottenham Court Road, was territory soon to be claimed by the visual arts counterpart to Bloomsbury, the Fitzroy Street Group, whose members converged on the Fitzroy Tavern, built in 1897 at the corner of Charlotte and Windmill streets, four blocks due west of the Crippens' new home. The group's most prominent and eventually most infamous member was the painter Walter Sickert, who from time to time in the years following his death would be considered a suspect in the Ripper murders. The Crippens shared the sidewalk with the brightest intellects of the day, including G. K. Chesterton, H. G. Wells, and Ford Madox Hueffer (later Ford Madox Ford), and the scholars of University College and the British Museum.

The neighborhood vibrated with sexual energy. Among the Bloomsbury Group, once it achieved full intellectual flower, conversation about sex flowed easily. The trigger, according to Virginia Woolf, was a moment when Lytton Strachey, the critic and biographer, walked into a drawing room where she and her sister Vanessa were seated.

Virginia wrote, "The door opened and the long and sinister figure of Mr. Lytton Strachey stood on the threshold. He pointed his finger at a stain on Vanessa's white dress.

" 'Semen?' he asked.

"Can one really say it? I thought and we burst out laughing. With that one word all barriers of reticence and reserve went down."

The dividing line between Bloomsbury and Fitzrovia, as the neighborhood around the Fitzroy Tavern eventually became known, was Tottenham Court Road, which happened also to be a fault line in the world's political crust and a part of London of no small interest to New Scotland Yard and the French Sûreté. For years the basement at No. 4 Tottenham had housed the Communist Working Men's Club, where firebrands of all stripe had spoken, raved, and cajoled. Nearby at No. 30 Charlotte was the equally notorious, though more radical, Epicerie Française, a center of the international anarchist movement, kept under periodic watch by French undercover detectives. Here men seethed at the rift between the poor and the rich that was then so glaring in Britain.

Each morning, as Crippen made his way to work at the sumptuous Munyon's office in Shaftesbury Avenue, he walked down Tottenham Court Road, past the notorious basement and past the Special Branch and Sûreté detectives who kept watch on the street and its surroundings.

None took the least notice of the little doctor, his eyes large behind his glasses, his feet thrown out to his sides as he walked, oblivious to the forces simmering around him.

Cora Crippen now launched her bid for fame in the variety halls of Britain. She had one significant advantage: British audiences loved acts from America. She resolved to make her debut in a brief musical of her own creation, in which of course she would play the leading role. She asked Crippen to pay the production costs, and he gladly assented, for the work seemed to improve Cora's outlook and her behavior toward him, though she remained prone to dramatic swings of mood, as if she believed volatility were as necessary to a diva as a good voice and an expensive dress, the purchase of which Crippen also cheerfully funded.

Cora drafted a libretto for her show but recognized that it needed work. She arranged a meeting with a woman named Adeline Harrison, a music hall actress and part-time journalist who also worked as an adviser helping other performers craft new acts and improve scripts. Crippen may have had something to do with recruiting Harrison, for the two women met at Munyon's suite of offices on Shaftesbury.

Harrison recalled her first glimpse of Cora. "Presently the green draperies parted and there entered a woman who suggested to me a brilliant, chattering bird of gorgeous plumage. She seemed to overflow the room with her personality. Her bright, dark eyes were twinkling with the joy of life. Her vivacious rounded face was radiant with smiles. She showed her teeth and there was a gleam of gold."

A photograph from about this time captured Cora in a pose for the stage. It shows her seated and singing from a songbook, beside a basket heaped with flowers of some lush species, possibly orchids or calla lilies, or both. She is on the far side of plump, with thick fingers and almost no neck. Her dress and the many layers underneath make her appear still larger, more weapon than woman. The dress is printed with daggerlike petals. Its billowing shoulders amplify the breadth of her bodice but also highlight the impossible narrowness of her abdomen, corseted perhaps in the famous "Patti" from the Y.C. Corset company, named for Adelina

Patti, one of the world's most beloved sopranos. Cora wears an expression that conveys both confidence and self-satisfaction. Not quite haughty, but vain and smug. Mighty.

Harrison read Cora's script. There wasn't much of it—"a few feeble lines of dialogue," Harrison wrote.

Cora told Harrison she wanted to make the act longer and asked how that might be achieved. Cora wanted it to be more of a freestanding operetta than a simple variety turn.

"I suggested that a little plot might improve matters," Harrison said.

The resulting show was called *The Unknown Quantity* and debuted at the Old Marylebone Music Hall, not quite the tier of theater Cora had hoped for. The Marylebone had developed something of a reputation for favoring melodramas that featured coffins, corpses, and blood, but it nonetheless was a known and credible venue that would give her an opportunity to show off her talents. That was all she wanted. Once London got a look at her, her future would be made.

A program from this period identified Cora as Macà Motzki—her maiden name divided in two—and as a principal in "Vio & Motzki's American Bright Lights Company, From the Principal American Theatres." Her foil was to be an Italian tenor named Sandro Vio, identified in the program as "General Manager and Sole Director." Crippen too was on the program, as "Acting Manager." The plot involved romance and extortion and required Cora at one point to hurl a fistful of banknotes at Vio. She insisted the cash be real, though the resulting first-night scramble by the audience caused the management to command that fake money be used in future performances. The show lasted one week. Cora demonstrated a lack of talent so complete that at least one critic mocked her as "the Brooklyn Matzos Ball."

The failure humiliated Cora and caused her to give up variety, at least for the time being.

THE CRIPPENS MOVED FROM South Crescent to Guildford Street, a block or so from where Dickens once lived, but soon afterward, around November 1899, Professor Munyon called Crippen back to America to run the company's Philadelphia headquarters for a few months. He left Cora in London.

Something happened during that stay, though exactly what isn't clear. When Crippen returned to London in June 1900, he was no longer employed by Munyon's. He took over management, instead, of another patent medicine company, the Sovereign Remedy Co., on nearby Newman Street. At about this time, he and Cora moved back to Bloomsbury, this time to Store Street, where a century earlier Mary Wollstonecraft had lived. The Crippens' new apartment was only half a block away from their old home on South Crescent and a brief walk to Crippen's new office.

Crippen learned to his displeasure that while he was in America Cora had begun singing again, at "smoking concerts for payment." She told him, moreover, that she intended to try once more to establish herself as a variety performer and had adopted a new stage name, Belle Elmore. And she had become even more ill tempered. "She was always finding fault with me," he complained, "and every night she took some opportunity of quarrelling with me, so that we went to bed in rather a temper with each other. A little later on, after I found that this continued and she apparently did not wish to be familiar with me, I asked her what the matter was."

And Cora—now Belle—told him. She revealed that during her husband's absence she had met a man named Bruce Miller and, Crippen said, "that this man visited her, had taken her about, and was very fond of her, and also she was fond of him."

THE GERMAN SPY

KAISER WILHELM II HAD INDEED taken notice of Marconi's achievements. He long had resented Britain's self-proclaimed superiority, despite the fact that he himself happened to be a nephew of Edward, the Prince of Wales, who would succeed Queen Victoria upon her death. He made no secret of his intention to build Germany into an imperial power and to hone his army and navy with the latest advances in science, including, if merited, wireless communication.

In the midst of a new series of tests at Salisbury Plain, during which Marconi set a new distance record of 6.8 miles, a German named Gilbert Kapp wrote to Preece to ask a favor. He was doing so, he stated, on behalf of a friend, whom he identified as "Privy Councillor Slaby." This was Adolf Slaby, a professor in Berlin's Technical High School. Kapp described him as "the private scientific adviser to the Emperor," and wrote: "Any new invention or discovery interests the Emperor and he always asks Slaby to explain it [to] him. Lately the Emperor has read of your and Marconi's experiments . . . and he wants Slaby to report on this invention."

Kapp had two questions:

"1) Is there anything in Marconi's invention?

"2) If yes, could you arrange for Slaby and myself to see the apparatus and witness experiments if we come over to London toward the end of next week?"

He added: "Please treat this letter as confidential and say nothing to Marconi about the Emperor."

Even though by now Marconi's fear of prying eyes was more acute than ever, Preece invited Slaby to come and observe a round of experiments set for mid-May 1897, during which Marconi would attempt for the first time to send signals across a body of water.

In the meantime Marconi contemplated a surprise of his own, for Preece.

Until now Marconi had considered that a contract with the post office might be the best way to profit from his invention and at the same time gain the resources to develop it into a practical means of telegraphy. But ever aware of each moment lost, Marconi had grown uneasy about the pace at which the post office made decisions. He told his father, "As far as the Government is concerned, I do not believe that they will decide very soon whether to acquire my rights or not. I also believe that they will not pay a great deal for them."

In the wake of Preece's lectures, investors had begun approaching Marconi with offers. Two Americans offered £10,000—equivalent to just over $1 million today—for his United States patent. Marconi evaluated this and other early proposals with the cold acuity of a lawyer and found none compelling enough to accept. In April, however, his cousin Henry Jameson Davis came to him with a proposal to form a company with a syndicate of investors linked to the Jameson family. The syndicate would pay Marconi £15,000 in cash—about $1.6 million today—and grant him controlling ownership of company stock, while also pledging £25,000 for future experiments.

Marconi gave the proposal the same scrutiny he had given all previous offers. The terms were generous. At the time £15,000 was a fortune. In H. G. Wells's novel *Tono-Bungay* one character exults in achieving a salary of £300 a year, because it was enough to provide a small house and a living for himself and a new wife. Best of all, Jameson Davis was family. Marconi knew him and trusted him. The investors, in turn, were known within the Jameson empire. It was too compelling to refuse, but Marconi understood that by accepting the offer he risked alienating Preece and the post office. The question was, could anything be done to keep Preece's sense of hurt from transforming the post office into a powerful enemy?

What followed was a carefully orchestrated campaign to depict this offer as something that Marconi himself had nothing to do with but that as a businessman he was obligated to take seriously, in the interests of his invention.

Marconi enlisted the help of a patent adviser, J. C. Graham, who knew Preece. On April 9, 1897, Graham wrote to Preece and told him the terms of the offer and added that Marconi had "some considerable doubts about closing with it lest apparently he should be doing anything which even appears to be ungrateful to you, for I understand from him that he is under a great debt of gratitude to you in more than one respect.

"As the matter is evidently weighing heavily on his mind, and as he appears to have only one object in view, viz., to do the right thing, I thought it was just possible that a letter from me might be of some use. I, of course, know nothing more of the case than I have set out above."

From the text alone, Graham's motive for writing the letter was far from obvious, and Preece must have read it several times. Was he asking Preece's opinion, or was his intent simply to notify Preece obliquely that Marconi intended to accept the offer and hoped there would be no hard feelings?

The next morning, Saturday, Marconi stopped by the post office building but found Preece gone. After returning to his home in Talbot Road, Westbourne Park, Marconi composed a letter to Preece.

He began it, "I am in difficulty."

The rest of the letter seemed structured to follow a choreography established by Marconi, Jameson Davis, and possibly Graham. It struck the same notes as Graham's letter and, like Graham's, said nothing about the fact that Jameson Davis happened to be Marconi's cousin.

Marconi referred to Jameson Davis and his syndicate as "those gentlemen" and couched the letter in such a way that anyone reading it would conclude that all of this was happening without his involvement, certainly without his encouragement—that this poor young man had suddenly found himself pressed to respond to an offer from the blue, one so generous that he found himself forced to consider it, though it gave him no joy to do so.

After setting out the details, Marconi added, "I beg to state, however, that I have never sought these offers, or given encouragement to the promoters."

Afterward he wrote to his father that he believed, based on what he had heard from Preece's associates, "that he will remain friends with me." In so doing, he revealed a trait of his character that throughout his life would color and often hamper his business and personal relationships: a social obtuseness that made him oblivious to how his actions affected others.

For in fact Preece felt deep personal hurt. Years later in a brief memoir, in which for some reason he described himself in the third person, Preece wrote, "Marconi at the end of 1897 naturally came under the influence of the business men who were financing his new Company, and it was no longer possible for Preece as a Government officer to maintain those cordial, and frequently almost parental, relations with the young inventor. No one regretted this more than Preece."

The depth of his hurt and its consequences would not become apparent for several months. For the moment Marconi's news did nothing to shake Preece's intention to make Marconi the centerpiece of his talk at the Royal Institution in June; nor did Preece immediately withdraw his support for Marconi's experiments. The new company had not yet formed, and Preece believed there was still a chance the government could acquire Marconi's patents. A decade later a select committee of Parliament would conclude that Preece should have tried harder. Had he done so, the committee reported, "an enterprise of national importance could have been prevented from passing into the hands of a private company and subsequent difficulties might have been avoided."

In April 1897, with Marconi's over-water tests still a month off, Britain was again wracked by a spasm of fear about the mounting danger of anarchists and immigrants. A bomb exploded on a train in the city's subterranean railway, killing one person and injuring others. The bomber was never caught, but most people blamed anarchists. Foreigners. Italians.

The world was growing more chaotic and speeding up. Rudyard Kipling could be spotted in his six-horsepower motorcar thundering around at fifteen miles an hour. The race among the great shipping companies to see whose liners could cross the Atlantic in the shortest time intensified and grew more and more costly as the size and speed of each ship increased and as the rivalry between British and German lines became freighted with an ever-heavier cargo of national pride. In April 1897 at the Vulcan shipyard in Stettin, Germany, thousands of workers raced to prepare the largest, grandest, fastest oceanliner yet for its launch on May 4, when it would join the stable of ships owned by North German Lloyd Line. Everything about this new liner breathed Germany's aspiration to become a world power, especially its name, *Kaiser Wilhelm der Grosse,* and its decor, which featured life-size portraits of its namesake and of Bismarck and Field Marshal Helmuth von Moltke, whose nephew all too soon would lead Germany into global war. The launch was to be overseen by Kaiser Wilhelm II himself.

In early May, Adolf Slaby sailed from Germany for Britain and made his way to the Bristol Channel, between England and Wales, where Marconi, with the help of a postal engineer named George Kemp, prepared for his next big demonstration.

MARCONI HOPED TO SEND messages across all nine miles of the Bristol Channel, but first he planned a more modest trial: to telegraph between Lavernock Point on the Wales side and a tiny island in the channel called Flat Holm, some 3.3 miles away. On Friday, May 7, Kemp traveled by tug to the island carrying a transmitter and took lodging "at a small house owned by the person in charge of the Cremation House."

Slaby arrived at Lavernock. Generously, if unwisely, Marconi sent a small transmitter to Slaby so that he could monitor the experiments firsthand.

On Thursday, May 13, one week after the tests began, Marconi keyed the message, "So be it, let it be so."

The sparks generated by his transmitter jabbed the air. Those present had to cover their ears against the miniature thunder of each blue discharge. The resulting chains of waves raced over the channel at the speed of light, from Flat Holm to Lavernock, where Marconi's main receiver now captured them without distortion.

Slaby realized how much the kaiser would value this new information. Slaby adored Wilhelm. In a letter to Preece, he would write, in unpolished English, "I can't love him more than I [do], he is the best and loveliest monarch who ever sit on a throne with the deepest understanding for the progress of his time. More than ever I regret, that those horrid politics have made him a stranger to your countrymen and to your whole country, that he is loving so very deeply."

But this adoration transformed Slaby from neutral academic into de facto spy. In Berlin Slaby had been experimenting on his own with coherers and induction coils to generate electromagnetic waves. He knew the fundamentals, but now he took detailed notes on how Marconi had designed, built, and assembled his apparatus. There can be little doubt that if Marconi had known just how much detail Slaby had harvested, he would have barred him from the tests, but apparently he was too deeply engrossed to notice.

More messages followed.

"It is cold here and the wind is up."

"How are you?"

"Go to bed."

"Go to tea."

And of course, this early example of wireless humor, possibly the first: "Go to Hull."

Next Marconi tried sending signals all the way across the channel. Though barely legible, they did reach the opposite bank nine miles away, a new record. To Slaby, such a distance seemed incomprehensible. "I had not been able to telegraph more than one hundred meters through the air," Slaby wrote. "It was at once clear to me that Marconi must have added something else—something new—to what was already known."

After the experiments Slaby returned quickly to Germany. He was back in Berlin within two days and immediately wrote to Preece to thank him for arranging his visit. "I came as a stranger and was received like a friend and experienced once more, that people may be separated by politics and newspapers but that science unites them."

Marconi did not share these warm feelings. Just as Preece felt betrayed by Marconi, Marconi now felt betrayed by Preece, for inviting Slaby to witness the experiments. Outwardly, however, he and Preece appeared still to be allies. With post office help Marconi continued his experiments, and Preece prepared for his talk at the Royal Institution, one of the most anticipated lectures in London.

In Berlin, Slaby immediately got to work replicating Marconi's equipment.

IN LIVERPOOL OLIVER LODGE roused himself from his dalliances with X-rays and ghosts. Angered by the attention being heaped upon Marconi, and by Preece's patronage, he hired his own attorney. On May 10, 1897, he filed an application to patent a means of tuning wireless transmissions so that signals sent from one transmitter would not interfere with those from another. In the same application he sought also to patent his own coherer and a tapping device that automatically thumped the coherer after each transmission to return its filings to their nonconductive state.

He had to withdraw these last two claims, however. Marconi's patent had priority.

This did nothing to quell Lodge's mounting resentment; nor did the news that Preece now planned a lecture on Marconi's wireless telegraphy at the Royal Institution. For Lodge, it was too much. On Saturday, May 29, 1897, he wrote to Preece to remind him of his own Royal Institution lecture three years earlier:

"The papers seem to treat the Marconi method as all new. Of course you know better, [and] so long as my scientific confreres are well informed it matters but little what the public press says.

"The stress of business may however have caused you to forget some of the details published by me in 1894. I used brass filings in vacuo then too. It could all have been done 3 years ago had I known that it was regarded as a commercially important desideratum. I had the automatic tapping-back [and] everything."

Preece went ahead with his lecture at the Royal Institution. He and Marconi included a demonstration similar to what they had done at Toynbee Hall, with bells "ringing merrily" from refuse cans, as *The Electrician* put it. The journal called the experiments "wizard-like."

Preece told the audience, "The distance to which signals have been sent is remarkable," and added, "we have by no means reached the limit." Here he aimed an attack at Oliver Lodge. Without identifying Lodge by name, Preece alluded to Lodge's declaration of three years earlier that Hertzian waves probably could not travel farther than half a mile. "It is interesting to read the surmises of others," Preece said. "Half a mile was the wildest dream."

Here, as *The Electrician* reported, Preece "scored an effective hit."

At the close of his lecture great applause rose from the audience. From Lodge and the Maxwellians came more fury. In a striking breach of the decorum that governed Victorian science, Lodge took his anger public. In a letter to *The Times* he wrote, "It appears that many persons suppose that the method of signaling across space by means of Hertzian waves received by a Branly tube of filings is a new discovery made by Signor Marconi. It is well known to physicists, and perhaps the public may be willing to share the information, that I myself showed what was essentially the same plan of signaling in 1894." He complained that "much of the language indulged in during the last few months by writers of popular articles on the subject about 'Marconi waves,' 'important discoveries' and 'brilliant novelties' has been more than usually absurd."

The attack startled even his friend and fellow physicist George FitzGerald, though FitzGerald shared Lodge's opinion. Shortly after the *Times* letter appeared, FitzGerald wrote to Lodge and cautioned, "It would be important to keep it from becoming a personal question between you and Marconi. The public don't care about that and will only say, 'This is a personal squabble: let them settle it amongst themselves.' "

FitzGerald did not blame Marconi. "This young chap himself, I understand he is merely 20"—actually, he was twenty-three—"deserves a great deal of credit for his persistency, enthusiasm, and pluck and must be really a very clever young fellow and it would be very hard to expect him to be quite judicial in his views as to everybody's credit in the matter." Marconi had not been "very open," he wrote, "but he is hardly to blame if his head is a bit swelled under the circumstances, and no Italian or other foreigner was ever really fair in their judgments so that it is quite unreasonable to expect them to be so."

The real problem was Preece, FitzGerald charged. He urged that Lodge focus his attacks on him, in particular on how Preece and the post office—"<u>absurdly</u> ignorant, as usual"—had ignored the scientific discoveries on which Marconi had based his apparatus and instead had been seduced by a "secret box."

He added, "Preece is, I think, distinctly and intentionally scoffing at scientific men and deserves severe rebuke."

~

ON JULY 2, 1897, MARCONI received his full, formal patent and, without Preece's knowledge, moved steadily closer to join with Jameson Davis to form a new company.

Preece may have believed he had stymied this plan. In a letter to superiors on July 15, in which he argued the time had come to consider acquiring the patent rights to Marconi's system, he wrote: "I have distinctly told him that as he has submitted his scheme to the consideration of the Post Office, the Admiralty and the War Department, he cannot morally enter into any negotiation with anyone else or listen to any financial proposals which might lead to a species of 'blackmailing' of his principal, if not his only, customers. He accepts and recognizes this position."

Preece recommended the government pay a mere £10,000 for the patent rights—about $1.1 million today—and doubted Marconi would feel himself in a position to argue. "It must be remembered that Mr. Marconi is a very young man. . . . He is a foreigner. He has proved himself to be open and candid and he has resisted very tempting offers. He has very little experience. On the other hand he cannot do much without our assistance and his system can scarcely be made practical for telegraphy by any one in this country but by ourselves."

But just five days later Marconi founded his new company. His representatives registered its name as the Wireless Telegraph and Signal Co. and identified its headquarters as being in London. Jameson Davis became managing director, with the understanding that once the enterprise was well established he would resign. Marconi received sixty thousand shares of stock valued at one pound each, representing 60 percent ownership of the company. He also received the £15,000 cash, and the company's pledge to spend another £25,000 developing the technology.

Within six months, the value of Marconi's stock tripled and suddenly his sixty thousand shares were worth £180,000 pounds, about $20 million today. At twenty-three years of age, he was both famous and rich.

IN BERLIN ADOLF SLABY had been busy. On June 17, one month after witnessing the Bristol Channel experiments, he wrote to Preece, "I have now constructed the whole apparatus of M[onsieur] Marconi and it works quite well. After returning from my holidays, which I intend to spend at the sea shore, I will try to signal through some distance. I feel always indebted to your extreme kindness in remembering those very pleasant and interesting days at Lavernock."

But Slaby's warm thanks belied grander ambitions, both for himself and for Germany. Soon he and two associates would begin marketing their own system and, with the enthusiastic backing of the kaiser and a cadre of powerful German investors, would become locked in a shadow war with Marconi that embodied the animosities then gaining sway in the larger world.

For the moment, however, Slaby pretended that all that mattered was science and knowledge. He wrote to Preece, "We are happy men, that we need not care for politics. The friendship that science had made cannot be disturbed and I wish to repeat to you the truest feelings of my heart."

Bruce Miller

Bruce Miller had once been a prizefighter and had the handsome but battered features to prove it. He had given up boxing for the stage and had come to England some months before meeting Belle in hopes of making a career in variety. He was, literally, a one-man band, playing drums, harmonica, and banjo all at the same time, and performed in London and in the provinces, at Southend-on-Sea, Weston-super-Mare, and elsewhere. When he met Belle, however, he was preparing to leave for Paris and the Paris Exposition of 1900, where he had entered into a partnership involving certain "attractions" at the exposition. He met her one evening in December 1899, about a month after Crippen's departure for Philadelphia. He was sharing an apartment with a male roommate, an American music teacher, on Torrington Square in Bloomsbury, adjacent to University College. That evening Belle came to the apartment to have dinner with his roommate, who introduced them. On that occasion, Miller said, "I merely shook hands with her and went away."

They met again, perhaps with Miller's roommate as intermediary, and became friends. Belle clearly was drawn to his size and rugged good looks. Miller was attracted by her energy and buoyancy and by her lush sexuality. He had a wife back in America, whom he had married in 1886, but as far as he was concerned the marriage had failed and he was married in name only.

"I cannot say that I told Belle Elmore that I was married," Miller conceded later, "but if I kept it from her it was not done intentionally. I never had anything to hide, or any object in keeping the information from her. When I first came to England I was separated from my wife. She wrote to me pleading to go back and live with her, and I showed Belle Elmore the letters." Belle agreed he should return to America and rejoin his wife.

Belle was not exactly forthcoming about her own marriage. "When I first met her, she was introduced to me as a Miss Belle Elmore," Miller said. "I met her several times before I knew that she was married. She frequently spoke of Dr. Crippen, and finally roused my curiosity, and I asked her who was Dr. Crippen?"

"That," she said, "is my husband. Didn't your friend tell you I was married?"

With Crippen away in America, Miller began coming to the apartment on Guildford Street two or three times a week, "sometimes in the afternoons," he said, "and sometimes in the evenings," though he contended later that the only room he entered was the front parlor.

He began calling Belle "brown eyes." He gave her photographs of himself, one of which she propped on the piano in the apartment. They went out together often, to restaurants popular with the theatrical crowd, like Jones's and Pinoli's, Kettner's in Soho, the Trocadero—the "Troc"—and most charismatic and infamous of all, the Café Royal on Regent near Piccadilly Circus, frequented by George Bernard Shaw, G. K. Chesterton, the sex researcher Havelock Ellis, the sex-obsessed Frank Harris, and before his fall the sexually indiscreet Oscar Wilde; here one Lady de Bathe, known best as Lillie Langtry, was said to have launched ice cream down the back of Edward, the future king. (Only partly true, as it happens—the incident did occur, but at a different place, and it involved a different actress.) Bookies mingled with barristers and ordered such drinks of the day as the Alabazan, the Bosom Caressa, Lemon Squash, and the Old Chum's Reviver.

For Bruce Miller and Belle, however, the drink of choice was champagne, and to commemorate their encounters they marked the date on each cork, until they had a string of them, which Belle kept in her possession. "Anything we do is always satisfactory to my husband," she told Miller. "I always tell him everything."

By the time Crippen returned, Miller was in Paris. He wrote to Belle "often enough to be sociable, to be friends." Crippen never met Miller, but Belle made sure he knew more than perhaps he wished. She continued to display at least one photograph of Miller in their home. In March 1901 she sent him an envelope containing six photographs of herself and told him they were taken by Crippen "with his Kodak." She hinted that Crippen knew she was sending them.

At one point, either by accident or by Belle's design, Crippen came across letters from Miller that closed with the line, "with love and kisses to Brown Eyes."

The letters induced in the little doctor a feeling akin to grief.

Later, Miller would be asked at length about these letters and about the true meaning of those closing words.

Enemies

DESPITE THE CITY'S ENDORSEMENT OF MARCONI, opposition elsewhere gained momentum, led as always by Oliver Lodge but now joined by new allies.

In September 1897 Britain's most influential electrical journal, *The Electrician,* came just short of accusing Marconi of stealing Lodge's work. "In fact Dr. Lodge published enough three years ago to enable the most simple-minded 'practician' to compound a system of practical telegraphy without deviating a single hair's breadth from Lodgian methods." Returning to Marconi's patent, the journal sniffed, "It is reputed to be easy enough for a clever lawyer to drive a coach and four through an Act of Parliament. If this patent be upheld in the courts of law it will be seen that it is equally easy for an eminent patent-counsel to compile a valid patent from the publicly described and exhibited products of another man's brain."

Meanwhile the public appeared to grow impatient with Marconi's secrecy and his failure to convert his technology into a practical system of telegraphy, despite reports of his successes at the post office, Salisbury Plain, and the Bristol Channel. This was an age that had come to expect progress. "What we want to know is the truth about all these questionable successes," one reader wrote in a letter published by *The Electrician.* "I say questionable, because this delay, this hanging fire, as regards practical results, raises in my mind certain doubts which in common probably with others I should like to have dispelled.

"Wherein are the present difficulties? Are they in the transmitter, in the receiver, or in the intervening and innocent ether, or do they exist in the financial syndicate, who upon the strength of hidden experiments and worthless newspaper reports, have embarked in this great and mysterious venture?"

Once Marconi could have counted on William Preece and the post office to come to his defense, but by now Preece had turned against him—though Marconi seemed oblivious to the change and to the danger it posed. In early September 1897, for example, the post office abruptly barred Marconi from tests it was conducting at Dover, even though the tests involved Marconi's equipment. Marconi complained to Preece that if he were not allowed to be present, the tests likely would fail. He feared that in the hands of others his wireless would not perform at its maximum; he also knew that the post office's engineers had not incorporated his latest improvements. He was twenty-three years old, Preece sixty-three, yet Marconi wrote as if chiding a schoolboy: "I hope this new attitude will not be continued, as otherwise very serious injury may be done to my Company in the event of the non success of the Dover experiments."

Shortly afterward he hired George Kemp away from the post office and made him his personal assistant, one of the most important hiring decisions he would make.

So far all this had taken place out of public view, but early in 1898 the post office exhibited what appeared to be the first official manifestation of Preece's disenchantment. The postmaster-general's annual report for the twelve months ended March 31, 1898, disclosed that tests of Marconi's apparatus had been conducted, "but no practical results have yet been achieved."

Marconi was stung. He believed he had demonstrated without doubt that wireless was a practical technology, ready for adoption. In December 1897 he and Kemp had erected a wireless mast on the Isle of Wight, on the grounds of the Needles Hotel at Alum Bay—the world's first permanent wireless station—and established communication between it and a coastal tugboat at a distance of eleven miles. In January 1898 they had erected a second station on the mainland, at another hotel, Madeira House in Bournemouth, fourteen and a half miles west along the coast. The two stations had been in communication ever since.

Seemingly blind to Preece's changed attitude, Marconi offered to sell the post office rights to use his technology within Britain for £30,000—an exorbitant price, equivalent to about $3 million today. The offer smacked of impudence. The government rejected the offer.

Now Preece struck again. In February 1899 he turned sixty-five, the post office's mandatory retirement age, but instead of retiring, he wrangled an appointment as Consulting Engineer to the Post Office, where circumstances contrived to make him an even more dangerous adversary. His superiors asked him to compile a report on Marconi's technology with an eye to determining whether the government ought to grant Marconi a license that would permit his stations to begin handling messages turned in at telegraph offices operated by the post office. Existing law, which gave the post office a monopoly over all telegraphy in the British Isles, forbade such use.

In his report of November 1899 Preece advised against granting the license. Marconi had yet to establish a viable commercial service anywhere, he argued. To grant a license now would merely enrich Marconi and his backers by causing an "ignorant excitement" among investors. "A new company would be formed with a large capital, the public would wildly subscribe to an undertaking endorsed by the imprimatur of the Postmaster-General and the Government would encourage another South Sea Bubble."

Later Preece wrote to Lodge, "I want to show you my Report. It is now with the Attorney General. It is very strong and dead against Marconi on all points."

LODGE WAS PLEASED. He wrote to Sylvanus Thompson about what he called "Preece's attempt to upset their applecart."

He wrote: "I can't help thinking it is a bit well deserved and just, though rather belated."

MARCONI CAME TO RECOGNIZE that he needed his own allies, both to neutralize the opposition of Lodge and to help dispel the still-pervasive skepticism that wireless telegraphy would ever be more than a novelty.

First he courted one of Britain's most revered men of science, Lord Kelvin. Early on Kelvin had declared himself a skeptic on the practical future of wireless, stating—famously—"Wireless is all very well but I'd rather send a message by a boy on a pony."

In May 1898 Kelvin stopped by Marconi's offices in London, where Marconi himself demonstrated his apparatus. Kelvin was impressed but remained skeptical about its future value. At this point Marconi and Lodge both were developing methods of tuning signals so that messages from one transmitter would not distort those from another, but Kelvin deemed interference a problem that would only grow worse as power and distance increased. Kelvin wrote Lodge, "The chief objection I see to much practical use at distances up to 15 miles is that two people speaking to one another would almost monopolize earth and air for miles around them. I don't think it would be possible to arrange for a dozen pairs of people to converse together by this method within a circle of 10 miles radius."

A month later Kelvin and his wife visited Marconi's station at the Needles Hotel on the Isle of Wight, where Marconi invited Kelvin to key in his own long-distance message. Now at last Kelvin seemed to awaken to the commercial potential of wireless. He insisted on paying Marconi for his message, the first paid wireless telegram and, incidentally, the first revenue for Marconi's company.

Marconi asked Kelvin to become a consulting engineer. On June 11 Kelvin agreed tentatively to do so, and the value of his support immediately became evident. That same day Kelvin wrote to Oliver Lodge, "I think it would be a very good thing if you would write direct to Marconi an olive branch letter." He told Lodge that after spending two days with Marconi, "I formed a very favorable opinion of him. He said I might write to you. . . . I know he would like your cooperation and I think it would be in every way right that you should in some way be connected with the work." He informed Lodge of his own decision to join with Marconi as consulting engineer and wrote, "I suggested that you also should be asked to act in the same capacity and he thoroughly approved of my suggestion. But before I had any idea of taking part myself I wished to promote the olive branch affair and I hope (indeed feel sure) you will take the same view of it as I do."

He added an enthusiastic postscript about his time at Marconi's Needles station: "I saw (and practiced!) telegraphing thence through ether to & from Bournemouth. Admirable. Quite practical!!!"

Kelvin seemed all but certain to join the company, when suddenly he expressed qualms that had nothing to do with Marconi or his technology. What troubled him was the idea that in allying himself with Marconi, he would be joining an enterprise devoted not just to exploring nature's secrets but to making as much profit as possible. On June 12, the day after his letter to Lodge, Kelvin wrote again. "In accepting to be consulting engineer, I am making a condition that no more money be asked from the public, for the present at all events; as it seems to me that the present Syndicate has as much capital as is needed for the work in prospect. . . . I am by no means confident that this condition will be acceptable to the promoters. But without it I cannot act."

For Marconi, this was an untenable condition, and Kelvin never did become consulting engineer.

Now Marconi concentrated on Lodge.

LODGE REVELED IN HIS NEW POWER to command Marconi's attention. For help in dealing with Marconi, Lodge recruited a friend, Alexander Muirhead, who ran a company that manufactured telegraphic instruments of high quality. Muirhead met Jameson Davis at the Reform Club in London and immediately afterward wrote to Lodge, "Today was only the beginning of the game. I feel sure now that they want to combine with us. Have patience it will come about."

In July Muirhead offered to sell Lodge's tuning technology to Marconi—for £30,000, the same steep price Marconi had quoted to the post office for his own patent rights. In a letter to Lodge dated July 29, 1898, Jameson Davis wrote, "This struck me and my directors as being exceedingly high, more especially, as we are without any information as to what the inventions may be." He wanted specifics, but so far none had been forthcoming. "As we are very anxious to have you with us, I should be very glad if we could have this matter cleared up, and come to some good business arrangement."

Now Marconi himself wrote to Lodge and, in an apparent attempt to advance the courtship by demonstrating his own rising prominence, included an intriguing return address:

Royal Yacht Osborne
Cowes
Isle of Wight

Anyone who could read a newspaper knew that at that moment Edward, the Prince of Wales, was aboard the royal yacht recuperating from injuries to his leg caused when he fell down a staircase during a ball in Paris. His mother, Queen Victoria, would have preferred that he spend this time at the royal estate, Osborne House, where she herself was staying, but Edward preferred the yacht, and a little distance. The vessel was moored about two miles away in the Solent, the channel between the Isle of Wight and the mainland. In any preceding age this distance would have assured Edward as much privacy as he wished, but his mother had read about Marconi and now asked him to establish a wireless link between the house and the *Osborne*.

Ever mindful of opportunities to draw the attention of the press, Marconi agreed. At his direction workers extended the *Osborne*'s mast and ran wire along its length to produce an antenna with a height above deck of eighty-three feet. He installed a transmitter whose spark flushed the sending cabin with light and caused a burst of miniature thunder that required him to fill his ears with cotton. On the grounds of Osborne House at an outbuilding called Ladywood Cottage, Marconi guided construction of another mast, this one reaching one hundred feet.

At one point, while adjusting his equipment, Marconi sought to cut through the Osborne House gardens, at a time when the queen herself was there in her wheeled chair. The queen valued her privacy and commanded her staff to guard against uninvited intrusion. A gardener stopped Marconi and told him to "go back and around."

Marconi, by now all of twenty-four years old, refused and told the gardener he would walk through the garden or abandon the project. He turned and went back to his hotel.

An attendant reported Marconi's response to the queen. In her mild but imperious way she said, "Get another electrician."

"Alas, Your Majesty," the attendant said, "England has no Marconi."

The queen considered this, then dispatched a royal carriage to Marconi's hotel to retrieve him. They met and talked. She was seventy-nine years old, he was barely out of boyhood, but he spoke with the confidence of a Lord Salisbury, and she was charmed. She praised his work and wished him well.

Soon the queen and Edward were in routine communication via wireless. Over the next two weeks Queen Victoria, Edward, and his doctor, Sir James Reid, traded 150 messages, the nature of which demonstrated yet again that no matter how innovative the means of communication, men and women will find a way to make the messages sent as tedious as possible.

August 4, 1898. Sir James to Victoria:

"H.R.H. the Prince of Wales has passed another excellent night, and is in very good spirits and health. The knee is most satisfactory."

August 5, 1898. Sir James to Victoria:

"H.R.H. the Prince of Wales has passed another excellent night, and the knee is in good condition."

There was this heated exchange, from a woman aboard the *Osborne* to another at the house, "Could you come to tea with us some day?"

A reply came rocketing back through the ether, "Very sorry cannot come to tea."

THIS NEW TECHNOLOGY intrigued Edward, and he was delighted to have Marconi aboard. By way of thanks he gave Marconi a royal tie pin.

While on the yacht, Marconi wrote to Lodge about his success in establishing communication between the queen and her son. "I am glad to say that everything has gone off first rate from the very start, having sent thousands of words both ways without having had once to repeat a word." He noted that although the distance was less than two miles, the two locations were "out of sight of each other, a hill intervening between."

He closed, "I am in great haste," and underlined his signature with a bold black slash of his pen, something he would do forever after.

In the midst of courting Lodge, Marconi's company experienced its first fatality, though the death had nothing to do with wireless itself.

One of Marconi's men, Edward Glanville, traveled to remote Rathlin Island, seven miles off the coast of Northern Ireland, to help conduct an experiment for Lloyd's of London, for which he was to help install wireless transmitters and receivers on Rathlin and on the mainland at Ballycastle, for use in reporting the passage of ships to Lloyd's central office in London. A tempestuous stretch of sea separated Rathlin and Ballycastle and up until now had made communication problematic.

In Ballycastle, where George Kemp managed the mainland portion of the work, the apparatus was placed in a child's bedroom in "a lady's house on the cliff," and the wires to the antenna were run out the child's window. If all went well, messages from Rathlin would be transmitted to Ballycastle by wireless, regardless of fog and storms, and relayed from there by conventional telegraph to Lloyd's.

One day Glanville disappeared. Searchers found his body at the base of a three-hundred-foot cliff.

Ever since joining Marconi, George Kemp had found himself called upon to perform diverse duties, but none so sad as now. On August 22, 1898, Kemp wrote in his diary, "I wired to London and arranged for the despatch of a coffin and the arrival of the coroner and steamer and, at 6 P.M. I went to Rathlin and examined the body with a doctor. I washed the body and placed it in the coffin. An inquest was held and the coroner returned 'death by accident.' It appeared that the people on the Island had often seen him climbing over the cliffs with a hammer with which he examined the various strata of the earth, and this was no doubt the cause of the accident."

Glanville's death briefly derailed the ongoing conversation between the Marconi company and Lodge, but now the courtship renewed. Lodge resisted, and declined even to give Marconi a demonstration of his technology.

Marconi grew impatient. On November 2, 1898, he wrote, "I sincerely hope you will be able to make us this exhibition or in some way arrange that we may work together rather than in opposition, which I am certain would be to the disadvantage of us both." In the meantime, he added, "It might . . . facilitate matters if you would kindly let us have an answer to the following questions." He then asked Lodge how many of his transmitters could operate in a given area without interference, what distance he so far had achieved, and what distance might be possible in the future.

Here again Marconi demonstrated his social obtuseness. He was asking Lodge to reveal his technology, yet just a couple of weeks earlier he himself had refused to do likewise for Lodge, stating: "I much regret that commercial considerations prevent me (at least at present) from mutually communicating the results we are obtaining."

This was exactly the wrong thing to say to Lodge, for whom the intrusion of commerce into science was so distasteful, but Marconi appeared not to register his antipathy.

In the same letter Marconi blithely asked Lodge if he would serve as one of the two sponsors required for his application to become a member of Britain's prestigious Institution of Electrical Engineers.

Lodge refused.

ANOTHER ENEMY NOW RAISED its standard—weather. Marconi saw that wireless likely would have its greatest value at sea, where it might end at last the isolation of ships, but achieving this goal required conducting experiments on the ocean and along coasts exposed to some of the most hostile weather the world had to offer. The more ambitious the experiment, the more weather became a factor, as proved the case at the end of 1898 when Trinity House, keeper of all lighthouses and lightships in Britain, agreed to let Marconi conduct tests involving the East Goodwin lightship, the same ship where William Preece's induction experiments had failed—a fact that could not have escaped Preece's increasingly jaundiced attention.

Marconi dispatched George Kemp to the ship to direct the installation of an antenna, transmitter, and receiver. Kemp chronicled the subsequent ordeal in his diary.

At nine A.M. on December 17, 1898, Kemp set out by boat from the beach at the village of Deal, notorious in the history of shipwreck both for the number of corpses routinely washed ashore after ships foundered on the Goodwin Sands and for the vocation of certain past residents who, as once chronicled by Daniel Defoe, had seen each new wreck as an opportunity for personal enrichment. Kemp's boat took three and a half hours to reach the lightship, which was moored at sea at a point roughly twelve miles northeast of the South Foreland Lighthouse, near Dover, where Marconi had erected a shore station for the trials.

Kemp arrived to find the lightship "pitching and rolling" in heavy seas. Nonetheless, Kemp and the lightship crew managed to erect a twenty-five-foot extension to one of the ship's tall masts, yielding an antenna that rose ninety feet above deck. "Beyond this," he wrote in his diary, "very little work was done as everyone appeared to be seasick." He left the lightship at four-thirty in the afternoon in an open boat, which he identified as a Life Boat Galley, and did not arrive at Deal until ten that night. He noted, with a good deal of understatement, "This was [a] rough experience in an open boat."

He returned to the ship on December 19, this time to stay awhile. He brought provisions for one week and immediately went to work installing equipment and running wire through a hole in a skylight. In his diary he noted that waves were crashing over the lightship's deck.

After a brief calm on December 21 and 22, the weather grew far worse. Late in the afternoon of December 23 "the wind increased and the Lightship began to toss about," he wrote. By evening "it was almost unbearable." He soldiered on and on Christmas Eve signaled greetings to Marconi, who was comfortably ensconced at South Foreland. That night Kemp volunteered to take watch over the ship's beacon so the crew could celebrate, which they did "until the early hours of the morning."

On Christmas Day, after Kemp and crew "managed to get over our Christmas dinner," a gale arose and the ship began to rise and plunge. Moored to the sea floor, it could not maneuver the way an ordinary ship could. "It was very miserable onboard," Kemp wrote, especially when the wind and tide conspired to hold the lightship broadside to the waves.

Over the next two days the sea continued to overwash the lightship. Water sluiced down hatchways. Kemp continued signaling. "Everything between decks was as wet as those on deck," he wrote on December 27.

The next day brought more of the same: "The weather was still bad and I told them at the Foreland that I was feeling ill, but I managed to send the 3 c.m. spark." Despite the increasingly awful conditions, Kemp recorded "splendid results," though it is hard to imagine how he achieved anything. "The waves were still breaking over the ship and I could not get on deck for fresh air. I was very cold, wet and miserable and had very little sleep."

By December 30 even stalwart Kemp began to crack. Conditions were so dangerous he could not venture topside for air. He used the ship's wireless to send a message of his own. "I told Mr. Marconi that I was not well enough to remain on board any longer and he must send for me when the wind dropped. . . . I told him that we wanted fresh meat, vegetables, bread and bacon but this was taken, it appeared, as a joke. The fact that I had come on board on Dec. 19, with provisions for one week had evidently been forgotten, also that I had been on board for 12 days living, the latter part, on quarter rations, consequently I had to beg, borrow and steal from the Lightshipmen."

New Year's Day 1899, his fourteenth day aboard, was cold and wet, with high seas, high winds, heavy rain. The next day he wrote, "I was so stiff and weak that I could scarcely move."

But at last, on January 4, the weather eased and became "quite calm." Supplies arrived for Kemp, "some mutton, a fowl, 2 bottles of Claret, 2 loaves, potatoes, a cabbage, sprouts and fruit." He added, with an underline for emphasis, "Had some good fresh food after 17 days."

Four months later the lightship's wireless provided a vivid demonstration of what Marconi long had hoped his technology would accomplish. In April, after heavy fog settled over the Goodwin Sands, a steamship named *R. F. Mathews,* 270 feet long and displacing nearly two thousand tons and carrying coal, yes, from Newcastle, rammed the lightship. The crew used Marconi's wireless transmitter to notify Trinity House and Lloyd's of the accident. Damage to both ships was minimal, and no one on either vessel was hurt.

DESPITE THIS AND OTHER SUCCESSES, including the first messages sent by wireless across the English Channel, the year 1899 proved a barren one for Marconi and his company, with no revenue from his invention and no prospect of any. Trinity House had been impressed with the Goodwin experiments but did not come forth with a contract. Nor did Lloyd's of London, though its representatives had been pleased with the Rathlin Island experiment and likely would have kept the system in operation if William Preece, citing the postal monopoly over telegraphy in Britain, had not stepped in and shut it down and replaced it with one of his own induction systems. Meanwhile the wall of skepticism that confronted Marconi seemed just as high and unbreachable as ever. Suspicion persisted about his motives and heritage and about the nature of the phenomena he had harnessed and its potential dangers. Rumor spread that wireless could be used to blow up warships.

In some people suspicion became outright fear, as evidenced by an incident that occurred at the station Marconi had built in France for his cross-channel experiments. The station was at Wimereux on the French coast, thirty-two miles across the English Channel from the South Foreland station that Marconi had erected for the lightship trials.

Anyone passing near the operator's room at night saw pulses of blue lightning and heard the loud crack of each spark, an eerie and disconcerting effect, especially on nights dusted with sea mist when the sparklight flared as a pale aurora. Inside, the juxtaposition of Marconi's apparatus against the decor of the room made things still more odd. The wallpaper and the rug and the tablecloth under the machine were all printed, dyed, or stitched with garish flowers.

One night, during a storm, an engineer named W. W. Bradfield was sitting at the Wimereux transmitter, when suddenly the door to the room crashed open. In the portal stood a man disheveled by the storm and apparently experiencing some form of internal agony. He blamed the transmissions and shouted that they must stop. The revolver in his hand imparted a certain added gravity.

Bradfield responded with the calm of a watchmaker. He told the intruder he understood his problem and that his experience was not unusual. He was in luck, however, Bradfield said, for he had "come to the only man alive who could cure him." This would require an "electrical inoculation," after which, Bradfield promised, he "would be immune to electro-magnetic waves for the rest of his life."

The man consented. Bradfield instructed him that for his own safety he must first remove from his person anything made of metal, including coins, timepieces, and of course the revolver in his hand. The intruder obliged, at which point Bradfield gave him a potent electrical shock, not so powerful as to kill him, but certainly enough to command his attention.

The man left, convinced that he was indeed cured.

MARCONI'S COMPETITORS FACED the same steadfast reluctance of the world to embrace wireless, but they nonetheless stepped up their own work. In America a new man, Reginald Fessenden, began drawing attention, and in France an inventor named Eugene Ducretet made news by transmitting messages from the Eiffel Tower in Paris to the Panthéon in the city's Latin Quarter. In Germany Slaby appeared to have joined forces with fellow countrymen Count Georg von Arco and Karl Ferdinand Braun, physicists also experimenting with wireless. Closer to home

Nevil Maskelyne, the magician who ran the Egyptian Hall, caused a stir when he placed a transmitter of his own design in a balloon and used it to ignite explosives on the ground below. And a newly roused Lodge, temporarily setting aside his hostility to the commercialization of science, began acting less like an academic and more like a man bent on establishing a business of his own. There were others working in the field as well, their progress revealed only in bits and pieces in the electrical press. No one doubted that even more contenders would emerge.

The race to build the first useful system of wireless telegraphy—the race, really, for distance—was well under way. Someone had to win, and timidity would not be an asset. It became clear to Marconi that he needed a demonstration grander and more daring than anything he so far had attempted. For several months he had been mulling an idea that would have fit well into a novel by H. G. Wells but that he knew was likely to spark nightmares, if not apoplexy, among his board of directors.

"Were You Her Lover, Sir?"

In time the precise character of the relationship between Bruce Miller and Crippen's wife would become a subject of interest to Scotland Yard and lead eventually to an interrogation by a barrister named Alfred A. Tobin, one of London's small cadre of impossibly articulate and learned lawyers who conducted the in-court portions of civil and criminal cases. Sometimes barristers served as prosecutors, assigned to trials by the director of public prosecutions; the rest of their cases came to them through a second tier of attorneys known as solicitors. The location of Miller's interrogation was the Old Bailey, London's Central Criminal Court. The subject: his letters to Belle Elmore.

"Were you writing to her as a lover?"

Miller: "No."

"Were you fond of her?"

"Yes."

"Did you ever tell her that you loved her?"

"Well, I do not know that I ever put it in that way."

"Did you indicate to her that you did love her?"

"She always understood it that way, I suppose."

"Then you did love her, I presume?"

"I do not mean to say that. I did not exactly love her; I thought a great deal of her as far as friendship was concerned. She was a married lady, and we will let it end at that. It was a platonic friendship. . . ."

"Do you know the difference between friendship and love?"

"Yes."

"Were you more than a friend?"

"I could not be more than a friend. She was a married lady and I was a married man."

"Were you more than a friend, sir?"

"I could not be more than a friend—I was not."

The presiding judge, Lord Chief Justice Alverstone, now joined in: "Answer the question whether you were or were not?"

"I was not more than a friend."

"Were there any improper relations between you?"

"No."

Now Tobin again:

"Did you ever write love letters to her?"

"I have written to her very nice letters perhaps."

"You know what a love letter is. Did you ever write a love letter to her?"

"Well, I do not remember that I ever put it just in that way. I often wrote to her very friendly letters; I might say they were affectionate letters."

"Then you wrote affectionate letters to her. Did you write love letters to her?"

"Affectionate letters."

"Ending 'Love and kisses to Brown Eyes'?"

"I have done so."

"Now, sir, do you think those are proper letters to write to a married woman?"

"Under the circumstances, yes. . . ."

"Do you agree now that those letters were most important letters to write to a married woman during her husband's absence?"

"I do not think they were, under the circumstances."

"Were you her lover, sir?"

"I was not."

"Have you been to any house in London with her for the purpose of illicit relationship?"

"I have not."

"Bloomsbury Street?"

"No place."

"Have you ever kissed her?"

"I have."

"Never done anything more than kiss her?"

"That is all."

"Why did you stop at that?"

"Because I always treated her as a gentleman, and never went any further."

The interrogation did nothing to clarify the relationship.

Fleming

One of the earliest messages Marconi sent across the English Channel was a brief telegram transmitted from his French station at Wimereux to the South Foreland station, where one of his men took it to an office of the post office telegraphs for relay by conventional land line to London. A messenger boy carried it to University College in Bloomsbury, near the British Museum, and there the telegram made its way to its intended recipient, John Ambrose Fleming, a professor of electrical engineering and friend of Oliver Lodge. On the date of the telegram, March 28, 1899, Fleming was forty-nine years old and possessed a degree of public fame and academic prominence just shy of what Lodge possessed. He was an expert in the amplification and distribution of electrical power.

The telegram read, simply,

> *Glad to send you greetings*
> *conveyed by electric waves through*
> *the ether from Boulogne*
> *to South Foreland twenty-eight*
> *miles [and] thence by*
> *postal telegraphs Marconi*

Though it seemed the most neutral of greetings, in fact it marked the start of Marconi's next attempt at seduction, and his most important, as future events would show.

Marconi recognized that with no revenue and no contracts and in the face of persistent skepticism, he needed more than ever to capture an ally of prominence and credibility. Through Fleming, however, Marconi also hoped to gain a benefit more tangible. His new idea, the feat he hoped would command the world's attention once and for all, would require more power and involve greater danger, physical and fiscal, than anything he had attempted before.

When it came to high-power engineering, he knew, Fleming was the man to consult.

UNLIKE LODGE OR KELVIN, Fleming was susceptible to flattery and needful of attention, as evidenced by the fact that upon receiving Marconi's telegram he made sure the London *Times* got a copy of it. *The Times* published it, as part of its coverage of Marconi's English Channel success. Next Fleming visited Marconi's station at South Foreland. He was deeply impressed, so much so that he wrote a long letter to *The Times* praising Marconi and his technology and acknowledging how the inventor had removed wireless from "the region of uncertain delicate laboratory experiments." Now, he wrote, it was a practical system marked by "certainty of action and ease of manipulation."

For Marconi, the letter was an affirmation of his strategy to gain credibility by refraction. Oliver Lodge too recognized that Fleming's letter had bestowed upon Marconi a new respectability and saw it as an act of betrayal. He wrote to Fleming, "My attention has been called to a letter of yours in the Times, in which it is suggested that you are attacking me and other scientific men who retain some jealousy for the memory of Hertz." He called the letter an "indictment against men of science" and asked Fleming for an explanation.

Fleming bristled. "I made no attack on you or any other scientific men in my letter to the Times," he replied. "I called attention to certain important achievements which in the public interest I thought should be noticed and described." He wrote that he was simply raising a point acknowledged by others, "that the time had arrived for a little more generous recognition of Signor Marconi's work as an original inventor. That after all is a matter on which different views may be held and if you dissent from it you are quite entitled to your opinion."

Soon afterward Marconi asked Fleming to become scientific adviser to his company. On May 2, 1899, Fleming wrote to Jameson Davis to set out certain conditions and to define "my position and views a little carefully."

If Jameson Davis cringed at this sentence, fearing a reprise of Lord Kelvin's qualms, he was soon reassured. Fleming wrote: "I have a strong conviction of the commercial possibilities of Mr. Marconi's inventions apart from their scientific interest provided they are properly handled. I should desire to see a genuine business of a solid character built up." He added a paragraph that later events would show to be in contradiction to elements of his character, especially his own deep need for recognition. Fleming wrote, "All that a scientific adviser does in the way of invention, suggestion or advice should be the sole property of those retaining him so far as their own affairs are concerned. I have noticed that any other course invariably leads sooner or later to difficulties and perhaps disputes."

Fleming agreed to work under a one-year contract, renewable at each party's discretion, for a fee of £300 a year. For the moment it seemed generous.

Now, Marconi revealed to him the nature of the grand experiment that had begun to occupy his thoughts. It would require two gigantic wireless stations and demand the production of more electrical energy than anything Marconi previously had attempted.

Though his maximum distance so far had been only thirty-two miles, what Marconi now proposed was to transmit messages across the full breadth of the Atlantic.

The Ladies' Guild Commences

Belle's avowed fondness for Bruce Miller had an important consequence. She told Crippen she no longer cared for him, and she threatened to leave him for Miller. Though they still slept in the same bed, they spent their nights without touch or warmth. They struck a bargain. Outwardly no one was to know of the strain in their marriage. "It was always agreed that we should treat each other as if there had never been any trouble," Crippen said.

He gave her money just as always, "with a free hand whatever she seemed to want at any time; if she asked me for money she always had it." She bought furs and jewelry and countless dresses. On one occasion he gave her £35—about $3,800 today—to buy an ermine cape. In public she always called him "dear."

In September 1903 Crippen went so far as to open a joint "current" account at Charing Cross Bank in the Strand. The account required his signature and Belle's but did not require both to be present when a check was presented for cashing. About three years later the Crippens opened a savings account at the same bank, with an initial deposit of £250—$26,000—under both their names.

Crippen paid for Belle's evenings out with friends and sometimes even came along, acting always the part of an affectionate and indulgent husband. He paid too for Belle's evenings with Miller.

Later Miller would contend that on some of his visits to the Crippens' flat he had the feeling that Crippen was at home, elsewhere among the rooms.

One evening Miller arrived at the Store Street apartment to find the table set for three. Belle held up dinner "until quite late," Miller recalled. Belle said, "I am getting uneasy; I expect another party to dinner."

The third party never arrived, and Belle told Miller, "I am often disappointed that way."

She never said who the third party was, but Miller "surmised that it was to be Dr. Crippen."

Crippen said, "I never interfered with her movements in any way. She went in and out just as she liked and did what she liked; it was of no interest to me."

He was not being entirely frank, however. "Of course, I hoped that she would give up this idea of hers at some time"—and by this he meant her idea of one day leaving with Bruce Miller.

Her other grand idea, of becoming a variety star, had been rekindled and now burned as brightly as ever. This time, however, she gave up trying to make her career in London and resolved instead to build a reputation at music halls in outlying towns and villages, known as "twice-nightlies" for the two variety programs performed each evening. "She got an engagement at the Town Hall, Teddington, to sing, and then from time to time she got engagements at music halls," Crippen said. She performed as a comedienne at a theater in Oxford, where she lasted about a week. She did turns at Camberwell, Balham, and Northampton.

Eventually she made it to the Palace, but not in London. In Swansea. Posters for a show there identified her as Miss B. Elmore and placed her performance between two musical groups, the Southern Belles and the Eclipse Trio.

"She would probably go away for about two weeks and return for about six weeks, but used to earn very little," Crippen said.

She began dying her hair a golden blond, at a time when dying hair was considered an act of suspect morality. "There was hardly any dyed hair," wrote W. Macqueen-Pope in his *Goodbye Piccadilly.* "It was considered 'fast' and the sign of prostitution." He recounted how a novelist, Marie Corelli, in one of her books described the owner of a fine country hotel as having dyed hair. The owner sued and won—even though she had indeed colored her hair. The court awarded only one farthing in damages. "She might have got more," Macqueen-Pope wrote, "but the dyed hair was most apparent."

Maintaining Belle's color took a lot of work. "When her hair was down in the morning one could see the original colour on the part nearest the roots," Belle's friend Adeline Harrison observed. Belle applied the bleaching chemicals to her hair every four or five days, and sometimes Crippen helped. "She was very anxious that nobody should ever know that she had any dark hair at all," he said. "She was a woman who was very particular about her hair. Only the tiniest portions of the hairs at the roots after they began to grow could be seen to be dark."

At one point Belle's travels brought her to a well-respected provincial theater in Dudley, the Empire, a theater with a sliding roof, where she wound up on a bill that included a much-loved comedian named George Formby. Another performer, Clarkson Rose, went to see Formby's act that night and happened also to catch Belle's performance. "She wasn't a top-rank artist, but, in her way, not bad—a blowsy, florid type of serio," meaning a seriocomic, a performer who mixed comedy and drama.

With so many turns a night, before boisterous audiences, it was never difficult to judge which performers the crowd favored. Belle was not one of them. Her singing was neither good enough nor sad enough to charm the crowd, and her comedy elicited only a halfhearted response from those accustomed to the likes of Formby and Dan Leno, one of the most popular comics of the day. She failed even in the halls of London's impoverished East End, considered one of the lowest tiers in the business. Robert Machray in his 1902 guide to the evening delights of the city, *The Night Side of London,* wrote, "To fail at even an East End hall must be a terrible business for an artiste. It means, if it means anything, the streets, starvation, death."

But not for Belle. She had Crippen, and she had his money. She did have one talent, however. She was gregarious and had a knack for making friends quickly. She gave up performing but lodged herself snugly among the theatrical crowd. Using Crippen's money, she continued her participation in the late-night revelries of actors and writers and their lovers and spouses. For appearances she sometimes brought Crippen along. Both kept to their bargain about keeping up the illusion of a happy marriage. They smiled at each other and told charming stories about their life together. Behind his thick glasses Crippen's magnified eyes seemed to glisten with genuine warmth and delight.

But not always. A photographer captured Crippen at a formal banquet. In the photograph he is wearing evening attire: black dinner jacket and pants, white bow tie, and a gleaming white shirtfront. He wears a flower in his lapel. He is surrounded by women in white, as if he is about to disappear in a cloud of taffeta, silk, and lace. Belle and two other women are seated on risers behind him. Two pretty young women sit on his right and left, so close to him that their dresses drape over his legs and thighs, meaning also that their bodies and his must be in contact, albeit with layers of cloth in between. The scene is faintly erotic. One woman's arm rests on his. The camera captures all five women in diverse expressions, odd for this era when one was never to move and above all not to smile. One woman stares into the middle distance, bored or sad or both. Another is smiling and glancing away. Belle, seated behind and above Crippen, has the pained expression of someone trying to get a room full of children to sit still. Only Crippen stares at the camera. His eyes, centered and magnified in the thick lenses of his glasses, are utterly without expression, as if he were a ventriloquist's dummy, momentarily inert.

Belle insinuated herself into a group of talented variety players and their spouses, among them Marie Lloyd, Lil Hawthorne of the Hawthorne Sisters, Paul Martinetti, a well-known "pantomimist," Eugene Stratton, a blackface singer, and others. At one gathering the women resolved to form a charity to provide for performers down on their luck and founded the Music Hall Ladies' Guild, a more subdued, women's counterpart to the Grand Order of Water Rats, founded in 1889, which Seymour Hicks, a performer and memoirist, called "the most distinguished brotherhood of world-famous music-hall artists."

The Water Rats' chairman was called the King Rat. The leader of the ladies' guild was merely its president: Its first was Marie Lloyd, its most famous member. Belle became treasurer.

The post gave Belle the kind of recognition she never got on stage. Her peers liked her and her unquenchable good spirits. Meetings were held every Wednesday afternoon, and Belle attended every one. Close friendships blossomed—close enough, certainly, that her friends knew about and had seen, even touched, the scar on her abdomen. Belle was proud of it. That long dark line gave her an element of mystery. When her friend and fellow guild member Clara Martinetti saw it, she was appalled. She had never seen a scar that size before. "Oh Belle does it hurt you?" she exclaimed.

"Oh no," Belle said, "it doesn't hurt me," and as she said it, she grabbed that portion of her abdomen and twisted.

"A Gigantic Experiment"

THE PLAN FLEW IN THE FACE of all that physicists believed about the optical character of electromagnetic waves. Like beams of light, waves traveled in a straight line. The earth was curved. Therefore, the physicists held, even if waves could travel thousands of miles—which they couldn't—they would continue in a straight line far out into space. Sending waves across the ocean was no more possible than casting a beam of light from London to New York. There was another question: Why bother at all? How could wireless improve on the transoceanic telegraphy already in place via undersea cable? In 1898 fourteen submarine cables draped the sea floor. The dozen in daily use carried 25 to 30 million words annually, only half their potential capacity. Transmission was expensive but fast and efficient.

Now Marconi proposed to set up a wireless service to do the same thing, with unproven technology, in the face of established physical law, and at the risk of destroying his company. The cost of a pair of wireless stations big enough and powerful enough for Marconi's plan to succeed would be immense and, if the effort failed, ruinous. And failure seemed a lot more likely than success. The scale of the stations Marconi now envisioned dwarfed anything he had built thus far. It was as if a carpenter, having erected his first house, set out next to construct St. Paul's Cathedral.

To Marconi, however, the greater risk lay in not making the attempt. He recognized that from a commercial standpoint his company was inert. He had amazed the world, but the world had not then come rushing to place orders for his apparatus. In the public view, wireless remained a novelty. Marconi saw that he had to do something big to jolt the world into at last recognizing the power and practicality of his technology.

That his plan might be impossible did not occur to him. He *saw* it in his mind. As far as he was concerned, he already had proven the physicists wrong. With each new experiment he had increased distance and clarity. If he could transmit across the English Channel, why not across the Atlantic? For him it came down to the height of his antenna and the intensity of charge that he was able to jolt into the sky.

He recognized, however, that to achieve his goal he needed help. Winding wire to produce an induction coil capable of signaling thirty feet was one thing, but building a power plant capable of sending a message thousands of miles was something else altogether. For this he needed Fleming.

At first Fleming was skeptical, but by August 1899, after studying the problems involved, he wrote to Marconi, "I have not the slightest doubt I can at once put up two masts 300 feet high and it is only a question of expense getting high enough to signal *to America*."

To better evaluate what it might entail, as well as to arrange another publicity event—coverage by wireless of the America's Cup race off New York at the request of the *New York Herald*—Marconi booked his first voyage to the United States. On September 11, 1899, accompanied by three assistants, including W. W. Bradfield, Marconi sailed for New York.

ON ARRIVAL MARCONI WAS THRONGED by reporters, who were startled by his youth—"a mere boy," the *Herald* observed—though at least one writer was struck by his alien appearance. "When you meet Marconi you're bound to notice that he's a 'for'ner.' The information is written all over him. His suit of clothes is English. In stature he is French. His boot heels are Spanish military. His hair and moustache are German. His mother is Irish. His father is Italian. And altogether, there's little doubt that Marconi is a thorough cosmopolitan." The passage was not meant as praise.

Marconi and his colleagues checked into the Hoffman House at Broadway and 24th Street in Manhattan, opposite a deepening triangular excavation that was soon to become the foundation of the Flatiron Building. They had just begun unpacking when the hotel's steam boiler, in the basement, exploded. A frightened guest blamed it on Marconi and his mysterious equipment. To quash the guest's concern, Marconi's men opened their trunks to reveal the quiescent apparatus within—and only then realized that the most important trunk was missing. Without the coherers it contained, Marconi would be forced to cancel his coverage of the America's Cup. His confident predictions of success had received a lot of attention from newspapers in America and abroad. His failure, with the weak excuse of lost luggage, would get at least as much publicity, perhaps even cause the price of his company's stock to slide and thereby eliminate any hope of paying for his transatlantic experiment.

Ordinarily Marconi's demeanor was cool and quiet. As the *Herald* noted, Marconi exuded the "peculiar semi-abstract air that characterizes men who devote their days to study and scientific experiment." The *New York Tribune* called him "a bit absent-minded." But now, upon finding the most important trunk missing, Marconi flew into a rage. With the petulance of a child, he proclaimed that he would leave for London aboard the next outbound ship.

His men calmed him. Bradfield and another assistant raced back to the wharf by horse-drawn cab to try to locate the trunk but failed. They returned to the hotel, no doubt fearing another outburst from their employer.

Now Bradfield remembered that on the day their ship left Liverpool, another liner also was scheduled to depart for America, but for Boston. He wondered if just possibly the trunk had gotten on the wrong ship. A reporter for the *Herald* headed north by train to check.

He found it, and Marconi's coverage of the yacht race, between the famed *Shamrock* owned by Sir Thomas Lipton and its American opponent, *Columbia II*, seized the world's attention. The *Columbia* won, and the *Herald* got the news first, by wireless.

DESPITE HIS SUCCESS, on November 8, 1899, when Marconi was scheduled to return to England, he had no new contracts to show for his effort. He had hoped to win the U.S. Navy as a customer, and while in America he had conducted a series of coastal trials, but the navy balked. Its report on the tests listed a host of speculative reasons to be wary of wireless, including this one: "The shock from the sending coil of wire may be quite severe and even dangerous to a person with a weak heart." Also, the navy's observers were peeved by Marconi's refusal to reveal his secrets. He allowed them to examine only certain components. Others, the navy complained, "were never dismantled, and these mechanics were explained in a general way. The exact dimensions of the parts were not divulged."

Far from being discouraged, Marconi arranged for yet another experiment, this one to take place during his voyage home aboard the *St. Paul,* a ship of great luxury and speed.

The ship's owner, the American Line, agreed to allow Marconi to equip the vessel with wireless and to rig an antenna high above deck. Marconi planned to begin transmitting from the ship to his stations at the Needles and Haven hotels as the liner approached England, to see how far from shore messages could be received.

As Marconi's assistants adjusted their shipboard equipment, Marconi demonstrated a paradox in his personality. Though he could be blind to the social needs of others, he also was able to command the allegiance of men older and younger and, as quickly became evident aboard ship, exuded a charm that women found compelling. One young woman, recalling the first time she met Marconi, said, "I noticed his peculiar, capable hands, and his rather sullen expression which would light up all at once in a wreath of smiles." He was said also to possess a dry humor, though occasionally it emerged heavily barbed. During one experiment, frustrated with the keying skills of an operator, Marconi asked via wireless if that was the best he could do. When the man replied that it was, Marconi fired back, "Well try using the other foot."

The *St. Paul* suited him. He had grown up amid luxury, conducted his first experiments amid luxury, and now, wealthy and famous, he did his traveling surrounded by something beyond luxury, for the designers of the great ships racing the Atlantic had sought to replicate in their first-class cabins and saloons the rich interiors of English country houses and Italian palazzi. Marconi associated with the wealthiest and most prominent of the ship's passengers, including Henry Herbert McClure, a well-known journalist. Marconi was the focus of attention and the subject of admiring, though discreet, observation by the women of the first-class deck. Always a connoisseur of beauty, Marconi returned the scrutiny.

As the *St. Paul* approached England, Marconi and his assistants stationed themselves at their wireless system, located in a first-class cabin, and began hailing the shore stations over and over. Taking turns, they kept at it through the night. They heard nothing in response, and indeed no one expected much this early in the voyage. The system had a maximum range under ideal conditions of perhaps fifty miles.

On Tuesday, November 14, 1899, the new managing director of Marconi's company, Maj. Samuel Flood Page, arrived at the Needles station on the Isle of Wight to observe the experiment. Jameson Davis, who several months earlier had retired from the post as planned, also came.

They calculated that the *St. Paul* would pass offshore at ten or eleven o'clock the next morning, Wednesday. Just in case, they assigned an operator to spend Tuesday night in the instrument room, where a bell rigged to the apparatus would announce the receipt of any incoming signals. No bells rang.

Flood Page returned to the instrument room at dawn as the sun began to bathe the Needles, a spine of chalk and flint sea-stacks from which the Needles Hotel took its name. "The Needles resembled pillars of salt as one after the other they were lighted up by the brilliant sunrise," Flood Page wrote. Marconi's men watched for ships to appear in the haze off the coast. "Breakfast over, the sun was delicious as we paced the lawn, but at sea the haze increased to fog; no ordinary signals"—meaning optical signals—"could have been read from any ship passing the place at which we were."

They saw no sign of the *St. Paul.* The hours dragged past. Flood Page claimed "the idea of failure never entered our minds," though this seems unlikely. Another hour passed, then another.

Then at 4:45 P.M. the bell rang.

The Needles operator signaled, "Is that you *St. Paul*?"

A moment later, an answer: "Yes."

"Where are you?"

"Sixty-six nautical miles away."

It was a new record. At Needles and aboard ship there was celebration, but soon the men at both nodes began running out of things to say. Pressed for fresh material, the Needles men began sending the latest news of the Boer War in South Africa, which had begun in mid-October and was now gaining ferocity. They sent other news as well.

Someone—it's not clear who—suggested publishing these dispatches in the form of a shipboard newspaper, the world's first. The captain granted Marconi use of the ship's print shop, which ordinarily had the more prosaic assignment of printing menus. The result was the *Transatlantic Times,* volume one, number one, a keepsake that passengers could purchase for a one-dollar contribution to the Seamen's Fund. "As all know," the newspaper's opening section stated, "this is the first time that such a venture as this has been undertaken. A Newspaper published at Sea with Wireless Telegraph messages received and printed on a ship going twenty knots an hour!"

Anyone reading closely would have found several passengers identified in the masthead of editors, including Marconi's assistant, Bradfield, as editor in chief and H. H. McClure as managing editor. But there was a third name as well, this one unfamiliar: J. B. Holman, treasurer.

As it happens, another momentous event had occurred during this crossing, albeit one of a rather more personal nature.

JOSEPHINE BOWEN HOLMAN was a young woman from Indianapolis—and from money. Though she lived now in New York with her mother, her roots lay in Woodruff Place, an enclave of some five hundred wealthy people incorporated as a distinct village within Indianapolis, occupying forested land once known locally as the Dark Woods.

Her hair was thick and dark, piled atop her head like a rich black turban. She had full lips and large eyes, and eyebrows that arced like gulls' wings over a gaze that was frank and direct. Marconi, now twenty-five years old, had always been drawn to feminine beauty, and he was drawn now to her.

They dined and danced and, despite the cold of a mid-November crossing, took long walks around the first-class deck. He taught her Morse code. He asked her to marry him, and she said yes. They kept their engagement a secret. When he disappeared into the wireless cabin for the last two days of the voyage, she was not troubled, though perhaps she should have been.

Once onshore, to evade discovery of the engagement by her mother, Holman inserted passages in her letters in Morse code.

Buoyed by love and by his success in signaling the Needles, Marconi prepared to reveal his idea to the company's directors and ask approval to build the two gigantic stations. By summer he was ready.

The directors balked. They considered it too risky and too expensive, and they doubted that apparatus capable of generating and managing the required power could even be built—and if so, whether the resulting station would smother every other Marconi station with interference.

Marconi countered that success in the venture would assert the company's dominance for once and for all. His confidence impressed the board, but so too did news from America that Nikola Tesla might be on the verge of attempting the same feat. In a much-read article in the June 1900 issue of *The Century Magazine,* Tesla alluded to things he had learned from experiments at his laboratory in Colorado Springs, Colorado, which he claimed could generate millions of volts of electricity, the equal of lightning. He wrote that in the course of his experiments he had found proof—"absolute certitude," as he put it—that "communication without wires to any point of the globe is practicable."

The article prompted J. P. Morgan to invite Tesla to his home, where Tesla revealed his idea for a "world system" of wireless that would transmit far more than just Morse code. "We shall be able to communicate with one another instantly irrespective of distance," Tesla wrote in the *Century* article. "Not only this, but through television and telephone we shall see and hear one another as perfectly as though we were face to face."

That word: *television*. In 1900.

In July Marconi's directors voted their approval.

The month provided another milestone as well. On July 4 Britain's Admiralty agreed to have Marconi's company supply and install wireless sets for twenty-six ships and six shore stations, at a cost of £3,200 pounds per installation—$350,000 today—with an additional annual royalty. The company would train the navy's men in how to use the apparatus. It was Marconi's first major order, but more important, it helped convince him and his directors of the need for a fundamental change in how the company operated.

As welcome as the contract was, providing first revenue just as Marconi's greatest quest was about to begin, it embodied the threat that the Royal Navy might now use Marconi's equipment to develop its own system, something it had the right to do under a British law that allowed the government to adopt any technology it wished, patented or not, in the interests of the empire's defense.

The contract caused Marconi to rethink the company's strategy for generating revenue through the manufacture and sale of apparatus to customers. As things stood, the post office monopoly barred the company from collecting fees for private telegrams sent by wireless and further blocked the automatic relay of telegrams from conventional land lines to Marconi's wireless stations. Thus the only likely customers were government agencies, of which only a few could be expected to see a need for wireless.

A loophole in Britain's telegraph laws suggested a possible new course. Instead of selling equipment, Marconi could provide customers with a wireless *service* that, if structured carefully, would skirt the postal monopoly. A shipping line, for example, would pay not for individual messages but for the rental of Marconi's apparatus and operators, who would receive their salaries from Marconi and communicate only with Marconi stations. Marconi argued successfully that the law allowed such an arrangement because all messages would be intracompany communications from one node of the Marconi company to another.

This change in strategy suited Marconi's personality. Ever since his days in the attic at the Villa Griffone, he had been worried about competitors. He saw this new approach as a means of erecting a bulwark against the competition. His new strategy included the requirement that any ship using his wireless service would communicate only with other ships likewise equipped, except in case of emergency. This meant that if a shipping line leased Marconi wireless for one vessel, it would have to do so for the rest of its fleet as well, if it wanted its ships to be able to communicate with one another. In theory the approach would also simplify things for shipowners, who would not have to pay for the construction and maintenance of their own networks of shore stations. Once fully equipped, a shipping line would be unlikely to switch to a competitor.

The new strategy seemed sound in principle, but now the question became, Would the change at last dispel the still-widespread reluctance to embrace wireless? Would customers come?

It seemed even more imperative now for Marconi to do something big to assert his dominance of the field and to publicize his technological prowess. If successful, the transatlantic bid would achieve both goals, as well as serve a third, more concrete purpose: It would demonstrate that his wireless could reach not only ships traveling near the coast but also liners far out in the deepest blue.

If successful. Marconi's certainty aside, the company was taking a grave gamble, betting its future and Marconi's reputation on a single experiment whose success nearly every established physicist believed to be impossible.

Late that summer Marconi and Flood Page and a recently hired engineer named Richard Vyvyan set out for Cornwall to search for a suitable location for the station that would serve as the English node of Marconi's transatlantic experiment. After tramping the coast, through fog and along paths that crossed mounds of heather and gorse and wildflowers, they settled on land atop Angrouse Cliff, near the village of Poldhu and adjacent to the large and comfortable Poldhu Hotel. Marconi did not mind remote locations, provided that a source of fine food and wine lay near at hand.

The first construction on the cliff began soon afterward, in October, directed by Vyvyan. Marconi planned the antenna; Fleming worked out details about how to amplify power to provide a spark intense enough to create waves capable of jumping the Atlantic, and how to do so safely, for with so much voltage coursing through the system even the act of keying a message could prove lethal. No ordinary Morse key could handle the power. This key would be a lever requiring muscle to operate, and courage as well, especially when sending Morse dashes—which required longer pulses of energy and increased the threat that uncontrolled sparks, or arcs, would be unleashed.

The extreme power of the station raised anew the board's concern about how its signals would affect transmissions from other, smaller wireless stations. Marconi by now had devised a means of tuning transmissions, for which he had received British patent no. 7777, often referred to as his "four sevens" patent. But the technology was fallible, as Fleming and Marconi well knew. In fact, they were sufficiently concerned that Marconi ordered George Kemp to build a second, far smaller station six miles away on a stretch of coast known as the Lizard, to gauge whatever interference might occur and to provide a receiver for trial messages once the new station began operation. Here Kemp directed the construction of an antenna consisting of three ships' masts secured end to end and stayed against the wind, rising to a height of 161 feet.

No reader of *The Times* would have guessed Fleming's concern, however, from reading his latest letter to the editor, published October 4, 1900, in which he praised a recent series of experiments that he claimed demonstrated Marconi's ability to tune transmissions to avoid interference.

Interestingly, Fleming at no point identified himself in the letter as Marconi's scientific adviser. He described how operators had sent messages simultaneously and how they had been captured on two receiving antennas "without delay or mistake."

"But greater wonders followed," Fleming wrote—at which point Oliver Lodge, reading *The Times* as he always did, must have spat his morning coffee onto the floor.

Fleming reported that the operators sent another round of simultaneous messages, one in English, one in French. This time both messages were received on a single antenna. Fleming gushed, "When it is realized that these visible dots and dashes are the results of trains of intermingled electric waves rushing with the speed of light across the intervening 30 miles, caught on one and the same short aerial wire and disentangled and sorted out automatically by the two machines into intelligible messages in different languages, the wonder of it all cannot but strike the mind."

Anyone reading the letter closely, however, would have seen that it presented anything but an objective, verifiable account of the experiments and instead derived its credibility entirely from the fact that its author was the great Ambrose Fleming. In effect, Fleming once again was asking the audience to trust him.

This letter—its glowful praise, its failure to note that Fleming was a paid employee—would prove costly. Not immediately, however. For now it merely kindled the curiosity, and professional skepticism, of Nevil Maskelyne, the magician.

Marconi shuttled between the Poldhu Hotel at Land's End and the Haven Hotel at Poole, though he spent most of his time at the latter. No railway ran directly from Poldhu to Poole, so Marconi had to travel first to London then catch another train south. This left him a lot of time for thinking and not a lot for his American beauty, Josephine Holman.

The engagement was still secret, and with nearly all Marconi's time consumed by travel and work, Holman must at times have wondered whether it was real or an artifact of imagination.

They wrote letters and sent telegrams. Marconi knew the news of their engagement would upset his mother, but he seemed not to realize that the longer he kept the engagement a secret, the more likely she was to feel hurt at his not sharing so important a part of his life. Annie Jameson had been his earliest and strongest ally, and she believed herself still to be his protector even in small things. Though he by now had turned twenty-six and was wealthy and famous the world over, she doted on him as if he were still a boy sequestered in his attic laboratory. She stayed often at the Haven Hotel. In one letter to him, written from there, she wrote, "After you left this morning I found you had not taken your rug with you. . . . I sent it to you at 3 o'clock today and hope you will get it all right by tomorrow." She urged him to keep "plenty of blankets" on his bed. "I have put all your things as tidy as possible in your room, and the key to your wardrobe I have put in one of the little drawers of your looking glass on the dressing table, but indeed there is little use in locking the wardrobe for all the keys are the same."

Later, from Bologna, she wrote, "I am thinking if it has got warmer at the Haven Hotel you will want your lighter flannels. Mrs. Woodward has the keys of your boxes. Your flannels are in the box with the two trays. Summer sleeping suits on the first tray. Summer vests under the two trays. Summer suit, jacket, waistcoat and trousers in the wardrobe (side of window)."

In London Ambrose Fleming awoke to the fact that he had taken on something far more involved and consuming than he had expected when he agreed to become scientific adviser. In a letter to Flood Page he complained that the company was making "extreme demands on my time" and cited by way of example a long letter from engineer Richard Vyvyan "which will take several hours to answer." His pay, he complained, was "in no way adequate."

He wrote, "I am willing to do this work on a scale of payment proportional to the responsibility. You are engaged in a gigantic experiment at Cornwell which if successful would revolutionize ocean telegraphy."

For him to continue, he wrote, his pay would have to be increased to £500 a year—more than $50,000 today. Further, he needed a promise of additional reward "if my work and inventions are of material assistance in getting across the Atlantic."

One week later, on December 1, 1900, Flood Page wrote back to notify Fleming that the directors had approved the increase. He added, however, that the board wanted assurance that Fleming understood a crucial point.

"I am desired to say," Flood Page wrote, "that while they recognize fully the great assistance you have given to Mr. Marconi with reference to the Cornwall Station, yet they cannot help feeling that if we get across the Atlantic, the main credit will be and must be Mr. Marconi's. As to any recognition in the future in the event of our getting successfully across the Atlantic, I do not think you will have cause to regret it, if you leave yourself in the hands of the Directors."

That Fleming truly understood the point—understood the lengths to which Marconi and his company would go to train the spotlight on Marconi alone—is doubtful. Fleming's roots lay in the loam of British academic science and in the British ideal of fair play. In his acceptance note, which he mailed to Flood Page two days later, Fleming wrote, "As regards any special recognition in the event of my services assisting in the accomplishment of transatlantic wireless telegraphy I can confidently leave this to be considered when the time arrives, assured that I shall meet with generous treatment."

As if on cue, the company now changed its name, from the Wireless Telegraph and Signal Co. to *Marconi's* Wireless Telegraph Co., though the name change would not become official until February.

SPEED WAS ESSENTIAL. Each week's issue of *The Electrician* brought some new and disturbing evidence of a groundswell of competition. Experiments involving wireless were occurring around the world, and in Britain there were troubling developments.

The Royal Navy installed thirty-one of its thirty-two new Marconi sets, but it shipped the last to an electrical equipment company where engineers, without authorization from Marconi, built fifty duplicates for the navy's use.

In December Nevil Maskelyne conducted tests in the Thames Estuary with his own wireless apparatus. The distance wasn't great—a few miles—but the customer who arranged the tests was impressive indeed, Col. Henry Montague Hozier, secretary of Lloyd's of London, a post he had held since 1874. He was the Lloyd's official who in 1898 had invited Marconi to conduct experiments on Rathlin Island, which, despite their success, failed to generate a Lloyd's contract. Now Hozier and Maskelyne formed an independent syndicate to develop and market Maskelyne's technology.

And there was Lodge: He continued to experiment with wireless and talked with his friend Muirhead, the instrument-maker, about possibly forming a new company to market the system. Happily for Marconi, however, Lodge became distracted once again. In 1900 he was appointed principal—the equivalent of president—of Birmingham University.

He accepted the position only after receiving assurance that he would be allowed to continue his investigations of the paranormal.

The End of the World

CRIPPEN BECAME CAUGHT UP in the social web of the Ladies' Guild. He attended parties thrown by music hall artists, visited their clubs, and when Belle felt it necessary to have a companion, dined out with other guild members and their husbands.

One afternoon Seymour Hicks, the performer and memoirist, encountered Crippen at London's Vaudeville Club. An acquaintance of Hicks's introduced them, and they spent half an hour together, over cocktails. Hicks knew something of Crippen and of the dynamics of his marriage to Belle. It was hard to understand—so large and robust a woman, exuding energy from every pore, coupled with so mild and self-erasing a man as Crippen.

"The most noticeable thing about him was his eyes," Hicks wrote. "They bulged considerably and appeared to be closely related to some kind of ophthalmic goiter. Added to this, as they were weak and watery he was obliged to wear spectacles with lenses of more than ordinary thickness, which so magnified his pupils that in looking at him I was by no means sure I was not talking to a bream or mullet or some other open-eyed and equally intelligent deep sea fish. He spoke with a slight American accent."

On this occasion Hicks's acquaintance complained of a toothache, and Crippen immediately handed him one of Munyon's remedies, "assuring him that it would instantly relieve the acute toothache from which he was suffering."

Hicks wrote, "There is no doubt that as the years rolled on, the home life of this little peddler of patent medicines must have been anything but a rest cure, and one for which even Dr. Munyon himself could not have found a remedy."

Hicks felt sympathy for Crippen. Looking back from the darkling year 1939, knowing by then all that had come to pass, Hicks mused, "Miserably unhappy, he would not have been human if he had not sought consolation elsewhere."

SO INTENSE WAS THE WORK and the fear of competitors—of Tesla, Lodge, Slaby, and now Maskelyne—that Marconi and his men seemed unaware of the passing of the nineteenth century, the age of Victoria. Out in the world beyond the windblown cliffs of Cornwall and the snug Christmas hearths of the Poldhu and Haven hotels, the first shadows of a long melancholy dusk had begun to gather.

In 1898, spurred by Admiral Alfred von Tirpitz and by widespread ill will among the German populace toward England, the German Reichstag passed the First Navy Law, which called for the production of seven new battleships. Two years later, in June 1900, much to the alarm of the British Admiralty, the Reichstag went further and passed its Second Navy Law, which doubled the number of battleships in the German navy and set in motion a cascade of events that over the next decade and a half would align the world for war.

There was wide agreement that some kind of war in Europe was inevitable, although no one could say when or between which nations; but there also was agreement that advances in science and in the power of weapons and ships would make the war mercifully short. The carnage would be too great, too vast, too sudden for the warring parties to endure. One voice dissented. In 1900 Ivan S. Bloch wrote, "At first there will be increased slaughter—increased slaughter on so terrible a scale as to render it impossible to get troops to push the battle to a decisive issue." At the onset of hostilities armies would try to fight under the old rules of warfare but quickly would find them no longer applicable. "The war, instead of being a hand-to-hand contest in which the combatants measure their physical and moral superiority, will become a kind of stalemate, in which neither army being able to get at the other, both armies will be maintained in opposition to each other, threatening each other, but never able to deliver a final and decisive attack."

They would dig in and hold their ground. "It will be a great war of entrenchments. The spade will be as indispensable to a soldier as his rifle."

In January 1901 something greater than fear of war settled over the nation. Twenty-three days into the new century, Queen Victoria died. Britain was cast into literal shadow as men and women donned black and thick black lines appeared along the edges of each page of *The Times*. A sense of dread shaded life. Henry James wrote, "I mourn the safe and motherly old middle-class queen, who held the nation warm under the fold of her big, hideous Scotch-plaid shawl and whose duration had been so extraordinarily convenient and beneficent. I felt her death much more than I should have expected; she was a sustaining symbol—the wild waters are upon us now."

Her son Edward, the Prince of Wales, would be king. James called him "Edward the Caresser" and feared his impending accession was "the worst omen for the dignity of things." In marked contrast to the old queen, the king-to-be was affable, indulgent, even funny. As Victoria lay dying, someone asked, not intending an answer, "I wonder if she will be happy in heaven?"

To which Edward replied, "I don't know. She will have to walk *behind* the angels—and she won't like that."

Marconi and his men mourned the queen but did not let her death interrupt their work. Kemp's diary makes no reference to her passing. On January 23, 1901, the day after her death, Marconi achieved his greatest distance yet, registered when the new test station at the Lizard on its first day of operation received messages sent from the Isle of Wight, 186 miles away.

The transatlantic station at Poldhu was well into the first phase of construction, and now Marconi turned to the matter of where to build its twin. He examined a map of the United States and began planning his second voyage to America.

Part III

Secrets

Marconi and associates launch a kite in Newfoundland.

Miss Le Neve

In 1901, at the age of seventeen, Ethel Clara Le Neve became an employee of the Drouet Institute for the Deaf, in Regent's Park, London, and soon afterward began working for another newcomer to the firm, Dr. Hawley Harvey Crippen.

Though its lofty name suggested otherwise, in fact the Drouet Institute was a seller of patent medicines, one of the wealthiest and most famous of the species. In time Britain's House of Commons Select Committee on Patent Medicines would expose the fraudulent and harmful practices of the industry and of Drouet in particular, but for now the company operated without scrutiny, its lavish offices reflecting the wealth accumulated in each day's harvest of mail. Drouet produced what it claimed was a cure for deafness, and its pitch was sufficiently compelling that an estimated one out of ten of Britain's deaf population bought one of its products.

Le Neve's true family name was Neave, but she had taken the name her father, once a singer, had used as a stage name. She was slender, about five feet five inches tall, and had full lips and large gray eyes. Her face formed a soft, pale V in which her cheekbones were clearly delineated, without appearing spare or gaunt. For the time, which tended to favor rounder faces and lush corseted bodies, her look was unusual but undeniably alluring. Her childhood peers would have been surprised at how she turned out. As a young girl she had prided herself on being a tomboy. "For dolls or other girlish toys I had no longing," she wrote. She loved climbing trees, playing marbles, and shooting her slingshot. "At that time my chief companion was my uncle, who was on the railway," she recalled. "Nothing delighted him more than to take me to see the trains, and even to this day." Even as an adult, she said, "there are few things which interest me more than an engine."

When she was seven years old, her family moved to London. She completed her schooling and resolved to make her own living. A family friend taught her and her older sister, Adine or more commonly Nina, how to type and take stenographic notes. Her sister achieved proficiency first and set out to find a job. The Drouet Institute hired her, and soon afterward Ethel joined the company, also as a stenographer and typist. "Very soon afterwards came Dr. Crippen, who was fated to influence my life so strangely."

Crippen came to the Drouet Institute when his previous employer, the Sovereign Remedy Co., went out of business. Drouet hired him to be a consulting physician, and in that capacity he soon encountered Ethel and her sister. "For some reason the doctor took kindly to us," Ethel wrote, "and almost from the first we were good friends. But really he was very considerate to everybody."

Nina became Crippen's private secretary, but Ethel too got to know the doctor. "I quickly discovered that Dr. Crippen was leading a somewhat isolated life. I did not know whether he was married or not. Certainly he never spoke about his wife."

Crippen and the sisters often took afternoon tea together. On one occasion, as Ethel and Nina prepared the tea and laid the service, a friend of Crippen's happened to come by the office. Seeing the preparations under way, the man sighed, "I wish I had someone to make tea for me."

With what Ethel termed his "customary geniality," Crippen urged the visitor to stay and join them. He did so, and during the conversation that followed, Ethel recalled, "mention was made of the doctor's wife." The sisters received the news in silence, though they found it both startling and intriguing. They said nothing to elicit further details.

At length the visitor left. After Ethel and Nina cleaned up the remains of the tea, Nina went to Crippen and asked if what the guest had said was true—that he really was married.

Crippen said only, "It would take the lawyers all their time to find out."

NINA BECAME ENGAGED, and as her wedding neared, she left her job at Drouet. Now Ethel became Crippen's private secretary. She missed her sister. "With her departure I felt very lonely," she recalled. "Dr. Crippen, too, was very lonely, and our friendship deepened almost inevitably. He used to come to see me at home. All this time his wife was shrouded in mystery."

One day a woman came to the office. She was large and energetic and had hair that clearly was dyed an amber blond. She wore a lot of jewelry and a dress that must have been expensive but was more flamboyant and gaudy than anything Ethel herself would have considered tasteful.

"Her coming was of a somewhat stormy character. I was leaving the office for lunch when I saw a woman come out of the doctor's room and bang the door behind her. She was obviously very angry about something."

Ethel turned to another employee, William Long, and whispered, "Who is that?"

"Don't you know?" he asked. "That's Mrs. Crippen."

"Oh," she said, startled. "Is it?"

Ethel needed a moment or two to absorb this revelation. Here was Crippen, so kind and soft-spoken, small in every way—one inch shorter, in fact, than she herself—married to this thunderhead of silk and diamond.

"After that," Ethel wrote, "I quickly realized Dr. Crippen's reluctance to speak about his wife."

AN EVEN STORMIER VISIT followed—a visit, Ethel wrote, "which might have ended tragically."

Belle was again in a fury and burst into the office in a cyclone of cloth and abraded corset. "There were more angry words, and just before she left I saw the doctor suddenly fall off his chair."

Belle roared out, slamming doors. Ethel ran to Crippen. "He was very ill, and I believed that he had taken poison. He told me that he could bear the ill-treatment of his wife no longer."

She found brandy and deployed it to revive him. Afterward, she wrote, "we did our best to forget the painful incident." But the violence of the encounter and Crippen's expression of such deep unhappiness caused a fundamental change in their relationship. Ethel wrote, "I think it was this, more than anything else, which served to draw us closer together."

SOON THE DROUET Institute also failed, helped along by a coroner's inquest that identified one of Drouet's cures—ear plasters—as a possible exacerbating factor in the death of a man whose ear infection spread to his brain with devastating consequences. Suddenly Drouet's advertising disappeared from the city's horse-drawn omnibuses. Though tolerance of patent medicine companies was beginning to wane, many companies continued to operate, and Crippen quickly found a new job as "Consulting Specialist" for the Aural Remedies Co., another firm that specialized in cures for deafness, though on his letterhead the only credential Crippen listed was, paradoxically, his degree in ophthalmology from New York.

The offices of Aural Remedies were in New Oxford Street, completed in 1847. It was a fitting location because the street had been built with the intent of eliminating one of the most crime-ridden parts of London, the Rookery, home previously to confidence men, pickpockets, and thieves. The construction cleared the neighborhood's worst precincts and triggered a lasting reformation, so that now only high-priced frauds such as Aural Remedies could afford the rents. Crippen brought with him the expertise he had gained at Drouet. He also brought Ethel, as his secretary.

In one letter, probably typed by Ethel, Crippen wrote to a reluctant customer about a special offer. "This places within your reach the possibility of being speedily . . . cured, and I hardly need point out that I could scarcely make such an offer, were I not convinced of the efficacy of my Treatment."

He proposed that the customer send him half the price quoted in an earlier letter, at which point Crippen would send him the treatment—"the complete Outfit"—on a trial basis. If after three weeks or so the treatment didn't accomplish anything, he wrote, the man would not have to pay another penny. "On the other hand, if you feel you have been benefited by the treatment, you can then remit the balance of the purchase price, namely 10s 6d." (Until 1971 British currency was configured in pounds, shillings, and pence. One pound equaled twenty shillings, written as 20 s., which in turn equaled 240 pence, or 240 d. A new pound is equal to 100 pennies, with one penny equal to 2.4 of the obsolete pence.)

Though the letter might seem to indicate otherwise, Aural Remedies was not going to take a loss if the customer never made the second payment. Patent medicines cost almost nothing to produce. Even the reduced price Crippen now offered would have yielded a substantial profit regardless of whether the customer paid another pence. The key point is that Crippen was *not* offering to return the initial payment.

In time Aural Remedies and Crippen would be identified by the muckraking magazine *Truth* in its "Cautionary List" of companies to avoid.

WITHIN THE CRIPPEN HOUSEHOLD the weather did not improve. They moved to another address on Store Street, No. 37, but this new apartment did not offer enough additional space to allow them to stay out of each other's way. They still had to sleep in the same bedroom. They could not afford anything larger, at least not in Bloomsbury. Crippen was only earning a fraction of the salary Munyon's had paid. Nonetheless he continued to allow Belle to spend heavily on clothing and jewelry. Crippen said, "although we apparently lived very happily together, as a matter of fact there were very frequent occasions when she got into most violent tempers, and often threatened she would leave me, saying she had a man she could go to, and she would end it all."

It was clear to Crippen that the man in question was Bruce Miller. In early April Miller came by the apartment for what would prove to be the last time. He wanted to say good-bye to Belle. He told her he was taking her advice and returning to Chicago to reunite with his wife. He sailed from England on April 21, 1904.

If Miller's departure reignited in Crippen any hope that his own marriage could now be restored, he immediately found those hopes dashed. Belle's temper worsened, and so too did the couple's financial situation, though he still made no effort to curb her expenditures. He began looking for another home that would be much larger but also cheaper, which meant necessarily that he would have to look outside the core of the city, at grave risk of annoying Belle even further.

The emotional landscape negotiated each day by the Crippens grew still more contorted. On her brief visits to Crippen's office, Belle had taken note of his typist, Ethel Le Neve. She was young and striking and slender. Her looks alone may have made Belle uneasy, or Belle may have sensed an unusual degree of warmth in the way Crippen and the young woman behaved toward each other, but there was indeed something about the typist that made Belle uneasy.

One morning a friend of Belle's named Maud Burroughs, who lived in the same building on Store Street, stopped by as Belle was getting dressed. In the course of their conversation, Belle mentioned her past surgery and asked Burroughs if she would like to see the scar.

Burroughs said no.

"Give me your hand," Belle said, "and you can feel where it was."

Belle took Burroughs's hand and, as Burroughs recalled, "placed it underneath her clothing upon her stomach. I felt what seemed to me to be a hole, so far as I remember, a little on one side of the lower part of her stomach."

The conversation shifted to Crippen, who by now, for reasons unclear, had taken to calling himself Peter. It was by this name that Belle and her friends addressed him.

Belle said, "I don't like the girl typist Peter has in his office."

"Why don't you ask Peter to get rid of her then?" Burroughs asked.

Belle replied that she already had asked him, but Crippen had told her the typist was "indispensable" to the company.

CRIPPEN'S RELATIONSHIP WITH ETHEL deepened. Later he would recall a particular Sunday in the summer of 1904 when "we had a whole day together, which meant so much to us then. A rainy day indeed, but how happy we were together, with all sunshine in our hearts." He recalled it as a time when he and she were in "perfect harmony with each other. Even without being wedded."

For her part Ethel came to view Crippen as "the only person in the world to whom I could go for help or comfort. There was a real love between us."

It was at about this point that Ethel, "by sheer accident," came across the letters Bruce Miller had sent to Belle. "This, I need hardly say, relieved me somewhat of any misgivings I had with regard to my relations with her husband."

WITHIN SIX MONTHS Aural Remedies also failed, and Crippen went back to work for Munyon's, this time out of a new location in a building called Albion House, also on New Oxford Street. Again he brought Ethel but also another past employee, William Long. Crippen returned not as a full-time employee but rather as an agent paid by commission. He made less than he had hoped. Finding a cheaper place to live became imperative, but here a challenge had no easy resolution: to find lodging that was not only less expensive but also much bigger and nice enough to keep Belle happy, or if not happy—which at this point must have seemed an impossible goal—at least to stop her behavior from degrading further. These clashing imperatives drove his search farther and farther from Bloomsbury.

"The Thunder Factory"

The hunt for land on which to build his first American station lasted far longer than Marconi had planned. Accompanied by Richard Vyvyan and an employee named John Bottomley, a nephew of Lord Kelvin, Marconi toured the coasts of New York, Connecticut, Rhode Island, and Massachusetts, taking trains as far as possible, then proceeding in wagons or on foot. Marconi before leaving Britain had appointed Vyvyan to build and run the new station.

Every piece of land seemed to harbor a critical flaw. No drinking water, no nearby town to supply labor and supplies, no rail line, and—a flaw that particularly rankled Marconi—no fine hotels of the kind that were common on the windswept coasts of Britain.

In February 1901 the group made its way to Cape Cod, landing at Provincetown. On a map the cape looked appealing, especially its midpoint where it hooked northward and the land rose to form oceanside cliffs over one hundred feet high.

In Provincetown Marconi hired a guide named Ed Cook, who was said to have a thorough knowledge of the cape's coastal lands. Cook did indeed know the beaches well. He was a "wrecker" who salvaged ships and boats wrecked off the cape on their way to Boston. In the previous century Henry David Thoreau had toured the cape and in his book *Cape Cod* described how wreckers descended on the remains of one ship, the *St. John,* even as grieving relatives came to the beach in search of lost loved ones. Cook used the profits of his salvage work to buy land.

Cook led Marconi the full length of the cape, traveling in Cook's wagon exposed to the frigid winds of February. When Cook led Marconi to the Highland Light, near the north end of the cape, opposite North Truro, Marconi believed he had found exactly the location he needed. The lighthouse stood atop a 125-foot cliff overlooking the shipping lanes that led to Boston harbor, which lay about fifty miles to the northwest. Here the keepers watched for inbound ships and consulted guidebooks to identify them, then telegraphed the news to the ships' owners in America and abroad, the latter messages traveling first by land line, then by cables laid under the Atlantic.

But the operators of the Highland Light did not trust Marconi. "They thought he was probably a charlatan," his daughter Degna wrote, "and they *knew* he was a foreigner. Not even Ed Cook was able to override their thorny New England resistance to strangers and new-fangled contraptions." They refused access.

Next Cook led him a few miles south to a parcel of land just outside South Wellfleet, consisting of eight acres atop a 130-foot cliff that overlooked the same beach along which Thoreau had walked half a century earlier. Buffeted by wind, now Marconi walked the ground. The land in all directions had been shaved to a stubble by gales and by loggers who over the previous century had stripped it to provide lumber for shipbuilders. Marconi knew he would have to import the tall masts necessary to hold his aerial aloft.

He liked this clifftop parcel. If he stood facing east, all he saw was the great spread of the Atlantic. As Thoreau observed, "There was nothing but that savage ocean between us and Europe."

When he faced the opposite direction, he saw the harbor at Wellfleet in clear view and very near. A railroad passed less than a mile away, and the nearest telegraph office, at Wellfleet Depot, was only four miles off. This meant lumber and machinery could be delivered to Wellfleet by ship or rail and hauled with relative ease overland to the cliff. A company report on Marconi's search states, "Plenty of water is available on the site and a very bad inn is situated about 3 miles away; there is, however, a residential house which we can rent on very moderate terms within 200 yards of the site." One bit of historical resonance was lost on Marconi.

During the eighteenth century Wellfleet had been named Poole, after a village in England—the same Poole whose Haven Hotel now served as Marconi's field headquarters.

Cook assured Marconi there would be no problem persuading the landowner to let Marconi build here. The landowner was Cook himself. He had acquired the land using the proceeds of his work as a wrecker. Whether either man recognized the paradox therein is unclear, but here was Marconi, whose technology promised to make the sea safer, acquiring land from a man who had made his living harvesting precisely the wrecks Marconi hoped to eliminate. In the future these eight acres of seaside land would be some of the most coveted terrain in the world, but at this time the stretch was considered worthless. Marconi bought it for next to nothing.

Marconi also hired Cook to be his general contractor, with a mandate to find workers, arrange living quarters and food supplies, and acquire necessary building materials. Marconi and his men took their initial meals at the nearby inn, but the food was so awful that he vowed never to eat there again. He arranged to have more elegant fare, and the wines to go with it, shipped from Boston and New York. Among the locals this caused a good deal of frowning and saddled Marconi with a lasting reputation as a culinary aesthete.

Soon Marconi headed back to England, leaving Vyvyan to face the true nature of the location.

IN LONDON COLONEL HOZIER of Lloyd's and Nevil Maskelyne of the Egyptian Hall, acting together as a syndicate, approached Marconi and offered to sell him Maskelyne's patents and apparatus. Marconi listened. As negotiations proceeded, Hozier somehow cut Maskelyne out of the syndicate and began negotiating on his own behalf, despite the fact that it was Maskelyne's technology upon which the syndicate was based. Hozier wanted £3,000—over $300,000 today—and a seat on Marconi's board. To make the arrangement more palatable, even irresistible,

Hozier promised that in return he would broker a deal between Marconi and Lloyd's itself.

Hozier's maneuver left Maskelyne embittered. But his anger, for the moment, seemed of no consequence.

THE LANDSCAPE THAT NOW confronted Vyvyan was lovely but spare. There were few trees, none tall enough to be worthy of the name, let alone to be useful for building houses or ships. Most of the surrounding flora hugged the ground. Hog cranberry coated the sand, tufted here and there by beach heather, also called "poverty grass," a name that captured the overall austerity of the terrain. There was crowberry, savory-leaved aster, mouse-ear, and goldenrod, as well as pitch pine planted during the previous century to keep wind-driven sand from overwhelming towns on the bay side of the cape. Everywhere the wind caught stalks of American beach grass and bent them until their tips scraped the sand, engraving precise circles and earning them the nickname "compass grass." Thoreau wrote, "The barren aspect of the land would hardly be believed if described."

Clouds often filled the sky. The Weather Bureau's Nantucket station, nearest the cape, reported for 1901 only 83 clear days, 101 days identified as partly cloudy, and 181 days where clouds reigned. On such days all color left the world. Sky, sea, and ground became as gray as shale, the color blue a memory. Frequent gales brought winds of fifty or sixty miles an hour and shot snow off the cliff edge in angry spirals. The boom of the sea paced the day like the tick of a gigantic clock.

The plan for the station called for the construction of living quarters for the staff, a boiler room to produce steam to generate electricity, a separate room full of equipment to concentrate power and produce a spark, and another room in which an operator would hammer out messages in Morse code. The most important structure was the aerial, and that was what most worried Vyvyan. In London Marconi had shown Vyvyan plans for a new antenna array to be built at Poldhu, and he ordered Vyvyan to build the same one in South Wellfleet. As soon as Vyvyan saw the plans, he grew concerned. He would have to erect twenty masts similar in design to the masts of sailing ships, complete with top gallants, royals, and yards. The finished masts would rise to 200 feet and stand in a circle about 200 feet in diameter, a Stonehenge of timber. The height of the masts, plus the 130-foot height of the cliff, would give Marconi's antenna an effective height of well over 300 feet, thus in theory—Marconi's theory—increasing the station's ability to send and receive signals over long distances. A complex series of guy wires and connectors was supposed to keep the masts from toppling. The masts in turn would support an aerial of wire. A heavy cable of twisted copper would connect the tops of all the masts, and from it would be strung hundreds of thinner wires, all converging to form a giant cone with its tip over the transmission building. A cable run through the roof would link the cone to the spark generator within.

What troubled Vyvyan most was the rigging. Each mast should have had its own array of guy wires, so that if one mast failed the others would remain standing. Instead, the top of each was connected to the tops of its neighbors with "triatic stays." Vyvyan realized that if one mast collapsed, these connections would cause the rest to fall as well. He told Marconi of his concerns, but Marconi overruled him and commanded that the station be built as designed. Vyvyan accepted his decision. "It was clear to me, however, that the mast system was distinctly unsafe."

Construction advanced slowly, hampered by what the Weather Bureau called a "period of exceptionally severe storms." April brought gales that scoured the coast with winds up to fifty-four miles an hour. May brought rainfall in quantities that broke all records for New England.

The men hired by Ed Cook lived in Wellfleet and adjacent communities, but Vyvyan, Bottomley, and the full-time Marconi employees lived on the grounds in a one-story residence with about two hundred feet of living space, a level of coziness that eventually prompted the station's chief engineer, W. W. Bradfield, to plead for an additional wing containing more sleeping space and a recreation room. He wrote, "In view of the isolation of the station, I regard it as almost necessary that this should be done in order that the men may be comfortable, contented, and that their best work may be got out of them."

The men did what they could to improve their living conditions. They dined on a table draped in white and spined with four candles jutting at odd angles from improvised holders. They read books, played the station's piano, and sang, and from time to time they hiked to the bay side to pick oysters at the mouth of Blackfish Creek, named for the herds of small whales called social whales that locals once drove onto the beach and butchered for oil. They went beachcombing, the sands below the Truro highlands being a lot more interesting in those days given the frequency of shipwrecks. One never knew what treasure might turn up, including crockery, luggage, fine soaps from a ship's cabin, and the occasional corpse, its cavities filled with sand. Thoreau called the beach "a vast *morgue*" for all the dead men and creatures the sea discharged. "There is naked Nature—inhumanly sincere, wasting no thought on man, nibbling at the cliffy shore where gulls wheel amid the spray."

This being the age of hired help, the station had a cook and employed two Wellfleet women who came each day to clean. They wore maid's caps and aprons. One of the women was Mable Tubman, daughter of a prominent Wellfleet resident, who caught the attention of one of Marconi's men, Carl Taylor. A photograph from the time shows Carl and Mable seated on the beach on a day bright with sun. What makes this photograph unusual is that it captures people actually having fun. Mable is wearing her apron and maid's cap and is turned away from the camera, watching the sea. Carl is wearing a light-colored suit and looks into the camera, a huge grin running from one earlobe to the other. He also is wearing a maid's cap.

THOUGH DEEPLY DISTRACTED by the work under way on both sides of the Atlantic and by myriad other developments, Marconi apparently believed that things were sufficiently under control that he and Josephine Holman could at last announce their engagement—though the major pressure to do so likely came not from him but from Josephine, who was growing increasingly concerned about just where she stood relative to his work. He still had not come to visit her family in Indianapolis.

Marconi's mother, Annie, had concerns about her own status in Marconi's life, now that he planned to marry. "To lose him to anyone, rich or poor, on his first flight from home was hard," Degna Marconi wrote years later.

Marconi's mother did her best to behave "self-effacingly," but she felt aggrieved when Josephine failed to write to her, and she complained to Marconi. Soon afterwards a letter did arrive, which Annie described as "very kind and sweet."

Now Annie wrote to Marconi, "I wish I had got this letter [from Josephine] before and I should not have said anything to you about her not writing. Now it is all right and I feel much happier and shall write to her soon." She added an odd line to this letter: "Our friends here think it is only right I should be at your wedding"—as if there had been serious consideration that she would not be present.

Spring brought successes and trials. On May 21, 1901, Oliver Lodge won a U.S. patent for "electrical telegraphy," and he and William Preece became de facto allies, more and more outspoken in their criticism of Marconi. Lodge also opened an attack on another front. He formed a new company with his friend Alexander Muirhead, the Lodge-Muirhead Syndicate, to begin selling Lodge's technology.

And Marconi endured a very public failure. He had agreed again to provide coverage by wireless of the America's Cup yacht races, now for the Associated Press, but this time he faced competition from two fledging American companies. The transmissions interfered with one another so much that Marconi could not send messages to his shore station for relay to the Associated Press—despite his claims to have perfected the technology for tuning transmissions to avoid interference. Afterward allegations arose that one of the competing companies had sought deliberately to create interference by transmitting exceptionally long dashes and, at one point, placing a weight on their transmission key and leaving it there, creating what one observer called "the longest dash ever sent by wireless."

In another realm, however, Marconi made progress. On May 21, 1901, the first British ship equipped with wireless, the *Lake Champlain,* departed Liverpool on a transatlantic voyage. At the same time Marconi's men also were installing his apparatus aboard Cunard's *Lucania.* During the *Lake Champlain*'s return voyage its Marconi operator got a surprise—a message from the *Lucania* at midocean. To mariners resigned to the isolation of the sea, the feat seemed a miracle.

Only years later would anyone take note of how strange it was that the second officer of the *Lake Champlain* at this pioneering moment was a young sailor by the name of Henry Kendall.

WITH WARMER WEATHER the work on Cape Cod proceeded quickly, though Vyvyan and Bottomley discovered that unlike Poldhu, where temperatures in summer remained cool, even cold, the cape often registered some of the hottest weather in New England, with temperatures in the nineties accompanied by wet-blanket humidity. At night thunderstorms arose often, shedding lightning that gave the terrain the pallor of a corpse. Fog would settle in for days, causing the edge of the cliff to look like the edge of the material world. At regular intervals the men heard the lost-calf moan of foghorns as steamships waited offshore for clarity.

As each mast rose, Vyvyan's concerns increased. Winds routinely blew at twenty to thirty miles an hour, sometimes more. By mid-June his workers had erected seventeen lower masts out of twenty. Fourteen of them now had top masts, and ten had a third stage, the top gallants. The plan called for each also to get a fourth stage, the royal masts, hair-raising work for the men who had to scale the masts and secure each portion to the next. A photograph shows these men, called riggers, at work—tiny figures alone at the tops of masts two hundred feet tall that swayed in even the slightest breeze.

By the end of the month the boiler house, generation equipment, and transmitter were in place and the circle of masts was complete. Photographs show a grove of two-hundred-foot masts linked and steadied with guy wires having all the substance of cobwebs draped on a candlestick.

Vyvyan tested the transmitter. At night the spark gap lit the sky with such intensity that it was visible and audible four miles down the beach. Up close it was deafening, like the crack of a starter pistol repeated over and over. An early employee, James Wilson, recalled, "If you opened the door and stepped out you had to hold your ears."

At times the wires of the antenna shimmered a cold blue. To keep electricity from flowing through the guy wires and distorting the signal, the men installed "deadeyes" of a very hard wood called lignum vitae at intervals along each wire. Shipwrights used dead eyes as connectors in rigging, but Marconi's men used them as insulators to break up potential paths for current. Still, current traveled to unexpected places. Through induction it charged drainpipes and stove flues. Even something as prosaic as hanging the wash became an electric experience. Mrs. Higgins, the station cook, reported feeling myriad electric shocks as she pinned clothing to the line.

August brought heat and fog, with Provincetown again recording the highest temperatures in New England, ninety-two degrees on August 12 and August 18, but storms were few and the winds reasonable, at no time exceeding thirty miles an hour. And yet, Vyvyan wrote, "In August, under the influence of nothing more than a stiff breeze, the heads of the masts on the windward side bent over to a dangerous degree."

The triatic stays that linked the tops of the masts ensured that when one mast swayed, they all swayed.

MARCONI MAINTAINED SECRECY. No one outside the company knew yet that he planned to try sending messages across the ocean. When an Admiralty official, G. C. Crowley, came to observe at Poldhu, Marconi enclosed the station's receiver in a box, just as he had done for his earliest demonstrations. Marconi willingly discussed his results, Crowley wrote, but would not let anyone look inside. "We used to call it 'the black box of Poldhu.'"

Marconi himself barely understood the nature of the phenomena he had marshaled, which made the process of preparing each station a matter of experiment. The huge cone-shaped aerial was the product of Marconi's instinctive sense of how Hertzian waves behaved. No established theory determined its shape. Its height reflected Marconi's conviction, partially confirmed by experimentation, that the distance a signal traveled varied in relation to the height of the antenna.

Beyond these assumptions lay an impossible array of other variables that likewise had to be resolved, any of which could affect overall performance. The subtlest of adjustments affected the nature and strength of the signal. Fleming, Marconi's scientific advisor, found that something as simple as polishing the metal balls of the spark gap greatly improved signal clarity. It was like playing chess with pieces ungoverned by rules, where a pawn might prove to be a queen for one turn, a knight for the next.

To make matters more difficult, the thing Marconi was trying to harness was invisible, and no means yet existed for measuring it. No one could say for sure even how Hertzian waves traveled or through what medium. Like Fleming and Lodge and other established physicists, Marconi believed that electromagnetic waves traveled through the ether, even though no one had been able to prove that this mysterious medium even existed.

Marconi and Fleming tried everything they could to boost the power and efficiency of the transmitters at Poldhu and Wellfleet, at times with surprising effect. As power increased, the ambient current became harder and harder to manage. At Poldhu the gutters on nearby homes sparked, and blue lightning suffused the Cornish mists. On August 9,

1901, George Kemp in Poldhu wrote in his diary, "We had an electric phenomenon—it was like a terrific clap of thunder over the top of the masts when every stay sparked to earth in spite of the insulated breaks. This caused the horses to stampede and the men to leave the ten acre enclosure in great haste."

The Poldhu station still was not finished when a period of extremely foul weather arrived and slowed things further. The masts were up, but high winds made it impossible for riggers to reach the mastheads. Squalls raked the cliff. On August 14, 1901, Kemp wrote, "The weather is still boisterous. The men could not work outside today."

The squalls and winds continued without ease for a full month, forcing Kemp to send workers home.

Signs went up beside the condensers that boosted the station's electrical power: "Caution. Very Dangerous. Stand Clear." At night the eruption of sparks could be seen for miles along the coast, followed by the crack of artificial thunder. One witness would call the Poldhu station "the thunder factory."

IN LONDON THERE WAS good news: Marconi's negotiations with Colonel Hozier of Lloyd's now yielded fruit. Hozier had given up trying to sell patents and technology to Marconi after realizing that the only aspect of the talks that interested him was the possibility of a contract with Lloyd's itself.

Hozier negotiated on behalf of Lloyd's but also on his own behalf, as a private individual, and on September 26, 1901, he struck a deal with Marconi. Hozier got his seat on the board and personally received £4,500 in cash and stock, half a million dollars today. Marconi got the right to build ten stations for Lloyd's, and Lloyd's agreed to use no other brand of wireless equipment for fourteen years.

The most important clause stipulated that the ten stations would be allowed to communicate only with ships carrying Marconi equipment, virtually ensuring that as shipping lines adopted wireless, they would choose Marconi's service. Shippers understood that wireless would make the process of reporting the arrivals and departures of ships to Lloyd's much safer and more efficient by eliminating the danger incurred when ships left course to venture close to shore for visual identification by Lloyd's agents. The agreement moved Marconi closer to achieving a monopoly over ship-to-shore communication, but it also fed resentment among governments, shipowners, and emerging competitors already unhappy with Marconi's policy barring his customers from communicating with any other wireless system.

Germany in particular took offense; and at the Egyptian Hall in Piccadilly, Nevil Maskelyne seethed.

Claustrophobia

Crippen's search for a new home took him into a leafy neighborhood in Kentish Town, north of London proper, on the upper edge of the Borough of Islington, where rents were lower than in Bloomsbury. Here he came across a house on a pleasant street called Hilldrop Crescent. The house seemed a solution to the challenge before him. On September 21, 1905, he signed a contract with the owner, Frederick Lown, under which he agreed to lease the place for three years for 52 pounds 10 shillings a year, about $5,500 today.

The details of the house would, in time, be of avid interest to countless millions around the world.

What the neighborhood lacked in glamour, it made up for by the fact that other people associated with the theater likewise had been drawn to it by its affordability, among them singers, acrobats, mimes, and magicians, some at the beginning of their careers, others at the end. It helped too that the house Crippen selected lay on a crescent, for the shape evoked some of London's nicest blocks and was reminiscent of the Crippens' past address in Bloomsbury. The new street, Hilldrop Crescent, formed a nearly perfect semicircle off the north side of Camden Road, the main thoroughfare in the district. The house Crippen chose was No. 39.

Trees lined the crescent and in summer presented an inviting arch of green to those passing in the trams, omnibuses, and hansoms that moved along Camden Road. The houses in Hilldrop Crescent all had the same design and were arrayed in connected pairs, every home sharing an interior wall with one neighbor but separated from the next by a narrow greensward. Residents did what they could to distinguish their homes from their attached twins, typically through gardening and the use of planters in shapes and colors that evoked Italy, Egypt, and India. Nonetheless, despite front stairwells that flared with cobalt and Tuscan rose, the neighborhood exuded an aura of failed ambition.

Like all the others on the crescent, the Crippens' house had four stories, including a basement level that, per custom, was used both for living space and for storage, with a coal cellar under the front steps and a kitchen and breakfast room toward the rear. The breakfast room was sunny and opened onto a long back garden surrounded by a brick wall.

At the front of the house a flight of steps led to a large door fitted with a knocker of substantial heft and a knob mounted at the door's center, behind which lay sitting rooms and a dining room. This was the formal entrance, reserved for princes and prime ministers, though guests of such stature never appeared on Hilldrop. When friends came for supper or cards, they typically entered through a door on the side. The next floor up had another sitting room and two bedrooms; the fourth and final level had a bathroom and three more bedrooms, one at the front, two at the back.

The walled garden behind the house was a long rectangle, in the midst of which lay two small glazed greenhouses overgrown with ivy. Behind one of the greenhouses and blocked from the view of anyone in the house stood a point where two walls intersected to form a secluded corner. Here lay a hundred crockery flower pots, empty and stacked neatly out of sight, as if someone once had planned a great garden but now had laid that dream aside.

Two decades earlier the crescent had been considered a good address, though with nothing like the prestige of a Mayfair or Belgravia. Since then it had begun a decline, which happened to be captured in one of the great works of nineteenth-century social reform.

In the 1880s a businessman named Charles Booth set out to conduct a block-by-block survey of London to determine the economic and social well-being of its residents as a means of contesting the outrageous claims of socialists that poverty in England was vast and deep. To conduct his survey, he employed a legion of investigators who gathered information by accompanying London School Board "visitors" on their rounds and Scotland Yard officers on their beats. For a time one of his investigators was Beatrice Potter (later Webb), soon to become a prominent social activist, not to be confused with Beatrix Potter, the creator of Peter Rabbit, though both women led nearly contemporaneous lives and would die in the same year, 1943, Beatrice at eighty-five, and Beatrix at seventy-seven. Booth was stunned to find that the incidence of poverty actually was greater than even the socialists had proclaimed. He found that just over 30 percent of the city's population lived below the "poverty line," a term he invented. He distilled his findings into seven groups from poor to rich and assigned a color to each, then applied these colors block by block to a map of 1889 London, thereby making the disparities in wealth vividly apparent. Hilldrop Crescent was colored red, for "Well-to-do. Middle class," Booth's second-wealthiest category.

A decade later Booth saw a need to revise his findings and again set out to tour the streets of London. At least three of Booth's walks with police officers took him into Hilldrop Crescent itself or adjacent streets. In his summary remarks for one walk, Booth wrote, "The best people are leaving." On another walk Booth followed Camden Road where it crossed the entrance to Hilldrop Crescent. Here he found "substantial houses" but "not such a good class of inhabitants as formerly." The decline continued a trend already well under way in neighborhoods just to the north. Booth wrote, "District is rapidly going down."

Various forces drove the trend, among them the increasing ease of transportation. A spreading network of suburban and subterranean railways enabled families of modest means to escape "darkest London" for distant suburbs. But the area around Hilldrop also happened to be blessed, or cursed, with the presence of three institutions unlikely to encourage housing values to soar. It is possible that Crippen persuaded himself that none of the three would have any direct effect on his life, but it is equally likely that he simply failed to notice their presence when making his choice. This state of ignorance could not have endured for long.

One of these institutions began its vast and fragrant sprawl about a minute's walk to the southeast of Crippen's house. Here lay the Metropolitan Cattle Market on Copenhagen Fields, opened in 1855 to replace the Smithfield Market where, as Charles Dickens observed in *Oliver Twist,* "the ground was covered nearly ankle deep with filth and mire" and the air was filled with a "hideous and discordant din." The new market covered thirty acres. Each year four million cattle, sheep, and pigs passed through its gates either for evisceration and dismemberment or to be sold on market days from its enclosed market stalls and "bullock lairs." There was less filth than at Smithfield, but the din was no less hideous, and when the weather was right and the market at its peak, typically Mondays—especially the Monday before Christmas, always the single busiest day—the chorus of lowing and bleating could be heard many blocks away, audible even to the residents of Hilldrop Crescent. Charles Booth found that animals were driven to the market through neighborhood streets, occasionally with comic effect. "Some go astray," he wrote. One bull remained loose for thirty-six hours. Another time a flock of sheep invaded a dress shop. "Loose pigs are about the worst to tackle," Booth noted: "they will spread so: attract a crowd in no time: make the police look ridiculous."

The market environs tended to draw a class of residents less savory than what the Crippens had encountered in Bloomsbury. "Very rough district," Booth observed, "many of the men working at the cattle market as drovers, slaughter-men, porters, &c; great deal of casual work. Some old cottage property partly in bad repair."

The two other institutions that tended to suppress the allure of Hilldrop Crescent were prisons. One was Holloway Gaol, also called City Prison, opened in 1852 to serve as the main prison for criminals accused or convicted of crimes within the City, London's financial district. Until 1902 Holloway had housed both men and women, including, briefly, Oscar Wilde, but by the time the Crippens moved to Hilldrop, it incarcerated only women—and soon would receive its first police van full of suffragettes arrested for seeking the right to vote. *Baedeker's Guide to London and its Environs* for 1900 describes the building as "rather handsome," an appraisal that could spring only from a predilection for gloom, for Holloway Gaol with its turrets, battlements, and jutting chimneys was the kind of place that stole the warmth from a sunny day.

But even Holloway's presence was benign compared to another prison, Pentonville, that fronted Caledonian Road, a short walk southeast of the Crippens' house. Its facade was no more or less dreary, but the frequency of executions conducted within imparted a black solemnity to its high, blank walls. Upon its opening in 1842, reformers called Pentonville a "model prison" in recognition of both its innovative design—four clean and bright cellblocks jutting from a central hub where guards had a ready view of every level—and its regimen, "the separate system," under which all of its initial 520 prisoners were kept in what later generations would call solitary confinement, each man alone in a cell and barred from ever speaking to his fellow inmates. The point was to compel prisoners to contemplate their behavior and—through solitary work, daily religious services, and the reading of soul-improving literature—to encourage them to shed their unhealthy behavior. In practice the separate system drove many insane and prompted a succession of suicides.

In 1902 the prison became a center for execution. This did not please the neighbors, although anyone with an appreciation of history would have seen this new role as a fitting return to the district's roots. In the 1700s an inn stood in Camden Town by the name of Mother Red Cap, a common stop for omnibuses but also the end of the line for many condemned prisoners, who were hung at a public gallows across the street. Public executions became great picnics of malice and drew increasing criticism until Parliament required that they be conducted within prison walls. By the time the Crippens arrived in Hilldrop Crescent, one remnant of prior practice remained enshrined in law and served as a persistent and depressing reminder for families living near Pentonville that they had a prison in their midst and that there were men within its walls who knew precisely the moment at which they would die. The law required that prison authorities ring the bell in the prison chapel fifteen times upon the completion of each hanging. This became wearing for neighbors and certainly for prisoners next in line for the gallows; for Belle it became another marker of her own social decline.

One immediate neighbor of the prison, Thomas Cole, grew sufficiently concerned about how the executions were affecting adjacent streets that he complained to the home secretary, a youngish man named Winston Churchill. "The Building," he wrote, "is surrounded by Houses occupied by respectable working people who find much difficulty in letting the apartments. It is also causing the Houses to remain empty and of course much loss to all concerned." He closed by asking, nicely, whether these executions couldn't take place elsewhere—in particular at a prison called Wormwood Scrubs, "which is, I believe in an isolated position?"

Cole's complaint made its way through the Home Office to the governor of Pentonville Prison, who responded promptly, in equally civil language and with empathy. After noting that indeed one of the major streets bordering the prison, Market Street, now contained several empty houses, he stated, "I do not think that the carrying out of executions here causes the difficulty in letting apartments, at least to any appreciable extent. The fact of a prison in itself being in a particular spot is bound to cause the neighborhood to be looked on as undesirable—However this agitation has been going on ever since executions were commenced in 1902."

But he argued that the prison was not the sole cause of neighborhood decay. Landlords were charging excessive rents, and "trams and tubes also tend to make the better class working people move farther away." Another problem was "the numbers of rats which I am informed infest the houses there." Decline clearly had occurred, he wrote. "I have observed that the people who reside in Market Street are not of as good a class as they were a few years ago. In fact every time places are to let they appear to be taken by people not so well off as those who lived there before them and so moves are more frequent."

In closing, he offered a suggestion: "that the tolling of the bell after an execution should be stopped—there appears to be no reason for it and if it were stopped I think very few people in the neighborhood would be aware of anything unusual taking place."

He was overruled. The bells had always rung. They would continue to ring. It was, after all, the law, and this was, after all, England.

BELLE DECORATED IN PINK. She wore pink dresses. She wore pink underclothes, including a pink silk ribbed undervest. She bought pink pillows with pink tassels, walled rooms with pink fabric, and hung pink velvet bows off the frames of paintings. She loathed green. It was unlucky, she believed. On seeing green wallpaper in the drawing room of a friend's home, she exclaimed, "Gee. You have got a hoo-doo here. Green paper! You'll have bad luck as sure as fate. When I have a house I won't have green in the house. It shall be pink right away through for luck." By that standard, she would now be very lucky indeed.

She exhibited an odd mix of frugality and extravagance, according to her friend Adeline Harrison. "Mrs. Crippen was strictly economical in small matters in connection with their private home living," Harrison wrote. "In fact, to such an extent did she carry it that it suggested parsimony. She would search out the cheapest shops for meat, and go to the Caledonian Market and buy cheap fowls. She was always trying to save the pence, but scattering the pounds."

She spent most heavily on clothing and jewelry for herself. For the house, she bought knickknacks and gewgaws and odd pieces of furniture. Given her passion for bargains, she no doubt frequented the famous Friday market of "miscellanies" at the Metropolitan Cattle Market. At first "miscellanies" was meant to describe livestock other than cattle, hogs, and sheep, such as donkeys and goats, but over the years the term had come to include anything that could be sold, whether animate or inanimate. On Fridays, as Charles Booth discovered when his survey of London took him to the cattle market, "nearly everything is sold & nearly everything finds a purchaser." Anyone browsing the stalls could find books, clothes, toys, locks, chains, rusty nails, and an array of worn and beaten wares that Booth described as "rubbish that one wd. think wd. not pay to move a yard." Another writer observed in 1891 that the "buyers and sellers are as miscellaneous, ragged and rusty as the articles in which they deal."

Belle adorned the house with ostrich feathers and in one room installed a pair of elephants' feet, a not uncommon decoration in middle-class homes. Her friends noted with cheery malignance that while Belle paid a lot of attention to how she looked and dressed, her housekeeping was haphazard, with the result that the atmosphere within the house was close and musty. "Mrs. Crippen disliked fresh air and open windows," Harrison wrote. "There was no regular house cleaning. It was done in spasms. The windows in all the rooms, including the basement, were rarely opened." Despite the size of the house—its three floors plus basement—Belle refused to spend money on a maid, even though servants could be hired for wages that later generations would consider laughably low. To reduce the amount of housework, she simply closed the top-floor bedrooms and regulated access to the rest of the house. "They lived practically in the kitchen, which was generally in a state of dirt and disorder," Harrison wrote. "The basement, owing to want of ventilation, smelt earthy and unpleasant. A strange 'creepy' feeling always came over me when I descended—it was so dark and dreary, although it was on a level with the back garden."

Harrison recalled visiting the house on a day when the contradictions of Belle's nature became strikingly evident. "I followed her into the kitchen one morning when she was busy. It was a warm, humid day, and the grimy windows were all tightly closed. On the dresser was a heterogeneous mass, consisting of dirty crockery, edibles, collars of the doctor's, false curls of her own, hairpins, brushes, letters, a gold jeweled purse, and other articles." In the kitchen the gas stove was brown with rust and stained from cooking. "The table was littered with packages, saucepans, dirty knives, plates, flat-irons, a washing basin, and a coffee pot." And yet in the midst of this clutter, a white chiffon gown with silk flowers lay draped "carelessly" across a chair.

At the window, which was closed, stood one of Belle's cats. "The little lady cat, who was a prisoner, was scratching wildly at a window in a vain attempt to attract the attention of a passing Don Juan."

OUTWARDLY, THE CRIPPENS seemed to have an idyllic marriage. Neighbors in the houses on either side and in back reported often seeing the couple at work together in the garden, and that Belle often sang. One neighbor, Jane Harrison, who lived next door at No. 38, reported, "They always appeared on very affectionate terms and I never heard them quarrel or have a cross word." On four occasions Harrison came over to help Belle prepare for parties, including one large affair that Belle hosted for the birthday of George Washington "Pony" Moore, manager of the blackface Moore and Burgess Christy Minstrels.

But those with a closer view saw a relationship that was not quite so idyllic. For one brief period Belle did try having a servant, a woman named Rhoda Ray. "Mr. and Mrs. Crippen were not altogether friendly to each other," Ray said, "and they spoke very little together." And a friend, John Burroughs, noticed that Belle could be "somewhat hasty" in her treatment of her husband. One change in how they configured their home appeared to cause no great concern among their friends, though within a few years it would take on great significance. For the first time in their marriage the Crippens occupied separate bedrooms.

What Belle's friends and neighbors did not seem to grasp was that Belle was lonely. She stayed at the house most of the time, though she often left for lunch, typically departing at about one o'clock and returning about three. She found comfort in pets, and soon the house was full of mewing and chirping and, eventually, barking. She acquired two cats, one an elegant white Persian; she bought seven canaries and installed them in a large gilt cage, another common feature of homes in the neighborhood. Later she and Crippen acquired a bull terrier.

At one point soon after moving into the house, she decided to take in boarders and placed an advertisement in the *Daily Telegraph*. Soon three young German men took up residence in the top-floor bedrooms. One of them, Karl Reinisch, later recalled that Belle had wanted more than just income.

He told his story in a letter that is now in the possession of Scotland Yard's Black Museum, accessible only to police officers and invited guests:

The house had a "beautiful garden," Reinisch wrote, and was situated on "a quiet, better street." He considered himself lucky to have been accepted as a tenant. "It counted at that time as a certain distinction to obtain board and lodging in the house of Dr. Crippen," he wrote. Crippen was "extremely quiet, gentlemanly, not only in thought but also in behaviour, not only towards his wife but also to me and everyone else. He idolized his wife, and sensed her every wish which he hastened to fulfill." That first Christmas, 1905, offered an example. "Dr. Crippen wished to give his wife a big surprise, one that would make her very happy, namely a gramophone. These were then very costly. Mrs. Crippen, a good piano player, was as pleased as a child at this attentiveness, and Dr. Crippen was even happier at the joy of his wife. He had gone to a great deal of trouble to procure the gramophone."

Crippen and Belle had opposite natures, Reinisch wrote. He found Crippen to be "extremely placid" and Belle "very high-spirited. Blonde, with a pretty face, of large, full, may I say of opulent figure." She was, he wrote, "a good housewife, unlike many other English women. She cooked herself, quite excellently." He noticed that despite the couple's "good financial circumstances," they had no servants.

Often Crippen and Belle recruited Reinisch and one of the other tenants for whist. "Mrs. Crippen could be extremely angry if she lost a halfpenny or a penny, and on the other hand extremely happy if she won a similar sum. The penny was not the most important thing here, but ambition. Merely so that his wife should not be angry, Dr. Crippen asked me . . . to play often intentionally badly, as he also often did, just to allow the mistress of the house to win and thus to make her happy."

Overall, however, the couple struck Reinisch as being reasonably content. "The marriage, at least during my time there, was very harmonious," he wrote. "I never once perceived any misunderstanding or bad feeling between the couple. I must mention that they lived a comparatively retiring life. It was only for this reason, so as not to be always alone together, that they took me into their household. I felt myself very much at home with this family, and never had the feeling that I was just an object to be made money out of, as was often the case elsewhere."

It was the absence of children, Reinisch believed, that had compelled Mrs. Crippen to seek lodgers.

"As a 'substitution' for offspring someone was to be in the house who was trustworthy and sociable," he wrote. "Thus the condition was made, on my being taken into the house, that I was not to go out every evening, but should rather stay in the house for the sake of the company. . . . It was not easy for me, as a young man who wished to enjoy himself in the big city, to agree to this. I had no need, however, to regret it, as the society of the two cultured people had only a good influence on me, and the frequent conversations beside the fire were very varied, stimulating and interesting."

Another tenant, however, had a different perception of the Crippens and told Belle's friend Adeline Harrison about a number of quarrels that always seemed one-sided, "Mrs. Crippen, excitable and irritable, chiding her husband; Crippen, pale, quiet, imperturbable."

THOUGH THE PRESENCE of Reinisch and the other tenants might have eased Belle's loneliness, it inserted extra tension into her relationship with Crippen. She made Crippen tend to their needs every day, and even on Sunday, which was Crippen's one full day off from work. "He had to rise at six o'clock in the morning to clean the boarders' boots, shovel up the coal, lay the breakfast, and help generally," Adeline Harrison wrote. He had to make beds, wash dishes, and on Sundays help prepare the tenants' midday dinner, all this without servants. "It was a trying time," Harrison wrote, "and quite unnecessary exertion for both, as Crippen was earning well, and gave his wife an ample supply of money." Belle used the income from the tenants to buy more clothing and jewelry.

In June 1906, after less than a year, Belle evicted the Germans. The work had become too much, a friend said, though it is possible too that the mounting fear of German spies influenced her decision. At nine-thirty on Saturday morning, June 23, Belle wrote, "As my sister is about to visit me, I regret exceedingly I shall want the house to ourselves, as I wish to do a great deal of entertaining and having Paying Guests in the house would interfere with my plans. I therefore hope you will find comfortable quarters elsewhere. Kindly do so at your own convenience as I do not want to rush you off and want you to feel thoroughly at home while you remain with us. I hope you will honor me with your presence at my weekly Receptions while my sister visits me."

The aspect of Belle's nature that most colored life at No. 39 Hilldrop Crescent was her need for dominion over Crippen. Placid and malleable, he was almost on a par with the household's other pets. He awaited definition. "He was a man with no apparent surface vices, or even the usual weaknesses or foibles of the ordinary man," Adeline Harrison wrote. "Restraint was the one and only evidence of firmness in his character. He was unable to smoke; it made him ill. He refrained from the consumption of alcoholic liquor in the form of wines and spirits, as it affected his heart and digestion. He drank light ale and stout, and that only sparingly. He was not a man's man. No man had ever known him to join in a convivial bout; he was always back to time, and never came home with a meaningless grin on his face at two o'clock in the morning attended by pals from a neighbouring club."

Soon after the move to Hilldrop, Belle insisted that Crippen convert to Roman Catholicism. She determined how he dressed. On January 5, 1909, she bought him three pairs of pajamas at the annual winter sale at Jones Brothers, a clothier, soon to prove among the most significant purchases of her life. She specified the color and cut of his suits. "His eccentric taste in the matter of neckties and dress generally may be attributed to the fact that it represented feminine taste," Harrison wrote. "His wife purchased his ties, and decided on the pattern of his clothing. She would discuss the colour of his trousers with the tailor, while he stood aside looking on, without venturing to give an opinion."

Her need for control extended as well to her cats. She never let them outside "for fear they should fall victims to the shafts of illicit love," Harrison wrote. Instead, she had Crippen build a cage for them in the garden.

Lest Crippen step out of line, there was always the threat that she would leave. She kept at least one photograph of Bruce Miller on display at all times. A reminder.

Later Crippen would tell a friend, "I have always hated that house."

IN 1907 A MAN calling himself Mr. Frankel rented a bedroom in a building on Wells Street a short walk west of Tottenham Court Road. He identified himself to his new landlord as an ear specialist. He was a small man with a large mustache and warm, if slightly protruding eyes, and when he walked, he tended to throw his feet out to the side. His manner was gentle. "The rooms which Frankel occupied were seldom used at night," his landlord said later, "but occasionally during the day I saw a girl coming down the stairs from the direction of Frankel's bedroom. I could not identify her."

Disaster

At Poldhu the weather continued rough. Kemp noted in his diary that on the morning of September 17 a gale from the southwest raked the station. Marconi was there. He, Kemp, and Fleming conducted experiments involving the generation of sparks. The wind intensified. "At 1 p.m.," Kemp wrote, "the wind suddenly changed to N.W. with a heavy squall which struck the circle [of masts] with increasing violence."

The masts rocked. The triatic stays that linked each mast to its neighbor caused them all to dance at once. Wind moaned through the wires.

One mast broke, but the triatics held—and transmitted the jolt of the collapse to the rest of the circle. All the masts failed. Half collapsed entirely, slamming onto the rain-soaked ground like great trees. The rest jutted from the ruin at haphazard angles.

No one was hurt, and somehow even the condensers, transformers, and generators in the buildings below escaped damage.

Marconi showed little emotion. Inside, however, he was impatient, nearly desperate. The disaster posed a wrenching challenge to his dream of transmitting across the Atlantic. He refused to postpone the attempt.

At his direction, the men at Poldhu erected two new masts, each 160 feet tall, and strung a thick cable across the top. From it they hung fifty-four bare copper wires, each 150 feet long, that converged over the condenser house and formed a giant fanshell in the sky. No particular law of physics dictated the design. It just struck Marconi as right.

Seven days after the disaster the new antenna was finished, and soon afterward, Marconi used it to make his first test transmissions to his station at the Lizard.

With this temporary antenna in place, he gave orders to build a new permanent station to consist of four towers each two hundred feet tall, constructed of cross-braced struts of pine. The four towers would anchor the corners of a piece of ground two hundred feet square. A thick cable of twisted wire would link the tops, and from it Marconi planned to string at least two hundred more wires to form a gigantic inverted pyramid reaching down to the roof of the condenser house. This time Marconi made sure the towers were designed to withstand the worst weather Cornwall was likely to deliver.

But construction of such an immense station would take months. He could not bear so long a delay in his transatlantic experiment. Partly his impatience was fueled by practical considerations. He worried that his board of directors would lose faith. So far the board, with reluctance, had allowed him to spend £50,000 on the stations at Poldhu and South Wellfleet—$5.4 million today. He needed to prove it was money well spent, though now with one station in ruins that proof would be harder to generate. He worried as always about the growing competition, especially from companies in America, and he still smarted from his failure at the last America's Cup. He knew also that the secrecy of his transatlantic plan could not be maintained much longer.

But the overriding motivation came from within. On an instinctive level he knew that his signals could cross the Atlantic, even though nothing in the laws of physics as then understood even hinted that such a feat might be possible.

The fact that his temporary antenna at Poldhu still allowed communication with a station he had built at Crookhaven, Ireland, 225 miles away bolstered his confidence. So too did a bit of new technology that had fallen into his hands. In August a friend and fellow countryman, Luigi Solari, an officer in the Italian Navy, had paid a visit to Poldhu and brought with him a new coherer developed by a navy signalman. Marconi tested it and found it to be far more sensitive to transmissions than even his own best receivers.

He devised a new plan. On November 4, 1901, he sent Kemp a conventional telegram: "PLEASE HOLD YOURSELF IN READINESS TO ACCOMPANY me to Newfoundland on the 16th inst. If you desire holidays you can have them now. Marconi."

He said nothing of the new plan to Fleming.

Marconi was enough of a realist to recognize that his temporary station was unlikely to be able to generate the power and the wavelengths that he believed would be necessary to produce waves capable of traveling all the way to Cape Cod. But Newfoundland was a different matter. It was far closer to Britain, yet still at the opposite side of the Atlantic. It was also well served by undersea cable, through the Anglo-American Telegraph Co., which held a monopoly on telegraphy between Britain and Newfoundland. This fact was vital to Marconi. He needed to be able to send and receive conventional telegrams in order to direct his operators in Poldhu and to gauge the progress of his experiments.

But now he faced the most important hurdle. Somehow, quickly, he had to erect a receiving station in Newfoundland with an antenna tall enough to receive signals from the temporary station at Poldhu. The antenna would have to be hundreds of feet high.

He came up with a novel solution. It was a good thing that he kept the whole effort secret, because if the shareholders had known of his plan, their confidence in him and his company likely would have plummeted.

Kemp packed his bags, as did another engineer, Percy Paget, and on Tuesday, November 26, 1901, at the wharves in Liverpool, they and Marconi boarded the Allan Line's *Sardinian*, bound for Newfoundland. Marconi carried his own coherers and versions of the Italian signalman's device given him by Solari. Tucked in the ship's baggage hold were two large balloons made of cotton and silk that, once inflated, would each have a diameter of fourteen feet. Here also the crew stowed a number of large canisters containing hydrogen to fill the balloons, and spools containing thousands of feet of copper wire, along with materials to assemble six kites measuring seven by nine feet, each capable of lifting a man.

The balloons and kites, Marconi wrote later, were a necessary concession to time and the elements, "as it was clearly impossible at that time of the year, owing to the inclement weather and especially in view of the shortness of the time at our disposal to erect high poles to support the aerial." Marconi envisioned using kites and balloons to loft a wire four hundred feet into the sky—twice the height of the masts at Cape Cod. At his instruction, the operators at Poldhu would send signals over and over at designated times until detected. Once he received a message, he planned then to go to his station at South Wellfleet to send a reply, thereby achieving at least a semblance of two-way communication across the Atlantic.

That evening, before the *Sardinian* sailed, Marconi, Kemp, and Paget sat down to dinner, their first meal onboard. It was a sumptuous affair, with excellent food and wine. The ship was warm and comfortable, the service attentive—not surprising, given that the three men comprised about half the ship's roster of passengers. Paget and Kemp shared a cabin; Marconi had his own.

In the middle of the meal a telegram arrived, addressed to Marconi.

AT SOUTH WELLFLEET, November was proving ferocious. The Weather Bureau called it the coldest November "for many years," with a mean temperature that was "phenomenally low." All month there was wind, rain, sleet, and snow, but the last week proved especially violent. On Saturday night, November 23, a nor'easter blew in and continued raging all the next day. Over the following two days the wind on Block Island reached eighty miles an hour, hurricane force. Storm flags went up and stayed up.

On Tuesday, November 26, the storm reached its peak. Powerful gusts of wind tore across the clifftop and caused the masts to undulate and twist. The triatic stays linking the tops of the masts caused them to move in unison, like dancers in some primitive ceremony.

The dance turned jagged. Eerily, the South Wellfleet station now experienced the same disaster that had destroyed its sister station in Poldhu. One mast failed; then all failed. A segment of mast the size of a tree trunk pierced the roof of the transmitting room. Another nearly struck Richard Vyvyan. It fell, he wrote, "within three feet of where I was standing at the time."

Now this station too lay shattered. Marconi's lavish investment had yielded only a dozen shipwrecks' worth of damaged spars, royals, and topgallants.

Vyvyan sent word of the disaster via undersea cable to the company's headquarters in London, which relayed the news to Marconi, now dining aboard the *Sardinian*. The telegram was concise: "Masts down Cape Cod."

The Poisons Book

In September 1908 Ethel Le Neve became a lodger in a house a few blocks south of Hampstead Heath and a mile or so west of Hilldrop Crescent. The house was occupied by Emily and Robert Jackson. Robert was a "traveler," or salesman, for a company that sold mineral water; his wife managed the letting of bedrooms in the house and provided the tenants with meals. Mrs. Jackson and Ethel took to each other immediately. Each evening when Ethel returned from work, Mrs. Jackson brought her a cup of tea in her room, where the two would spend a few moments catching up on the day's events. Soon Ethel was calling Mrs. Jackson "Mum" and "Ma."

What Mrs. Jackson did not know was that Ethel was four months pregnant, but this became apparent two weeks later, when Ethel had what Mrs. Jackson called a "miscarriage," though that could have been a euphemism. Female doctors were rare, but one such physician, Ethel Vernon, came to the house to care for Le Neve. "I never saw the baby," Mrs. Jackson said, later, "and I was present in the room when Miss Vernon asked her where it was." Le Neve said she did not know, "but eventually said she had been to the lavatory and whilst there felt something come from her."

The doctor and Mrs. Jackson questioned Ethel "closely" for the name of the father, but she would not reveal his name.

Ethel became ill and Mrs. Jackson tended to her as if she were her daughter. Two or three days later Crippen came to the house and asked to see Ethel, giving Mrs. Jackson his card. He stayed only a few minutes. A week later he returned, but this visit was just as brief as the last.

Mrs. Jackson said of him later, "I thought him quite the nicest man I had ever met."

Ethel remained in bed about two weeks, then returned to work.

Crippen retained a vague connection to Munyon but threw most of his energy into founding a new business, a dental practice, with a New Zealand dentist named Gilbert Mervin Rylance. They called their new venture the Yale Tooth Specialists. "He was the financier," Rylance said, "and I was the dental partner." Crippen managed the company and produced the necessary anesthetics. They agreed to split all profits evenly. The practice occupied an office in the building where Crippen already had been working, Albion House on New Oxford Street, and where the Ladies' Guild maintained its headquarters. Crippen continued to concoct and sell medicines of his own design, including a treatment for deafness called Horsorl.

The miscarriage changed the tenor of Ethel's relationship with Crippen. Where once the affair had been carefree and daring, especially given the proximity of the Ladies' Guild, now there was loss and along with it a realization on Ethel's part that her love for Crippen had grown deeper. She found it increasingly difficult to endure the fact that each night he returned to his home and to his wife, she to a single room in Hampstead, alone.

The Music Hall Ladies' Guild continued its good works. Its members grew fond of Belle Elmore and her energy, and Belle returned their affection. Though she herself was not performing, she daily encountered those who were, and at least for the time being this seemed to be enough. The one stubbornly dreary part of her life was her husband. She assured him time and again that there were many men who would have her in a heartbeat. With increasing frequency she reiterated her threat to leave.

She appeared not to realize, however, that her threat had lost a good deal of its power. Crippen was in love with Ethel Le Neve and promised her that one day he would make her his legal wife. She was, he believed, the woman who should have shared his bed all along. Belle's departure would be a blessing, for desertion was one of the very few grounds that British law accepted as cause for divorce.

In turn, Crippen did not realize that Belle had become increasingly serious about her threat and had begun planning ahead. Their savings account at Charing Cross Bank in the Strand now contained £600 (more than $60,000 today). Under bank rules, Belle and Crippen each had the right to withdraw money, without need of the other's signature. There was a catch, however. Only the interest could be withdrawn on demand. Closing the account or withdrawing any of the principal required advance notice of one full year.

On December 15, 1909, the bank received a notice of intent to withdraw the entire amount. It was signed by Belle alone.

BELLE WAS GENEROUS with her new friends at the guild. On Friday, January 7, she and Crippen went together to the guild offices, and there Belle gave a birthday gift, a coral necklace, to her friend and fellow member Louie Davis. Belle was troubled by something that had happened the night before, and now, as she handed the present to Davis, she said, "I didn't think I should be able to come to give it to you as I woke up in the night stifling and I wanted to send Peter for the priest. I was stifling and it was so dark."

Belle turned to Crippen. "Didn't I dear?"

"Yes," Crippen agreed, "but you are alright now."

The three then left and walked together to a nearby Lyons & Co. teahouse, crowded as always, and there Belle repeated her story, with still more drama.

"I shall never forget it," she said. "It was terrible."

She put her hand to her throat.

Davis found it odd to hear Belle recount such a story, for one of

Belle's most salient characteristics was her robust good health. As one friend said, Belle "did not seem to know what an ache or pain was."

Over tea Crippen blamed the incident on anxiety generated by Belle's work for the guild. He urged her to resign, advice that he had given before and that she had ignored, just as she ignored it now. One reason he wanted her to quit was to remove her from Albion House, where she had become a near-constant presence, forcing Crippen and Ethel to maintain a level of circumspection that both found cumbersome and inhibiting.

At Lyons the conversation moved on.

On Saturday, January 15, 1910, Crippen left his office and walked along New Oxford Street to the nearby shop of Mssrs. Lewis & Burrows, Chemists, where he always bought the compounds he used in his medicines and anesthetics. Over the previous year he had acquired hydrochloric acid, hydrogen peroxide, morphine salts, and—his highest-volume purchase—cocaine, which he bought on nine occasions throughout the preceding year, for a total of 170 grains. Today, however, he wanted something different. He asked the clerk, Charles Hetherington, for five grains of hyoscine hydrobromide.

The order did not surprise Hetherington. He knew Crippen and liked him. Crippen always smiled and exuded an aura of kindness. Part of it was the way he looked—the now-graying mustache and beard made him seem approachable, and his eyes, magnified by the lenses of his spectacles, made him seem somehow vulnerable. Hetherington knew also that Crippen made homeopathic medicines and dental anesthetics, and that hyoscine was sometimes used in drugs meant to have a tranquilizing effect on patients.

But Hetherington could not fill the order. Hyoscine was an exceedingly dangerous poison and was rarely used, and as a consequence he did not have it in stock. Indeed, in his three years working for Lewis & Burrows, he had never known the shop to have that large a quantity on hand at any one time. He told Crippen he would have to order it and that it ought to arrive within a few days.

HETHERINGTON RELAYED THE ORDER by telephone to a drug wholesaler, British Drug Houses Ltd., "the largest firm of Druggists in London, and probably in England," according to its managing director, Charles Alexander Hill.

His company had no problem filling the order, as it typically had about two hundred grains on hand, supplied by Merck & Co. of Darmstadt, Germany. Hyoscine had "very limited demand," he said. Ordinarily his company supplied chemists with a maximum of one grain at a time, though a wholesale drug firm once ordered three grains and a hospital fifteen, the largest single order he could recall.

The company shipped the five grains of hyoscine to Lewis & Burrows on January 18, along with other compounds the shop had ordered.

THE NEXT DAY, Wednesday, January 19, Crippen again walked to Lewis & Burrows's shop and asked for his order.

In the past whenever Crippen picked up his poisons—his morphine and cocaine—the clerks on duty did not make him sign the poisons book, in which they recorded purchases of "scheduled" poisons. "We did not require him to do so," said Harold Kirby, an assistant at the shop, "because we knew him, and knew him to be a medical man."

On this occasion, however, the shop did ask him to fill out an entry in the book and sign it, because of the unusual nature of the order and the potency of the drug. Crippen "did not raise the slightest objection," Kirby said.

First the form asked for the "Name of Purchaser," and here Crippen wrote, "Munyons per H. H. Crippen," though by now he had only a slight connection to Munyon. Where the form asked, "Purpose for which it is required," he filled in, "homœopathic preparations." He signed his name.

Kirby handed him a small container that enclosed tiny crystals that weighed only one-hundredth of an ounce yet were capable of killing twenty men. He put the container in his pocket and returned to Albion House.

The Secret of the Kites

Despite the news about South Wellfleet, Marconi, Kemp, and Paget sailed for Newfoundland. The crossing took ten days and was marred by what Kemp called a "terrific gale" and a blizzard at sea. On Friday, December 6, 1901, they entered the harbor at St. John's and docked at Shea's Wharf. Snow bearded the *Sardinian*'s hull and lay in drifts on deck.

A throng of reporters and dignitaries met Marconi as he disembarked. To mask the true purpose of his mission, he hinted that he had come to Newfoundland to explore aspects of ship-to-shore communication. He reinforced the ruse by cabling the Cunard Line in Liverpool to inquire about the locations of the wireless-equipped *Lucania* and the more recently outfitted *Campania*. "He reasoned," Vyvyan wrote, "that if he stated his purpose beforehand and failed, it would throw some discredit on his system . . . whereas if he succeeded the success would be all the greater by reason of its total unexpectedness."

Soon after landing, Marconi began scouting for a site to launch his kites and balloons and settled on a "lofty eminence" that he had spotted from the ship, which bore the apt name Signal Hill, for the fact that it previously had been used for visual communication. It rose three hundred feet over the harbor and had a two-acre plateau at its top. Marconi and Kemp decided to set up their receiver and other equipment in a building on the plateau that previously had been a fever hospital.

On Monday, December 9, three days after their arrival, they began their work in earnest. They buried twenty sheets of zinc to provide a ground, assembled two kites, and oiled the skin of one balloon so that it would retain hydrogen.

Before leaving England, Marconi had given his operators at Poldhu instructions to wait for a cable from him specifying a day to begin signaling. Here too he was concerned about secrecy, for he knew that information leaked from telegraph offices as readily as water from a colander. By establishing the protocol in advance, all he now had to do was send a cable stating a date, with no further instructions. On that date, at three o'clock Greenwich Mean Time, his Poldhu station was to begin sending the letter S, three dots, over and over. Marconi had chosen S not out of nostalgia for his first great success on the lawn at Villa Griffone, but because the transmitter at Poldhu channeled so much power, he feared that repeatedly holding down the key for a dash might cause an electric arc to span the spark gap and damage his equipment. His operators were to send ten-minute volleys of 250 S's spaced by five-minute intervals of rest. The pauses were important, for the key required to manage so much power had more in common with the lever on a water pump than the key typically found in conventional telegraph bureaus—it required strength and stamina to operate.

That Monday Marconi sent Poldhu his cable. The message read, simply, "BEGIN WEDNESDAY 11TH."

So far, he had succeeded in keeping his real goal a secret. Only the *New York Herald* had bothered to send a correspondent to Signal Hill, and now the newspaper reported that Marconi "hopes to have everything completed by Thursday or Friday, when he will try and communicate with the Cunard steamer *Lucania*, which left Liverpool on Saturday."

On Tuesday Kemp and Paget conducted a test flight of one kite, which rose into the sky trailing an antenna of five hundred feet of wire. The weather was fair, and the kite flew nicely. The next day, Wednesday, when the signaling was to begin, the weather changed.

Of course.

A strong wind huffed across the clifftop and raised the hems of the men's coats. They decided to try a balloon first, thinking it would have more stability in the rough air. They filled a balloon with a thousand cubic feet of hydrogen and, with Kemp holding tight to a mooring line, sent it aloft. This time the wire was six hundred feet long. The balloon's silk and cotton sleeve, expanded by gas to fourteen feet in diameter, acted now as a giant sail far overhead. In his diary Kemp wrote that he "had great trouble with it."

Abruptly, the wind intensified. The balloon rose to about one hundred feet when Marconi decided the weather was too turbulent. The men began hauling it back down.

The balloon tore free. Had the balloon moved in a different direction, Kemp noted, "I should have gone with it as its speed was like a shot out of a gun."

With six hundred feet of wire following in a graceful arc, the balloon, Marconi wrote, "disappeared to parts unknown."

MARCONI TOLD THE *Herald*'s man, "Today's accident will delay us for a few days and it will not be possible to communicate with a Cunarder this week. I hope, however, to do so next week, possibly with the steamer leaving New York on Saturday."

THE NEXT MORNING, Thursday, December 12, the plateau atop Signal Hill was engulfed in what Marconi called a "furious gale."

"I came to the conclusion that perhaps the kites would answer better," he wrote, and so despite the storm Kemp and Paget readied one for launch. This time they attached two wires, each 510 feet long. Coats flapping, they launched the kite into the gale. It dipped and heaved but rose quickly to about four hundred feet.

"It was a bluff, raw day," Marconi wrote: "at the base of the cliff, three hundred feet below us, thundered a cold sea. Oceanward, through the mist I could discern dimly the outlines of Cape Spear, the easternmost reach of the North American continent, while beyond that rolled the unbroken ocean, nearly two thousand miles of which stretched between me and the British coast. Across the harbor the city of St. John's lay on its hillside, wrapped in fog."

Once the kite was airborne, Marconi, Kemp, and Paget retreated from the weather into the transmitter room. "In view of the importance of all that was at stake," Marconi wrote, "I had decided not to trust to the usual arrangement of having the coherer signals recorded automatically through a relay and a Morse instrument on a paper tape." Instead, he connected his receiver to the handset of a telephone, "the human ear being far more sensitive than the recorder."

It seemed at the time a prudent decision.

Not only the press was kept in the dark. Ambrose Fleming had left Poldhu on September 2 and soon afterward departed for his first vacation in years. Despite his crucial role in designing and adjusting Poldhu's transmitter and power supply, he knew nothing about the attempt then under way in Newfoundland. On returning from his holiday, he occupied himself with his teaching duties at University College in Bloomsbury and worked on an important upcoming talk, a Christmas lecture at the Royal Institution.

For reasons that remain unclear, Marconi had excluded Fleming from the very thing that he had hired him to achieve. It may simply have been an oversight, owing to the turmoil raised by the destruction of the stations at Poldhu and South Wellfleet. It may, however, have been another example of Marconi's periodic lapse into social blindness with its attendant disregard for the needs of others.

The kite shuddered through the sky and strained at the line that tethered it to the plateau. At the appointed time Marconi held the telephone receiver to his ear. He heard nothing but static and the noise of wind. Each new gust stabbed the room with the scent of winter.

In Poldhu the operator began slamming down the key to make each dot.

To anyone watching, the whole quest would have seemed utterly hopeless, *deserving* of ridicule—three men huddled around a crude electrical device as a gigantic kite stumbled through the sky four hundred feet overhead. If not for the atmosphere of sober concentration that suffused the room, the scene would have served well as a *Punch* parody of Marconi's quest.

Wretched Love

Toward the end of January 1910 Ethel's friend and landlady, Mrs. Jackson, began to notice a change in her behavior. Ordinarily Ethel left the house at ten o'clock in the morning for work, then returned around six or seven in the evening, when she would greet Mrs. Jackson warmly. In practice if not blood they were mother and daughter. But now Ethel's demeanor changed. She was, Jackson said, "rather strange in manner, sometimes she would speak to me, sometimes not and was depressed. People noticed it."

After several nights of this Mrs. Jackson resolved to ask Ethel why she seemed so depressed, though indeed London in deep winter—so dark, cold, and wet, the streets conduits of black water and manure—was enough to depress anyone.

As was her habit, Mrs. Jackson followed Ethel into her bedroom, where in more pleasant times they would spend the evening talking about work or the day's news. The thing most on people's minds was the rising power of Germany and the near-certainty of invasion, raised anew by a terrifying but popular play, *An Englishman's Home,* by Guy du Maurier.

At first Ethel said nothing. She undressed and changed into bedclothes, then undid her hair and let it fall to her shoulders. Her cheeks still red from the cold, her hair dark and loose, she really was lovely, though sad. She put curlers in her hair, Jackson recalled—and probably these were examples of the latest in grooming technology, the Hinde's Patent Brevetee, about three inches long, with a Vulcanite central core and two parallel metal bands.

She had difficulty. Her hands moved in clumsy jolts. Her fingers twitched. She did her hair, then undid it, "pulled and clawed it, looked straight into a recess in the corner of the room and shuddered violently," Jackson said. Ethel had a "horrible staring look in her eyes."

Jackson was too worried to leave and stayed with Ethel until nearly two o'clock in the morning. She begged Ethel to tell her what was wrong, but Ethel would only say that the cause had nothing to do with Mrs. Jackson. "Go to bed," Ethel said, "I shall be alright in the morning."

Ethel lay back in bed and turned her face to the wall. Mrs. Jackson sat beside her awhile longer, then left.

IN THE MORNING ETHEL was no better. Mrs. Jackson brought her a cup of tea in her room. Later, after Mrs. Jackson's husband had left—his liking for Ethel had waned after her miscarriage—Ethel came into the kitchen for breakfast. She ate nothing. She rose and put on her coat, preparing to leave for work. Mrs. Jackson stopped her. "I can't let you go out of the house like this," she said.

It was clear to Mrs. Jackson that Ethel was too ill to leave. She telephoned Crippen at Albion House, then returned to Ethel. "For the love of God," she now said, "tell me what is the matter, are you in the family way again?"

Ethel said no.

Mrs. Jackson persisted: "I told her she must have something on her mind, and that it must be something awful or she would not be in that state." She told Ethel, "You must relieve your mind or you will go absolutely mad."

Ethel said she would tell her the story later in the day, after dinner, but within two hours she came to Mrs. Jackson and said, "Would you be surprised if I told you it was doctor?"

Mrs. Jackson assumed that Ethel was now revealing for the first time that Crippen had been the father of her lost baby, and that for some reason the whole incident had come back to cause her renewed grief.

Mrs. Jackson said, "Why worry about that now its all past and gone?"

Ethel burst into tears. "Its Miss Elmore."

This perplexed Mrs. Jackson. The name was new to her. Ethel had never mentioned anyone named Elmore—she was sure of it. "Who's that?" she asked.

"She is his wife you know, and I feel it very much, when I see the Doctor go off with her after the other affair." Ethel added, "it makes me realize my position, what she is and what I am."

On this score Jackson had little sympathy. "What's the use worrying about another woman's husband?"

Ethel told her that Crippen's wife had threatened to leave with another man and that he hoped to divorce her.

"Don't you think he is asking rather a lot of you?" Mrs. Jackson asked. "At your age it seems to me to be most unfair. Tell him what you have told me, as regards feeling your position. Tell him that you have told me."

Ethel remained in her room the rest of the day. The next day, however, she returned to work and spoke to Crippen just as Mrs. Jackson had advised. Crippen assured her that he had every intention of marrying her someday.

That night Ethel told Mrs. Jackson how thankful she was that she had confessed her troubles. From then on her mood improved. Said Mrs. Jackson, "she seemed very much more cheerful." Their evening conversations resumed, though now a new and compelling topic had been added to the palette already available for discussion.

Despite their address in the northern reaches of London, the Crippens took full advantage of the city's gleaming nightlife. Electric trams, motorized buses, and a rapid shift from steam to electric locomotives in the subterranean railways had made travel within the city a fluid, easy thing. Starting in about 1907 a new term had entered the language, *taximeter,* for a device invented in Germany that allowed cab drivers to know at a glance how much to charge their customers. In short order the term was reduced to *taxi* and applied to any kind of cab, be it growler, a hansom, or one of the new motorized variety.

The Crippens also often invited friends to their home, typically for casual dinners followed by whist, though occasionally Belle threw parties of a more boisterous nature to which she invited some of London's most prominent variety performers. For Crippen, these occasions became ordeals of labor and hectoring, for the house invariably was a shambles and had to be cleaned and neatened while Belle prepared the food.

Two friends were regular visitors to the house, Paul and Clara Martinetti, who lived in a flat on Shaftesbury Avenue, an easy walk from Crippen's office. Paul had once been a prominent variety performer, a pantomime sketch artist, but he had retired from the stage and lately had been in poor health from a chronic illness that required weekly visits to a physician. The Martinettis first encountered the Crippens at a party at the home of Pony Moore, the minstrel director. At Belle's suggestion, Clara joined the Music Hall Ladies' Guild and became a member of its executive committee. They saw each other every Wednesday at guild meetings and became friends. Soon the couples began visiting each other's homes and, as a foursome, going to the theater and then out to dinner in Piccadilly and Bloomsbury. The Martinettis were unaware of the tensions that suffused their friends' marriage. "I would describe Dr. Crippen as an amiable kind-hearted man," Clara said, "and it always looked to me as if he and his wife were on the best of terms." Belle, she said, "always appeared to be very happy and jolly and to get on very well with Dr. Crippen."

Late in the afternoon of January 31, 1910, a Monday, Crippen left his office at Yale Tooth and walked to the Martinettis' flat to invite them to Hilldrop Crescent that evening for supper and cards. Clara at first demurred. Paul was at his doctor's office, and she knew from past experience that when he returned, he would be tired and feeling poorly.

"Oh make him," Crippen said, "we'll cheer him up, and after dinner we'll have a game of whist."

Crippen left.

Paul returned from his appointment at about six o'clock. Soon afterward Crippen also returned and, exhibiting an unusual degree of insistence, repeated his invitation directly to Paul. His friend looked tired and pale and told Crippen, "I feel rather queer." Nonetheless Paul agreed to come. He and his wife said they could be at the Crippens' house by seven o'clock.

Despite the ease of transportation, the journey proved something of an ordeal. The Martinettis encountered an age-old problem—they could not find a taxi. They walked instead to Tottenham Court Road, where they caught one of the new motorized buses, then rode it north through congested streets to Hampstead Road, where they got off and caught an electric tram that took them to Hilldrop Crescent. It was now about eight o'clock, one hour later than they had intended. As they walked to No. 39, they saw Crippen at the door, watching for them. Now Belle too came barreling out, jerking her head backward as was her custom, smiling, and calling out, "You call that seven o'clock?"

For Paul, the trip had been exhausting. He did not look well. As always, there were no servants, so Clara took off her own coat and hat and took them to a spare bedroom. Belle went down to the kitchen on the basement level and continued preparing dinner. She called up to Crippen to take care of the Martinettis. Paul had two whiskeys.

At length, dinner was ready, and Crippen and the Martinettis descended to the breakfast room, where for these casual suppers the couples always converged. Belle greeted them by first showing off a new addition to the family, "a funny little bull terrier," Clara recalled, "and she tried to show us how funny he was." Belle clearly was delighted with the dog but complained about his lack of cleanliness, though she just as quickly excused his condition on grounds he was after all only a puppy.

Dinner consisted of several salads and "a joint" of roast beef. Crippen carved.

It was about eleven o'clock when Belle brought out dessert—two or three sweets, what E. M. Forster called "the little deadlies"—and served them with liqueurs and coffee. Real coffee, at eleven at night. Belle offered cigarettes, but only Paul accepted, then began to smoke. He and

Crippen went upstairs to the first-floor parlor, while Belle and Clara stayed behind to clean up. Belle told Clara to remove only the "necessary" things from the table; she and Crippen would finish in the morning.

They chose partners for whist, Belle with Paul, Crippen with Clara. As the game progressed, the room grew warm and soon was stifling. Crippen left the table and turned down the gas. Paul became quiet. "I had got a chill while playing cards and was not feeling well," he said.

One day soon, great importance would be assigned to every detail of what happened next. At the time, however, it all seemed utterly without significance.

PAUL EXCUSED HIMSELF and left the room, heading for the bathroom. "Mr. Martinetti wanted to go upstairs," Crippen said later, "and, as I thought he knew the house perfectly well, having been there many times during eighteen months, I thought it was quite all right that he should go up himself."

When Paul came back, he looked worse than ever. "He returned looking white," his wife said. He took his place at the card table, but his hands were cold and he began to tremble.

Belle poured him a brandy, but Clara protested. "Oh no, Belle, that is too much," she said. "I don't think he ought to have brandy after the whiskey he had."

Belle insisted. "Let him have it."

Clara: "No, Belle, I rather not, you know I have to take Paul home."

"You let him drink it, I take the responsibility."

They struck a compromise. "Give him some pure whiskey," Clara said. "I really don't care for him to mix his drinks."

Belle poured him a whiskey, straight, then commanded Crippen to find a taxi. He put on his coat and left. He found nothing, no two-wheeled hansom, no four-wheeled growler, none of the new motorized taximeter cabs. Repeatedly Belle glanced out the front window to look for Crippen. "It seemed to us," said Clara, "that he would never return."

At last he did come back, but without a cab. Belle sent him out again. This time he returned within a few moments with a growler.

Crippen helped Paul down the front steps and into the cab. Belle and Clara kissed, and Belle too started down the steps, but without a coat. Clara stopped her. "Don't come down, Belle, you'll catch a cold."

The cab rumbled off into the night. Later Clara would recall that Crippen and Belle "were certainly on affectionate terms" and that apart from Paul's discomfort the evening had been a pleasure. Belle as always had been warm and jolly, Crippen self-effacing and solicitous of her needs. "On the night of the party," Clara said, "we were the happiest party imaginable."

But when Crippen walked back up the stairs to the house after saying a last good-bye to the Martinettis, he found that Belle had undergone a transformation.

"Immediately after they had left my wife got into a very great rage with me, and blamed me for not having gone upstairs with Mr. Martinetti," Crippen said, referring to Paul's exit to use the bathroom. "She said a great many things—I do not recollect them all—she abused me, and said some pretty strong words to me; she said she had had about enough of this—that if I could not be a gentleman she would not stand it any longer, and she was going to leave me." He quoted her as shouting, "This is the finish of it I won't stand it any longer. I shall leave you tomorrow, and you will never hear of me again."

So far none of this was novel. "She had said this so often that I did not take much notice of it," Crippen said.

But now she went one step further and said something she had never said before—"that I was to arrange to cover up any scandal with our mutual friends and the Guild the best way I could."

Belle retired to her bedroom, while Crippen retreated to his. "I did not even see her the next morning," he said. "We retired very late, and it was the usual thing that I was the first one up and out of the house before she was ever up at all."

THAT MORNING, TUESDAY, February 1, Crippen went to his office at Yale Tooth as usual and was, according to Ethel Le Neve, "his own calm self." She wrote, "Surely we, who knew him so well and every expression of his face, would have noticed at once if he had shown the slightest agitation."

At midday Crippen left Albion House and walked to the Martinettis' flat on Shaftesbury to check on Paul. Clara greeted him at the door and told him Paul was sleeping. Crippen was pleased to learn that Paul had gotten no worse during the night. They chatted a few moments longer, then Crippen turned to leave.

Clara asked, "How's Belle?"

"Oh, she is all right."

"Give her my love."

"Yes," Crippen said, "I will."

When he returned to Hilldrop Crescent at seven-thirty that evening, he found the house empty, save for the cats, the canaries, and the bull terrier.

Belle had gone.

The main question that now occupied him, he said, was how to avoid the scandal that would arise if the true reason for Belle's departure ever got out.

The Fatal Obstacle

On Signal Hill the weather worsened. Marconi listened hard for the sound of three snaps in the static mist that filled his telephone receiver, but he detected nothing. Outside, his men struggled to keep the kite aloft and stable. Each time it bobbed and dipped, its two trailing wires grew longer or shorter. Marconi still had only a vague understanding of how electromagnetic waves traveled and how the length of his antennas affected transmission and reception, but he did recognize that this constant rising and falling could not be helpful.

Trying to send signals to a wildly shifting kite was a bit like trying to catch a fish in a whirlpool.

In Poldhu Marconi's operators fired chains of S's into the sky over Cornwall. Thousands of watts of power pulsed through the spark gap. Lightning cracked, and pipes tingled. Electromagnetic waves coursed in all directions at the speed of light. Receivers at the Lizard, at Niton, and at Crookhaven instantly detected their presence. The signals were likely received aboard at least one of the increasing number of ocean liners equipped with wireless, perhaps the *Kaiser Wilhelm der Grosse* or the *Lake Champlain* or one of Cunard's grand ships, depending on their locations. But the receiver on Signal Hill remained inert.

At about twelve-thirty the receiver issued a sharp click, the sound of the tapper striking the coherer. It meant the receiver had detected waves.

The tension in the room increased. Marconi's face bore its usual sober expression. As was so often the case, his lips conveyed distaste, as if he had scented an unpleasant odor.

"Unmistakably," he wrote, "the three sharp little clicks corresponding to three dots, sounded several times in my ear."

He was excited but skeptical. He wanted to hear the clicks so badly that he felt he could not trust his own judgment. He passed the telephone receiver to Kemp.

"Can you hear anything Mr. Kemp?" he asked.

Kemp listened, and he too heard, or claimed to hear, sequences of three dots. They passed the telephone receiver to Paget as well, who listened but heard nothing. He, however, had grown increasingly hard of hearing.

"Kemp heard the same thing as I," Marconi wrote, "and I knew then that I had been absolutely right in my calculation. The electric waves which were being sent out from Poldhu had traversed the Atlantic, serenely ignoring the curvature of the earth which so many doubters considered would be a fatal obstacle, and they were now affecting my receiver in Newfoundland."

There was no serenity on Signal Hill. A burst of wind tore the kite free. The men lofted a second one, now with a single wire of five hundred feet. This configuration, Kemp wrote in his diary, "appeared more in harmony with the earth's electric medium and the signals from Poldhu station. We were able to keep this kite up for three hours and it appeared to give good signals." In all, they picked up twenty-five of the three-dot sequences.

Marconi wrote a draft of a telegram to Managing-Director Flood Page in London to announce his success but held it back. He wanted to hear more signals before notifying his board and especially before the news became public.

He tried again the next day, Friday, December 13, 1901. The weather grew more ferocious. There was snow, rain, hail, and wind—great gasps of it. Three times they launched kites, and three times the weather drove the kites to ground. During the brief periods the kites were in the air, however, Marconi claimed that he again heard three-dot sequences from Poldhu, though these signals were even less distinct than what he had heard the day before.

Frustrated by the lack of clarity, Marconi still did not send his cable to headquarters. He resolved to wait one more day, until Saturday, to allow time for more trials.

The wind accelerated. On Saturday it reached a point where an attempt to fly anything, balloon or kite, was out of the question. In desperation, Kemp and his helpers began constructing a very different kind of antenna. They began stringing a wire from the top of Signal Hill to an iceberg marooned in St. John's harbor.

To Kemp's regret, he never got the chance to test it.

Marconi debated what to do next. He could wait and hope that the weather would improve or that Kemp's iceberg antenna would work, or he could simply trust that he had indeed heard signals from Poldhu and go ahead and notify Flood Page in London.

He sent the cable. "SIGNALS ARE BEING RECEIVED," it read. "WEATHER MAKES CONTINUOUS TESTS VERY DIFFICULT." That night, he released a statement to *The Times* of London.

For one so aware of the "doubters" arrayed against him and of the hostility in particular of Lodge, Preece, and the electrical press, Marconi in orchestrating this transatlantic experiment and in now revealing it to the world had made errors of fundamental importance.

Once again he had failed to provide an independent witness to observe and confirm his tests. Moreover, in choosing to listen for the signals with a telephone receiver instead of recording their receipt automatically with his usual Morse inker, he had eliminated the one bit of physical evidence—the tapes from the inker—that could have corroborated his account. He had to have recognized that his claims of so wondrous an accomplishment, deemed impossible by the world's greatest scientists, would pique skepticism and draw scrutiny, but apparently he believed that his own credibility would be sufficient to put all doubts to rest. This belief was a miscalculation that would prove costly.

THAT SUNDAY, DECEMBER 15, the governor of Newfoundland, Sir Cavendish Boyle, held a celebratory lunch for Marconi at which, as Kemp recalled, the governor served champagne that had been retrieved from a shipwreck after years underwater. The *New York Times* called Marconi's feat "the most wonderful scientific development in modern times."

Over the next few days the stock prices of the transatlantic cable companies began to fall. Within a week the Anglo-American Telegraph Co. saw its preferred stock drop seven points and its common stock four. Shares of Eastern Telegraph Co. lost five and a half.

AMBROSE FLEMING LEARNED of Marconi's feat only by reading a newspaper. He wrote later that he had been "left in ignorance of this success" until he opened the December 16 edition of the *Daily Mail,* where he saw the headline, "MR. MARCONI'S TRIUMPH."

He had been left out of the whole affair, yet it was he who had designed and configured the power system at Poldhu and who during his many grueling trips to the station had made it all work.

He was hurt and angry.

JOSEPHINE HOLMAN PROFESSED to be delighted. In an interview she revealed that she had known all along about Marconi's plan to span the Atlantic. "It has been a terrible state secret with me for more than a year," she said. She omitted the fact that during that year she had seen him only rarely. She hoped that she would see more of him, now that his great goal had been achieved, and that he might even pay a visit at last to Indianapolis to meet her family.

"I would rather marry that kind of man than a king," she said, and pronounced herself "the happiest woman in the world."

Several days later her grandmother hosted an engagement party at her home in Woodruff Place in Indianapolis—without Marconi. Josephine now lived in New York with her mother but had come down for a six-week visit. The party occurred at its end, during a spell of deep cold that raised fears of what the newspapers called a "coal famine." After the party, she set out to return to New York to rejoin her mother and, most important, to be reunited with her fiancé, who was headed there now, booked to stay at his favorite hotel, the Hoffman House.

It seemed a delightful prospect: Christmas in New York, with her husband-to-be, now more famous than ever.

To the Ball

When Ethel Le Neve arrived at work on the morning of Wednesday, February 2, 1910, she found a packet on her desk with a note on top that caused a soaring of spirit. Written in Crippen's hand, the text was simple and direct: "B.E. has gone to America." The note asked Ethel to deliver the packet to Melinda May, secretary of the Music Hall Ladies' Guild.

"Shall be in later," Crippen wrote, "when we can arrange for a pleasant little evening."

So Belle was gone. "I was, of course, immensely excited at this disappearance of Dr. Crippen's mysterious wife," Ethel wrote. "I knew well enough that they had been on bad terms together. I knew that she had often threatened to go away and leave him. I knew also that she had a secret affection for Mr. Bruce Miller, who lived in New York." Ethel assumed that Belle had at last made good on her threat and had run off to join the ex-prizefighter. If true, if really true, it meant that Crippen now would be free to seek divorce and, despite the strictures of British law, likely would prevail. It was, as she put it, "amazing news."

Ethel took the packet down the hall to the offices of the guild, which was due to meet that day, then returned to Yale Tooth to await her lover. She had many questions.

At noon he still had not appeared. She believed he was conducting business at nearby Craven House, on Kingsway. She busied herself with the work of the office, though Crippen's news made it hard for her to concentrate.

Crippen did not come back until four o'clock that afternoon. "He was not in a mood then for a long conversation on the subject," she recalled, "and his reticence I readily understood." But she had to speak with him.

"Has Belle Elmore really gone away?"

"Yes," Crippen said. "She has left me."

"Did you see her go?"

"No. I found her gone when I got home last night."

"Do you think she will come back?"

Crippen shook his head. "No," he said, "I don't."

On this score Ethel needed reassurance: "Did she take any luggage with her?"

"I don't know what luggage she had, because I did not see her go. I daresay she took what she wanted. She always said that the things I gave her were not good enough, so I suppose she thinks she can get better elsewhere."

Though Crippen seemed downcast, Ethel offered neither condolence nor sympathy. "I could not pretend to commiserate with him," she wrote. "He had led me into the secret of his unhappy married life, and now that his wife had disappeared it seemed to me best for him, perhaps also best for her."

Now Crippen surprised her. He reached into his pocket and pulled out a handful of jewels that Belle had left behind. "Look here," he said. "You had better have those." He held them out. "These are good, and I should like to know you had some good jewelry. They will be useful when we are dining out, and you will please me if you will accept them."

"If you really wish it," Ethel said, "I will have one or two. Pick out what you like. You know my tastes."

He chose several diamond rings; a more elaborate ring with four diamonds and a ruby; and a brooch in a pattern that evoked a rising sun, with a diamond at its center and pearls radiating outward in zigzag fashion.

The jewels were lovely, and Ethel believed them to be of the finest quality, for Crippen, as she put it, "was a real expert in diamonds." Previously he had shown her how to judge a diamond by color and clarity, and how to tell at a glance whether a diamond had been set in New York or London.

She suggested he pawn the remaining jewels—a dozen rings and a large brooch inlaid with rows of diamonds in the shape of a tiara. The idea of doing so had not struck Crippen, but now he told Ethel it was a good plan. He walked to a pawnshop on the same street as his office, Mssrs. Jay & Attenborough.

He showed a clerk named Ernest Stuart three diamond rings. After examining them closely, Stuart agreed to lend Crippen £80. Crippen returned a few days later with the rest of the jewels, and got another £115 pounds, for a total of £195—nearly $20,000 today.

That night Ethel Le Neve slept in Crippen's bed at Hilldrop Crescent for the first time.

FOR THE LADIES of the guild, the news was equally amazing. The packet delivered to the guild office that morning contained two letters—one for Melinda May, and one for the guild's executive committee. It also contained the guild's ledger and checkbook, which Belle in her role as treasurer had kept at home.

The letters were dated that same day, February 2, and were from Belle Elmore. A notation after the closing of May's letter indicated it had been prepared by Crippen at Belle's request.

"Dear Miss May," it began, "Illness of a near relative has called me to America on only a few hours' notice, so I must ask you to bring my resignation as treasurer before the meeting to-day, so that a new treasurer can be elected at once. You will appreciate my haste when I tell you that I have not been to bed all night packing, and getting ready to go. I shall hope to see you again a few months later, but I cannot spare a moment to call on you before I go. I wish you everything nice till I return to London again."

The letter to the executive committee repeated the news and noted the enclosure of the checkbook and ledger. It urged the committee to suspend the usual rules and appoint a new treasurer immediately. "I hope some months later to be with you again, and in meantime wish the Guild every success and ask my good friends and pals to accept my sincere and loving wishes for their own personal welfare."

The news of Belle's departure and the selection of her replacement consumed most of that day's meeting, though no one thought to walk the short distance to Crippen's office to ask for a fuller explanation.

A FEW DAYS LATER—MOST LIKELY it was Saturday, February 5—Ethel and Crippen arranged to spend an evening together at the theater. "He thought it would cheer us both up," Ethel said, though she herself needed no cheering. She reveled in her new status. No longer would she have to endure the sight of Crippen going off with his wife to some evening function, when rightfully it should have been she, Ethel, who accompanied him.

They were both in the office, Saturday being a workday, when Crippen remembered that he had forgotten to leave out food for his pets—the seven canaries, two cats, and bull terrier. He could not get away to feed them, but the prospect of leaving them so long without food troubled him.

Lest this problem destroy the evening and their first opportunity to go out together in public without fear of discovery, Ethel volunteered to go to Hilldrop Crescent and feed the animals. Crippen offered his keys. She left after lunch.

Ethel entered the house through the side door and found herself alone in the place for the first time. She had seen little of it so far, only the kitchen, the parlor, the bathroom, and of course Crippen's bedroom. She made her way to the kitchen, where she found most of the pets. She went to the pantry, near the door to the coal cellar, to get some milk for the cats, but as she did so, one of the cats, a beautiful white Persian—Belle's favorite—escaped and dashed upstairs. Ethel gave chase.

The cat led her throughout the house. "The faster I ran the faster went the cat," she recalled. At last she cornered it and brought it back downstairs to the kitchen.

Her tour had taken her through rooms she had never seen before, giving her a new sense of what life had been like for Crippen—nothing "uncanny," as she put it, just a sense of loneliness and what she termed a "strange untidiness."

"Rich gowns lay about the bedrooms, creased and tumbled in disorder," Ethel wrote. "Lengths of silk which had never been made into frocks were piled up, and on the pegs was a regular wardrobe, like part of a dressmaker's show-room." There were piles of clothes and "cheap stuff" that appeared never to have been worn or used. "I was struck," she wrote, "by this extraordinary litter." That Belle had left so much jewelry and clothing behind, even a number of gorgeous and expensive furs, seemed to Ethel a measure of how thoroughly her marriage to Crippen had failed. "I did not question the fact that she had walked straight out of the house, abandoning her old home life, and relinquishing everything it had contained."

What did surprise Ethel was the decor, especially in light of Belle's obvious attention to her own appearance. The house had been furnished "in a higgledy-piggledy way," Ethel wrote. "There was scarcely anything which matched. The only thing in the house which I liked was the ebony piano. All the other things had been picked up at sales by the doctor and his wife, and were of the most miscellaneous description. There was a tremendous number of trumpery knickknacks, cheap vases, china dogs, and occasional tables. There were lots of pictures—small oil and water-colour paintings by unknown artists—with bows of velvet on them to add to their beauty."

The air was stale, the rooms dark. Overall a sense of loneliness and gloom suffused the place. "From the first," Ethel said, "I took a dislike to the house."

THAT MONDAY CRIPPEN stopped in at the Martinettis' flat on Shaftesbury Avenue. Clara asked, "What is all this about Belle? She has gone to America and you said nothing about it."

"We were busy packing the whole night the cable came," Crippen said.

Clara asked why Belle had not sent her a message; Crippen replied they had been too busy getting Belle ready for departure.

"Packing and crying?" Clara asked.

"No," Crippen said, "we have got over all that."

The next week he told Clara that he had received disturbing news from Belle, by telegram. She was ill, a pulmonary ailment. Nothing to worry about, but troubling all the same.

WITH EACH DAY that Belle did not return, Ethel Le Neve found her confidence growing. She began wearing the jewelry Crippen had given her and allowed herself to be seen with him on the street, at the theater, and at restaurants. Her landlady, Mrs. Jackson, noticed that Ethel seemed to be in fine spirits almost all the time, noticed too that she had begun wearing new clothes and jewelry, including a brooch with a central diamond and radiating beams of pearls, and a trio of bracelets, though one of the bracelets, set with amethyst stones, seemed far too big for Ethel's tiny wrist. Ethel also showed off two new gold watches. One evening, beaming, she showed Mrs. Jackson a diamond solitaire ring and called it her "proper engagement ring." A few nights later Ethel displayed yet another ring. She flashed the diamond in the light. "Do you know what this cost?" she exclaimed.

"I have no idea," Jackson said.

"*Twenty pounds.*" More than $2,000 today.

One night, playfully, Mrs. Jackson asked Ethel if someone had died and left her a lot of money.

No, Ethel replied with delight. "Somebody has gone to America."

ETHEL BEGAN SPENDING NIGHTS away from Mrs. Jackson's house. In the first week of February she was gone only one or two nights, but soon she was spending nearly every night away. She told Mrs. Jackson she was staying with friends and was helping Crippen search the house for certain papers and belongings of Belle's, and she mentioned too that he had been teaching her how to shoot a revolver, a small nickel-plated weapon that he kept in a wardrobe in his bedroom.

Soon Ethel began giving gifts of clothing to her friends and to Mrs. Jackson. A widow with two daughters roomed at Constantine Road, and Ethel now gave the children an imitation pearl necklace, a piece of white lace, an imitation diamond tiara, two spray scent bottles, a pink waistband, two pairs of shoes with stockings to match, and four pairs of stockings—white, pink, and black—all of which became the daughters' most-loved possessions. To her sister Nina she gave a black silk petticoat, a dress of gold Shantung silk, a black coat, "a very big cream coloured curly cape with long stole ends," a white ostrich neck-wrapper, and two hats, one of gold silk, the other saxe blue with two pink roses.

At the time Nina said, "Fancy anyone going away and leaving such lovely clothes behind."

Yes, Ethel agreed, "that Mrs. Crippen must have been wonderfully extravagant."

But it was Mrs. Jackson who received the greatest windfall. She later had occasion to make a precise list:

1 outfit of mole skin trimmed in black
1 long coat, brown
1 long coat, black
1 coat and skirt, dark gray, striped
1 fur coat
1 coat, cream-colored
1 voile blouse and skirt, black
2 blouses, black (old)
2 blouses, one blue silk and lace, the other cream lace (new)
1 pair slippers
11 pairs stockings, brown, black, blue, white, pink, and black-and-white-striped
1 felt hat, brown, trimmed
1 lace hat, brown, trimmed with flowers
1 mole hat, pink, covered in sateen
1 imitation diamond

1 lizard-shaped diamond
1 harp-shaped brooch
2 hair stones, paste
3 night dresses, white (new)
1 skirt, yellow
1 outfit, heliotrope (new)

Ethel and Crippen grew more and more bold about declaring their romance to the world. Ethel wore Belle's furs on the street and to work at Albion House, despite the proximity of the ladies of the guild, to whom Belle's clothing was nearly as familiar and recognizable as their own. Crippen bought two tickets to one of the most important social events of the variety world, the annual banquet of the Music Hall Artists Benevolent Fund, set to take place on Sunday, February 20, at the much-loved Criterion Restaurant in Piccadilly.

"Neither of us was very anxious to go," Ethel wrote. "The doctor had bought a couple of tickets, and naturally he wanted to use them. He asked me if I would go with him. I said that I was not very keen, as I had not danced for some years, and I had not a suitable dress." Ethel ordered a new one, in pale pink, from Swan and Edgar, a prominent draper.

This decision to attend the ball was the couple's most daring declaration yet and, as it happened, most unwise.

Built in 1873, the Criterion combined glamour and raffishness, especially its Long Bar, for men only, where a Scotland Yard inspector might find himself in amiable conversation with a former convict. In its dining rooms painters, writers, judges, and barristers gathered for lunch and dinner. Later, after the theaters of the Strand and Shaftesbury Avenue closed for the night, the city's population of actors, comedians, and magicians thronged the "Cri" and its bar and its Grand Hall and its East Room and West.

Crippen wore an evening coat, Ethel wore her new dress, and as a further touch, she pinned to her bodice the rising sun brooch that Belle had left behind. Men watched her and admired the way her dress set off her slender figure. The ladies of the guild watched too, but what most caught their attention was the brooch. They knew it well—it had been a favorite of Belle's. Louise Smythson saw it. Clara Martinetti saw it, and later noted that the typist "wore it without any attempt at concealment." Annie Stratton saw it, as did her husband, Eugene, who sang in blackface with Pony Moore's minstrels. Lil Hawthorne, attending with her husband and manager John Nash, sat opposite Crippen and the typist, and they too noticed the brooch. John Nash said, "it impressed me." Maud Burroughs saw it: "I know [Belle] was very particular whenever she went away to have all her jewelry, except what she took with her, placed in a safe deposit, and this is why it struck me as so strange that the typist was seen wearing a brooch of hers."

The atmosphere shimmered with hostility. Crippen sat between Clara Martinetti and Ethel. The two women did not speak, but at one point their eyes met. Mrs. Martinetti nodded. She recalled that Ethel seemed "very quiet." John Nash said, "I noticed that Crippen and the girl were drinking very freely of wine."

Mrs. Louise Smythson approached Crippen and asked for Belle's address in America and said how strange it was that Belle had not yet written, to anyone.

"She is away up the mountains in the wilds of California," he said.

"Has she no settled address?"

"No," Crippen said, but then offered to forward anything that Smythson wanted to send.

For the moment, Mrs. Smythson let the matter drop.

"AFTER THIS," ETHEL WROTE, "I noticed that the members of the Music Hall Ladies Guild were showing marked curiosity in my movements." Her sense of being spied upon and gossiped about became acute. She could not help but run into the ladies of the guild when she entered and left the building and walked the hall to Crippen's office. Nothing was said directly, but much was communicated by glance and rigid cordiality, deadly for its iciness. "Often when I went along the street with Dr. Crippen," she wrote, "I remarked people staring at me in a curious way."

It made her uncomfortable. She wished the ladies could just accept the fact of her relationship with Crippen and be done with it.

But she had made the mistake of allowing the affair to become public: This was the England of Edward VII, but it was also the England that served as the setting for *Howards End,* to be published later that year, in which E. M. Forster plunged one of his heroines, Helen Schlegel, into an illicit pregnancy. He wrote, "The pack was turning on Helen, to deny her human rights."

On March 12 Crippen took a cab to Mrs. Jackson's house on Constantine Road and thanked her for all she had done for his "little girl," but now, he said, he was taking her away. They loaded all her things into a cab, then went to a nearby public house to celebrate. Even Mrs. Jackson's husband came along, though he did not approve of Crippen and did not consider Le Neve's recent behavior at all ladylike. Crippen bought champagne. They all drank.

Then Crippen took Ethel home.

"I Don't Believe It"

Marconi had expected some skepticism about his Newfoundland success, but he was dismayed to find himself now confronting a barrage of incredulous commentary.

"I doubt this story," Thomas Edison told the Associated Press. "I don't believe it." He said, "That letter 'S' with the three dots is a very simple one, but I have been fooled myself. Until the published reports are verified I shall doubt the accuracy of the account."

In London that same day the *Daily Telegraph* reported, "Skepticism prevailed in the city. . . . The view generally held was that electric strays and not rays were responsible for activating the delicate instruments recording the 'S's' supposed to have been transmitted from near the Lizard to Newfoundland on Thursday or Friday." The paper cited one widely held theory making the rounds that the signals had come from a "Cunarder fitted with the Marconi apparatus, which was, or should have been, within 200 miles of the receiving station at St. John's on the day of the experiment." It also quoted William Preece as stating that "the letters S and R are just the letters most frequently signaled as the result of disturbance in the earth or atmosphere."

Two days later *The Electrical Review* called Marconi's claim "so sensational that we are inclined for the present to think that his enthusiasm has got the better of his scientific caution." The *Review* proposed that the signals most likely came from a station in America. "A practical joker who had learned when the signals were expected, might easily have fulfilled the expectations of the watchers at the Newfoundland station."

The Times of London published a letter from Oliver Lodge that was a model of artful damnation. "It is rash to express an opinion either way as to the probability of the correctness of Mr. Marconi's evidently genuine impression that he has obtained evidence on the other side of the Atlantic of electrical disturbances purposely made on this side, but I sincerely trust he is not deceived." Acknowledging that he had been critical of Marconi in the past, Lodge wrote, "I should not like to be behindhand in welcoming, even prematurely, the possibility of so immense and barely expected an increase of range as now appears to be foreshadowed. Proof, of course, is still absent, but by making the announcement in an incautious and enthusiastic manner Mr. Marconi has awakened sympathy and a hope that his energy and enterprise may not turn out to have been deceived by the unwonted electrical dryness of the atmosphere on that wintry shore."

But at least one longtime skeptic took Marconi at his word, and saw in his achievement a glimmer of threat.

ON THE EVENING OF MONDAY, December 16, 1901, as he dined at his hotel in St. John's, Marconi was approached by a young man bearing a letter addressed to him. Marconi's dinner companion was a Canadian postal official named William Smith, who was staying at the same hotel and had a room just off the dining room. As the young man crossed the room toward the table, Marconi was telling Smith that he now planned to build a permanent station on Newfoundland, most likely at Cape Spear, a spit of land that jutted into the sea four miles southeast of Signal Hill.

Smith watched as Marconi opened the letter. As Marconi read, he became distraught. When Smith expressed concern, Marconi passed him the letter.

Smith too found it appalling. The letter was from a law firm representing the Anglo-American Telegraph Co., the big undersea cable company that provided telegraph service between Britain and Newfoundland.

The letter was brief, a single long paragraph that charged Marconi with violating Anglo-American's legal monopoly over telegraphic communication between Britain and Newfoundland. "Unless we receive an intimation from you during the day that you will not proceed any further with the work you are engaged in and remove the appliances erected for the purpose of telegraph communication legal proceedings will be instituted to restrain you from the further prosecution of your work and for any damages which our clients may sustain or have sustained; and we further give you notice that our clients will hold you responsible for any loss or damage sustained by reason of [your] trespass on their rights."

Marconi was furious, but he took Anglo-American's threat seriously. He knew his own company could not withstand litigation with so powerful a foe, and he recognized too that harm had indeed been done to Anglo-American, because of the decline in the price of its stock.

Smith asked him into his room, calmed him, and on impulse invited him—"begged him," Smith recalled—to bring his experiments to Canada. (At this point Newfoundland was a colony of Britain; it did not join Canada until 1949.) Over the next few days Smith arranged a formal invitation from the Canadian government. Marconi relented and set off for Nova Scotia, part of Canada since 1867, to scout a new location.

A party of dignitaries met him at the wharf in North Sydney, at the eastern tip of Nova Scotia, and whisked him into a train for a brief trip south to Glace Bay to show him a spot called Table Head. Aptly named, it was a flat plateau of ice and blown snow atop cliffs striated with bands of blue-gray and rust that fell a hundred feet straight down to the sea. "The site," Smith said, "delighted Marconi."

He set off for Ottawa to negotiate a formal agreement with the government.

ON CHRISTMAS DAY two operators with Anglo-American Cable exchanged salvos of doggerel. One in Nova Scotia tapped out,

Best Christmas greetings from North Sydney,
Hope you are sound in heart and kidney.

Next year will find us quite unable
To exchange over the cable.
Marconi will our finish see,
The Cable Co's have ceased to be.
No further need of automatics
Retards, resistances and statics.
I'll then across the ether sea
Waft Christmas greetings unto thee.

His counterpart in Liverpool responded,

Don't be alarmed, the Cable Co's
Will not be dead as you suppose.
Marconi may have been deceived,
In what he firmly has believed.
But be it so, or be it not,
The cable routes won't be forgot.
His speed will never equal ours,
Where we take minutes, he'll want hours.
Besides, his poor weak undulations
Must be confined to their own stations.
This is for him to overcome,
Before we're sent to our long home.
Don't be alarmed, my worthy friend.
Full many a year precedes our end.

North Sydney ended the exchange:

Thanks old man, for the soothing balm,
Which makes me resolute and calm.
I do not feel the least alarm,
The signal S can do no harm.
It might mean sell to anxious sellers,
It may mean sold to other fellers.

Whether it is sold or simply sell,
Marconi's S may go to—well!

IN NEW YORK JOSEPHINE Holman spent Christmas without her fiancé. She was coming to see that being pledged to a man so obsessed with work brought with it certain disadvantages, one of them being loneliness.

News from America

Letter,
Sunday, March 20, 1910
To Clara and Paul Martinetti

Dear Clara and Paul,

Please forgive me for not running in during the week, but I have really been so upset by very bad news from Belle that I did not feel equal to talking about anything, and now I have had a cable saying she is so dangerously ill with double pleuropneumonia that I am considering if I had better not go over at once. I don't want to worry you with my troubles, but I felt that I must explain why I had not been to see you. I will try and run in during the week and have a chat. Hope both of you are well, with love and good wishes.

Yours very sincerely,
Peter

Telegram,
Thursday, March 24, 1910
To Paul and Clara Martinetti

Belle died yesterday at 6 o'clock.

Part IV

An Inspector Calls

Chief Inspector Walter Dew.

"Damn the Sun!"

On his way back to London, with a formal offer from Canada in hand, Marconi stopped off in New York and attended a January 13, 1902, banquet of the American Institute of Electrical Engineers, where he was to be the guest of honor. Unknown to him, the affair nearly proved to be a disaster.

At first a number of prominent scientists declined to attend, expressing doubt as to whether Marconi really had sent signals across the Atlantic, but by the night of January 13 the institute's leaders had managed to recruit a ballroom full of believers. They held an elaborate banquet. Black signs at three points in the room bore the names Marconi, Poldhu, and St. John's, with strings of lamps hung between them. At intervals, the lamps flashed three dots. The menus were printed with ink made from Italian olive oil, and the soup for the evening was "Potage Electrolytique." Bowls of sorbet emerged, decorated with telegraph poles and wireless masts.

Thomas Edison had been invited but could not attend. Instead he sent a telegram, which the master of ceremonies read aloud. Clearly Edison had changed his mind and now accepted Marconi's claims. His telegram read, "I am sorry that I am prevented from attending your dinner tonight especially as I should like to pay my respects to Marconi, the young man who had the monumental audacity to attempt and succeed in jumping an electric wave clear across the Atlantic Ocean."

Cheers and applause rose from the audience. For Marconi, it was a rare moment of adulation, but he understood that his achievement in Newfoundland, though striking, was only the beginning of a long struggle. What he did not recognize was the extent to which the applause masked a deep and pervasive skepticism toward him and his claims of success.

In London Ambrose Fleming sulked. After learning of remarks Marconi had made in Canada and at the banquet, he felt doubly hurt. He believed that he deserved a big share of the credit for Marconi's success, yet when the great moment had arrived, he had been frozen out. In his own account of events, Fleming wrote that during Marconi's celebratory lunch with the governor of Newfoundland, Marconi had "made no frank acknowledgement . . . of the names of those who had assisted him but spoke continuously of 'my system' and 'my work.'" At the New York banquet, Fleming wrote, Marconi "pursued the same policy."

Josephine Holman too grew disenchanted. If she had expected to be the center of Marconi's attention during his stay in New York, she now found that she was mistaken. Marconi attended luncheons and dinners and kept busy in between by overseeing the installation of wireless aboard the SS *Philadelphia,* the ship that would take him and Kemp back home.

Josephine conceded defeat. On January 21, 1902, her mother, Mrs. H. B. Holman, issued an announcement to the press: Her daughter had asked Marconi to release her from the engagement, and Marconi had done so.

It made the front page of the *Indianapolis News* in an article just three paragraphs long under the headline, "ENGAGEMENT IS BROKEN." The item offered few details.

Later, a *News* reporter managed to catch up with Marconi at the Hoffman House in New York and asked if he had anything more to say.

"No, except that I am sorry."

The reporter asked, "Have your feelings in any way changed toward Miss Holman?"

"I don't think I can answer that—just say simply, please, that I am sorry."

The reporter probed further: "Had your experiments reached the point where you were at liberty to be married?"

"Well, hardly," Marconi said, "but if other things had not occurred things might have been arranged." He continued: "I have not one word of criticism to make on Miss Holman's notion. She concluded, I suppose, that her future happiness did not rest in my keeping, and the letter of request followed. I had reason to believe that our relations were quite happy and mutual until lately, and it is only natural that I should feel a little depressed at the result."

He added a tincture of mystery when he told another reporter that while delays in his work had indeed been a factor, "there was also a very delicate question involved." He gave no further explanation.

Miss Holman said little but did tell one newspaper, "There have been disasters on both sides." She was *not* referring to the collapse of the masts at Poldhu and South Wellfleet.

By the end of the day, Wednesday, January 22, 1902, as gossip about the breakup became the opening course at dinner tables in Indianapolis, New York, and London, both Marconi and Holman were at sea, Marconi aboard the *Philadelphia* bound for Southampton, Holman aboard the *Kaiser Wilhelm der Grosse,* one of the few German liners afloat that was equipped with her ex-lover's apparatus.

Holman sought escape to the Continent, hoping that travel would prove a salve for her broken heart; Marconi got back to work.

Loath to let a day pass without further experimentation, Marconi installed himself in the *Philadelphia*'s wireless cabin. As the liner approached the English coast, he made contact with Poldhu and set a new record for ship-to-shore communications: 150 miles.

Despite his failed romance, Marconi arrived in London feeling more confident than he had in a long time—a good thing, for he faced a year that would prove especially trying and raise a grave new threat from Germany.

IN LONDON MARCONI explained the details of the new Canadian arrangement to his board of directors. Much to the directors' delight, Canada had agreed to pay for the construction of the Nova Scotia station. Less delightful was Marconi's promise to provide transatlantic wireless service for 60 percent less than the rate charged by the cable companies, a maximum of ten cents a word. This was a bold commitment, given that all Marconi had sent thus far was a couple of dozen three-dot signals. Nonetheless, the board gave its approval.

Next Marconi addressed the annual meeting of his company's shareholders and for the first time in public launched into a direct attack against William Preece and Oliver Lodge and their much-publicized harping about flaws in his system. A man more able to sense the subtler bounds of accepted scientific behavior might have omitted this attack or at least phrased it differently, with the kind of oblique but slashing wit at which British parliamentarians seemed so adept, but Marconi was about to cross a dangerous invisible line—especially in touching on that most sensitive of subjects, Lodge's interest in ghosts.

First Marconi took on Preece. "Sir William Preece is, I believe, a gentleman with various claims to scientific distinction; but, whatever his attainments in other walks of science, I regret to say that the most careful examination reveals absolutely no testimonial to his competency for this most recent of his undertakings. Such knowledge of my work as he may possess is at least three years old—a very long period, I would remind you, in the brief history of my system. . . . Of the conditions under which the system is now worked Sir William Preece is, in fact, wholly ignorant."

Now he addressed Lodge's criticisms. "I regret to say that, distinguished as Dr. Lodge may be as a professor of physics or as a student of psychical phenomena, the same statement applies also in his case, so far as my present system or wireless telegraphy is concerned."

Marconi declared that his tuning technology allowed him to send messages across the Atlantic "without interfering with, or, under ordinary conditions, being interfered with, by any ship working its own wireless installations." He then challenged Preece and Lodge to attempt to interfere with his transmissions and even offered them the use of his own stations for the experiment.

His shareholders applauded, but to others outside the company, his remarks, published in the press, smacked of impudence and mockery.

The *Westminster Gazette* suggested that "Signor Marconi would have done better if he had spared his sneers at the capacity of the more important of his critics. . . . Bitter retorts and jeers at the intelligence of opponents are not the marks of the scientific spirit. There would seem to be no a priori reason why the student of psychic phenomena should not be permitted to express an opinion upon the future of wireless telegraphy."

The *Electrical Times* condemned Marconi for speaking "with scarcely veiled contempt" of Lodge and Preece. "Had it not been for the scientific work of the former it is doubtful whether Mr. Marconi would have had any wireless telegraphy to boast about, while to the latter he is indebted for help and encouragement when he first came to England. . . . But, apart from that, the tone Mr. Marconi adopts is hardly decent in so young a man towards one so much his senior and of so high a standing in the engineering and scientific world."

The journal further charged that if no one knew much about the current state of Marconi's technology, it was Marconi's own fault. "If Mr. Marconi would but describe his methods and apparatus openly and fully, as scientific men are accustomed to do, he would find no lack of sympathy and appreciation."

Far from ending here, the battle was about to get a lot uglier.

TWO DAYS LATER, on Saturday, February 22, 1902, Marconi once again boarded the *Philadelphia*. The main purpose of this voyage was to return to Canada to close the agreement with the government, but he also saw an opportunity to counter the skepticism confronting his Newfoundland achievement. He installed a new and taller antenna on the *Philadelphia* to attempt to increase the range at which signals could be received from Poldhu, and invited the ship's captain, A. R. Mills, to witness the tests. He abandoned the telephone receiver he had used in Newfoundland and attached his usual Morse inker, so that at least there would be a physical record of whatever signals came through.

Everyone by now accepted that Marconi's system worked well over short distances, so the first messages exchanged with his shore stations caused little stir. It was on the morning of the second day, when the ship was precisely 464.5 miles from Poldhu, that things got interesting.

The equipment snapped to life. The receiver captured the message, "All in order. V.E.," with *V.E.* being code for "Do you understand?"

Messages and S's continued to arrive as per Marconi's schedule.

At 1,032.3 miles the ship received this message: "Thanks for telegram. Hope all are still well. Good luck."

Five hundred miles later the last message containing complete words arrived. "All in order. Do you understand?" But even at 2,099 miles from Poldhu the ship's receiver continued to pick up distinct three-dot patterns.

Captain Mills saw the blue dots as they emerged from the inker. Marconi turned to him. "Is that proof enough, Captain?"

It was. The captain agreed to stand witness and signed the tape and a brief affidavit reading, "Received on S. S. 'Philadelphia,' Lat. 42.1 N., Long. 47.23 W., distance 2,099 (two thousand and ninety-nine) statute miles from Poldhu."

On landing in New York, Marconi told a gathering of reporters, "This merely confirms what I have previously done in Newfoundland. There is no longer any question about the ability of wireless telegraphy to transmit messages across the Atlantic." In an interview with H. H. McClure, Marconi said, "Why, I can sit down now and figure out just how much power, and what equipment would be required to send messages from Cornwall to the Cape of Good Hope or to Australia. I cannot understand why the scientists do not see this thing as I do."

But the voyage had brought forth a troubling revelation, which Marconi for now kept secret. He had discovered that during daylight hours, once the ship was more than seven hundred miles out, it received no signals at all, though reception resumed after dark. He called this the "daylight effect." It seemed, he said, that "clear sunlight and blue skies, though transparent, act as a kind of fog to powerful Hertzian rays."

A couple of months later, still mystified and frustrated by the effect, Marconi was less judicious in his choice of words. "Damn the sun!" he shouted. "How long will it torment us?"

THAT SAME SPRING Marconi discovered that he had made a personal enemy of Kaiser Wilhelm.

It was a minor incident and very likely did not happen in the way the kaiser believed, but it occurred against a backdrop of degrading relations between Germany and Britain. Wilhelm's drive to strengthen the German Navy had prompted Britain's leaders to reconsider the merits of "splendid isolation" and to contemplate alliances with Russia and the once-feared French. That summer the *Daily Mail* would go so far as to recommend a preemptive strike at the German fleet, expressing in print an idea already in private circulation in the clubs of London and among some military planners.

The growing discord had its private analog in the long-standing animosity between Marconi and Adolf Slaby, and between Marconi's company and its German opponent, Telefunken, which had begun marketing the Slaby-Arco-Braun apparatus around the world. Even the U.S. Navy was a customer. For Kaiser Wilhelm and Telefunken officials, Marconi's policy that ships equipped with Marconi apparatus communicate only with other Marconi stations had become a source of rising irritation.

So things stood when, early in 1902, Prince Heinrich of Prussia, Kaiser Wilhelm's younger brother, set out for New York aboard the German liner *Kronprinz Wilhelm,* equipped with Marconi's tunable wireless. As the ship came within range of the Lizard and Poldhu, the prince observed a demonstration of how messages from both stations could be received simultaneously through the ship's one antenna. As the liner approached New York, the prince discovered to his surprise that communication between ships and a new Marconi station on Nantucket had become almost routine. (The new South Wellfleet station, with its four giant towers, was under construction.)

During his voyage back home, Prince Heinrich sailed aboard another German liner, the *Deutschland,* but this ship was equipped with Telefunken apparatus. The prince expected once again to experience the miracle of wireless conversation, but heard nothing from Nantucket, the

Lizard, or Poldhu. Charges arose that Marconi's men had chosen to snub the *Deutschland,* and by proxy the prince himself, and might even have jammed her wireless. The kaiser was furious, as was the German public. A wave of what one journal called "malignant Marconiphobia" swept across Germany.

But the Marconi company had not jammed the German ships wireless. Out of respect for the prince, it had ordered its operators to suspend temporarily the prohibition against conversing with alien apparatus. The cause of the silence encountered by the *Deutschland* cannot be known, but may have been a technical fault in the Telefunken apparatus.

Kaiser Wilhelm chose to see it as a deliberate affront and demanded that an international conference be convened to establish rules for wireless at sea. Marconi understood that his true intent was to seek an agreement requiring that all wireless systems communicate with one another. Marconi saw this proposal as a serious threat and condemned it. His company had built the world's most elaborate and efficient network of wireless stations. To allow others now to use this network, Marconi argued, was simply unfair.

To Lodge and other Marconi critics, Kaiser Wilhelm's campaign promised a comeuppance for Marconi that was long overdue. On April 2, 1902, Sylvanus Thompson wrote to Lodge, "Marconi's whining about others coming in to rob him of the fruits of his work is too funny—a mere adventurer like him with his pinchbeck claims to be an original inventor!" (The word *pinchbeck,* from the name of an eighteenth-century watchmaker, is an archaic term for a goldlike alloy used in cheap jewelry. It served as a synonym for such words as *counterfeit, fake,* and *sham.*)

Relations with Germany degraded further. At Glace Bay Richard Vyvyan and his men got an unexpected, and unwelcome, visit from the Imperial German Navy. As they worked atop the cliffs at Table Head, they caught sight in the distance of a fleet of ships, which anchored off Glace Bay. Vyvyan immediately guessed their purpose, for the station was the only thing likely to draw the Germans to this desolate and dangerous roadstead.

A party came ashore that included an admiral and thirty officers. The day was hot, the walk a long one. Vyvyan met them at the gate to the station and offered refreshments.

The admiral declined. He and his men, he said, had come to see the station.

Vyvyan told him he would be delighted to show him around, provided of course that the admiral possessed written authorization from Marconi or the directors of the company.

The admiral had neither.

Vyvyan expressed his deepest regret. Without such authorization, he said, it simply was impossible to admit the admiral and his landing party.

The admiral bristled. He declared that His Imperial Majesty, Kaiser Wilhelm, would hear of the incident and be furious.

Vyvyan was very sorry to hear it but was helpless to do anything further in the matter. He again offered his regrets. The admiral and his staff trudged off.

But the fleet remained at anchor. Vyvyan posted a sentry in one of the new towers.

His instincts proved correct. The next day the sentry spotted boats pulling away from the fleet with about 150 men on board. They landed and gathered at the gate. This time, Vyvyan noticed, no officers accompanied them.

The men attempted to push past him "in an unruly mob."

Vyvyan stood his ground. "I informed them admission was forbidden and if they persisted I would use force to prevent them entering the station."

The compound behind him was full of workmen, who sensed trouble and began to converge on the gate. Tension mounted.

But then, unexpectedly, one of the Germans blew a whistle. The sailors formed ranks and departed, transformed suddenly into "a disciplined force, and no longer an unruly crowd of men."

The fleet departed.

Frustrating failures in Marconi's long-range system continued to haunt him.

In June 1902 Edward was to have his coronation but was felled by appendicitis. At first the likelihood of his survival seemed slim, but he underwent surgery and survived, and once again he retreated to a royal yacht, the *Victoria and Albert,* to recover. Meanwhile the dignitaries dispatched to attend the coronation abruptly found themselves without a mission. Italy had sent a warship, the *Carlo Alberto,* and now loaned the ship and its six-hundred-man crew to Marconi to use as a floating laboratory until Edward's recovery was advanced enough for the coronation to take place.

Italy's King Victor Emmanuel III decided that in the interim he would pay a visit to Tsar Nicholas II of Russia. He ordered the *Carlo Alberto* to meet him in Kronstadt, the Russian naval base, where he and the tsar would come aboard for a demonstration of Marconi's wireless. En route, during a stop at the German naval port of Kiel, Marconi was able to receive signals at six hundred miles, and on the night of July 15, 1902, while in Kronstadt harbor, at sixteen hundred miles. But he found again that sunlight played havoc with daytime reception, and he heard nothing from Poldhu between sunrise and sunset. Which now posed a problem, what with King Victor Emmanuel and Tsar Nicholas about to visit. Marconi wanted to demonstrate the receipt of a message to his royal visitors but knew that it would be awkward to insist that they visit after dark. Luigi Solari proposed that Marconi install a wireless transmitter elsewhere on the ship and send a message from there. He intended no deception, he claimed, merely to demonstrate by day what could easily be achieved at night.

On July 17 the king and tsar came aboard and proceeded to Marconi's wireless cabin, where Marconi showed off tapes of the messages received from Poldhu. Suddenly the receiver came to life and the Morse inker printed out a message of welcome and congratulations for Nicholas.

Startled and impressed, the tsar asked where the message had originated. Marconi confessed and disclosed the hidden transmitter. The tsar took no offense, apparently, for he asked to meet Solari and applauded his ingenuity.

The next month, while still engaged in experiments aboard the *Carlo Alberto,* Marconi confronted an inexplicable failure of his system. In one experiment he planned to receive messages for King Victor Emmanuel sent via Poldhu, but no messages came through. Nothing he tried improved reception, and he could find no good reason for the failure. He once had told Solari, "I am never emotional." But now Solari watched as he smashed the receiver to pieces.

Marconi blamed Fleming. Without consulting Marconi, Fleming had altered a key component of the Poldhu station, thereby reversing a previous change ordered by Marconi himself. Fleming had also installed a new spark device of his own design.

Marconi complained to his new managing director, Cuthbert Hall, who had been the company's second-ranked manager until the resignation a year earlier of Major Flood Page. Fleming's device, Marconi wrote, had "proved in practical working to be unsatisfactory."

Marconi ordered his men at Poldhu to replace Fleming's invention with one of his own design—and now Fleming felt slighted. He objected that he ought to be consulted before changes of that magnitude were made.

Which only annoyed Marconi further.

In another letter to Cuthbert Hall, Marconi wrote, "It should be explained to [Fleming] that his function as Consulting Engineer is simply to advise upon points which may be expressly referred to him and in no way places upon the Company any obligation to seek his advice upon any matters in which it is deemed unnecessary. . . . I do not wish to inflict any unnecessary wound on Dr. Fleming's susceptibilities, but, unless you are able to put the matter before him effectively in a right light, I shall feel bound to make a formal communication to the Board with reference to his general position."

None of this, however, made it into a report by Luigi Solari on the *Carlo Alberto* experiments, published in the October 24, 1902, edition of *The Electrician.* His account made it seem as if everything had gone exactly as planned. Ordinarily readers would have had to accept Solari's report at face value, for once again Marconi had made no provision for an impartial observer to vouch for his results.

In this case, however, someone else happened to have been listening in, without Marconi's knowledge.

That summer the Eastern Telegraph Co., an undersea cable concern, had decided to install a wireless station of its own, at its cablehead at Porthcurno in Cornwall, about eighteen miles from Poldhu. The transatlantic cable industry still did not expect much competition from wireless but did see that it might have value as a source of additional traffic to be fed into their cables and for communicating with cable-repair ships. Eastern Telegraph hired Nevil Maskelyne for the job, and in August 1902 the magician erected a temporary antenna twenty-five feet tall. Immediately Maskelyne began picking up Morse signals from Poldhu, something the Marconi company had touted as being next to impossible given its tuning technology.

Maskelyne picked up a repeated signal, the letters CBCB. "Knowing that experiments were in progress between Poldhu and the Carlo Alberto," Maskelyne wrote, "it did not take a Sherlock Holmes to discover that 'CBCB' was the call signal for the Carlo Alberto." He and Eastern's men nicknamed the ship the *Carlo Bertie.*

Maskelyne not only listened but kept copies of the tapes that emerged from his own Morse inker. Their true significance was not yet clear to him.

The Ladies Investigate

First she disappeared, allegedly to America, and now she was dead. None of it made sense; all of it stretched credibility. It was wonderful, in the Edwardian sense of the word, yet here was Crippen, the very soul of credibility, telling them it was so. He was, according to Maud and John Burroughs, "a model husband"; so "kind and attentive," said Clara Martinetti; a "kind-hearted humane man," said Adeline Harrison.

And yet.

There was the rising sun brooch worn so brazenly by the typist, and the fact that Belle had neither written nor cabled her friends since her departure and had not thought to send a wireless message—by now a "Marconigram"—from her ship, the kind of thing she would have delighted in doing for the surprise of it. There was the fact too that Crippen all along had seemed unsure of Belle's exact whereabouts and was unable to produce an address. She was in the "wilds of California," as he had put it, yet Belle never had mentioned relatives in California, let alone in the state's nether portions.

Even before word arrived of Belle's death, her successor as guild treasurer, Lottie Albert, asked a friend, Michael Bernstein, to make inquiries about Belle on behalf of the guild.

Crippen had said Belle had sailed aboard a ship of French registry and that it had sailed out of Le Havre. The name, he thought, was something like *La Touee* or *Touvee.* Bernstein searched the passenger lists of French ships for a passenger named Crippen or Elmore but found nothing.

On March 30, a Wednesday and thus a day when the Ladies' Guild met, Clara Martinetti and Louise Smythson walked down the hall to Crippen's office ostensibly to offer condolences. In fact, they intended to perform a kind of interrogation.

Mrs. Martinetti asked him for the address of the person who had nursed Belle in her last moments, but Crippen said he did not know who it was.

She asked how long Belle had been ill. Crippen said she had become ill on the ship and failed to look after herself and as a consequence contracted pneumonia.

Mrs. Martinetti asked where Belle was buried and explained that the guild wanted to send an "everlasting wreath" to place on her grave. Crippen said she had not been buried—she had been cremated, and that soon her ashes would arrive by post.

Cremated.

Belle had never once mentioned a wish to be cremated after death. She was so forthcoming about everything in her life, to the point of having friends touch her scar, that surely she would have mentioned something as novel as cremation.

Mrs. Martinetti asked where Belle had died. Crippen did not answer directly. He said only, "I will give you my son's address."

"Did she die with him, and did he see her die?" Mrs. Martinetti asked.

Crippen answered yes, but in a confused manner, then gave her Otto's address in Los Angeles.

She and Mrs. Smythson left, their suspicions aflame. Mrs. Martinetti immediately wrote a postcard to Otto asking for details of Belle's death.

It took him a month to reply. He apologized for the delay but explained that he had been distracted by the illness and death of his own son.

Turning to the subject at hand, he wrote, "The death of my stepmother was as great a surprise to me as to anyone. She died at San Francisco and the first I heard of it was through my father, who wrote to me immediately afterwards. He asked me to forward all letters to him and he would make the necessary explanations. He said he had through a mistake given out my name and address as my step-mother's death-place. I would be very glad if you find out any particulars of her death if you would let me know of them as I know as a fact that she died at San Francisco."

AT NO. 39 HILLDROP CRESCENT Ethel Le Neve cleaned house. This proved a challenge. To begin with, the place smelled terrible, especially downstairs in the vicinity of the kitchen, though the odor to a degree had permeated the entire house. Mrs. Jackson sensed it immediately on her first visit and mentioned it to Ethel. "The smell," Jackson said, "was a damp frowsy one, and might have resulted from the damp and dirt. It was a stuffy sort of smell."

"Yes," Ethel told her, "the place is very damp and in a filthy condition. This is how Belle Elmore left it before she went away to America."

Ethel opened windows and cleared away clothing and excess furniture, piling much of it in the kitchen. Crippen invited his longtime employee William Long to come over and see if there was anything he wanted. "A night or two after this," Long said, "I went there and in the kitchen he shewed me a pile of woman's clothing such as stockings, underclothing, shoes and a lot of old theatrical skirts, and old window curtains, table cloths, rugs, etc."

Over several evenings Long took it all. Crippen also gave him the gilt cage and the seven canaries.

Ethel hired a servant, a French girl named Valentine Lecocq. "I have at last got a girl which I am thankful for," Ethel wrote to Mrs. Jackson. "She is only 18 yrs. but seems anxious to learn & willing enough. The poor girl however hasn't hardly a rag to her back, not a black blouse or anything & as Dr. is asking some friends to Dinner next Sunday, I feel I must rig her out nice & tidy."

With the French girl's help, Ethel made progress. In another letter to Mrs. Jackson she wrote, "Have been ever so busy with that wretched house and think you would hardly recognize same." She found it hard to keep the house "anywhere near clean" and at the same time attend to her duties at Crippen's office. "It gives me little time to myself," she complained.

But the housework soon would end, she knew. Crippen's lease was to expire on August 11, at which point they planned to move to a flat in Shaftesbury Avenue.

"Still," she told Mrs. Jackson, "notwithstanding the hard work [I] am indeed happy."

She delighted in the little moments with Crippen. In her memoir she wrote, "He used to come with me to the coal cellar, scuttle in hand, and while he was shoveling up the coals I would lean up against the door holding a lighted candle and chatting with him."

The house grew brighter and more welcoming, and the awful scent dissipated. Crippen helped whenever he could and every day held her, kissed her, talked with her. They were not yet married in the eyes of the law and could not be married until Belle's death in America was duly certified, but they were as much husband and wife as could be.

"So time slipped along," Ethel wrote, "—both of us extremely happy and contented, working each of us hard in different ways."

The ladies watched.

They saw Crippen leave with the typist and arrive with the typist. They saw them walking together. The typist wore furs that looked very much like Belle's, but of course one could never be sure, as furs were hard to tell apart. They saw them together at the theater and at restaurants. One day Annie Stratton and Clara Martinetti ran into Crippen on New Oxford Street. "Whilst we were talking to him," Mrs. Martinetti said, "he seemed anxious to get away, and after he left us I saw him joined by the typist, and both got into a bus."

And the ladies learned a troubling fact: Only one French liner had been scheduled to sail for America on the day of Belle's departure, a steamer called *La Touraine.*

The ship had never left port, however. It was under repair.

Strange news, but then these were strange times. On May 6, 1910, at 11:45 P.M. King Edward VII died, casting the nation into mourning. For the first time in England's history the directors of Ascot ruled that all in attendance must wear black, a moment known ever after as "Black Ascot" and familiar in future generations to anyone who saw *My Fair Lady.*

As if the world really were coming to an end, Halley's comet appeared in the skies overhead, raising fears of a collision and prompting rumors of dire events yet to come.

A Duty to Be Wicked

Marconi's long voyage of experiment aboard the *Carlo Alberto* ended on Halloween morning 1902, when the ship arrived at Nova Scotia. Marconi's goal—his hope—was now to move beyond mere three-dot signals and send the first complete messages from England to North America. It was imperative that he succeed. Skepticism about his transmission to Newfoundland had continued to deepen. Success would not only counter the doubters but also ease growing worries among his board of directors about whether all this costly experimenting would ever yield a financial return.

By now Marconi had completed construction of the new stations at South Wellfleet and Poldhu, and on Table Head at Glace Bay, the most powerful of all. Each station had more or less the same design: four strong towers of cross-braced wood, each 210 feet tall, supporting an inverted pyramid of four hundred wires. Each station had a power house nearby, where steam engines drove generators to produce electricity, which then entered an array of transformers and condensers. At South Wellfleet the process yielded 30,000 watts of power, at Glace Bay 75,000. At South Wellfleet a thick glass porthole and soundproof door had to be installed between the sending room and the sparking apparatus to prevent injury to the operator's eyes and ears.

Marconi began his new attempt the day after his arrival, coordinating each step with his operators at Poldhu through telegrams sent by conventional undersea cable. The first signals to arrive "were very weak and unintelligible," according to Richard Vyvyan. But they did arrive. Heartened by the fact that Poldhu had been operating only at half power, Marconi ordered his engineers there to increase wattage to maximum, expecting it would resolve the problem. It didn't. Now he heard nothing at all.

The hundreds of wires that comprised the Glace Bay aerial could be used all at once or in segments. Marconi and Vyvyan tried different combinations. Again, nothing. Night after night they worked to find the magic junction with only trial and error as their guide. To attempt to receive by day seemed hopeless, so they often worked the entire night. Over eighteen consecutive nights they received no signals. Tensions grew, especially in the Vyvyan household. He had brought his new wife, Jane, to live with him at Glace Bay, and now she was pregnant, hugely so, the baby due any day.

Snow began to fall and soon covered the clifftop. At night the sparks from the transmitter lit the descending flakes. With each concussion a pale blue aura burst across the landscape, as if the transmission house were a factory stamping out ghosts for dispersal into the ether. Three-foot daggers of ice draped wires.

In the midst of it all Marconi received a telegram from headquarters, stating that the price of his company's stock was falling. Though Marconi did not yet know it, the decline was conjured by a magician.

NEVIL MASKELYNE LOATHED FRAUD but loved to mislead and mystify audiences. His base was the Egyptian Hall in Piccadilly, one of London's most popular venues for entertainment and one of the city's strangest buildings. "It is beyond the powers of delineation to attempt any thing in the shape of a description of the front of this most singular piece of architecture," wrote one early visitor. Built in 1812, its facade mimicked the entrance to an Egyptian temple. Two huge figures jutted from its yellow cladding, and hieroglyphs covered its pilasters and sills. The building originally served as a museum of natural history but failed to draw many visitors and became instead a venue for the display of a succession of oddities, including an entire family of Laplanders, an eighty-pound man called the Living Skeleton, and in 1829 the original Siamese twins. Its most famous living exhibit was one of its smallest, a native of Bridgeport, Connecticut, named Charles S. Stratton, placed on display here in 1844 by Phineas T. Barnum. By then Stratton was best known by his stage name, General Tom Thumb.

Nevil's father, also named John Nevil Maskelyne, took over the Egyptian Hall with a partner, George A. Cooke, and by 1896 turned it into "England's House of Mystery," where twice a day audiences watched magic shows and encountered illusions and mechanical chimeras. By then Maskelyne and Cooke, as they were known, had achieved fame by exposing two celebrated American spirit mediums, the Davenport brothers. The magicians billed themselves as "The Royal Illusionists and anti-Spiritualists." One of their most popular attractions was an automaton named Psycho, an oriental mystic whose robe and turban disguised internal mechanical devices that enabled him to solve math problems, spell words, and most famously, play whist with members of the audience. Nevil Jr. took over from his father, and when he was not dabbling in wireless, he performed in shows alongside his own partner, a magician named David Devant. Together Maskelyne and Devant revealed to audiences the tricks deployed by mediums, with such aplomb that some Spiritualists believed they really did have psychic powers and merely pretended disbelief in a cynical drive to make profits.

Maskelyne did not trust Marconi. The Italian claimed to have performed amazing feats but provided little hard proof beyond the testimonials of such allies as Ambrose Fleming and Luigi Solari. The latest example was Solari's glowing report in *The Electrician* about Marconi's *Carlo Alberto* experiments.

Maskelyne read it with distaste, then delight. Suddenly he realized that the tapes he had collected while eavesdropping on Marconi's transmissions included some of the messages Solari described. These tapes showed that Marconi's system was more flawed than he was letting on.

Maskelyne decided to reveal his findings. In an article published by *The Electrician* on November 7, 1902, he disclosed that using his own apparatus at the Porthcurno station near Poldhu, he had intercepted Marconi's signals and that the tapes from his Morse inker proved that Solari's account had been less than accurate. He stopped short, however, of accusing Solari and Marconi of fraud.

The tapes, he wrote, showed that errors due to atmospheric distortions were common and that transmissions from some other station had interfered with communication between Poldhu and the *Carlo Alberto*. Maskelyne also challenged Solari's claim that the Poldhu station could transmit at a rate of fifteen words a minute. By his own count, he wrote, the rate was closer to five.

He also addressed a claim by Solari that a message from the Italian Embassy in London, transmitted by wireless from Poldhu, had been received without flaw aboard the ship at precisely four-thirty P.M. on September 9, 1902. In fact, Maskelyne found, transmission of the message had begun several nights earlier, on September 6 at nine o'clock. (This may have been the message that drove Marconi to smash his equipment.)

One thing was certain: Maskelyne had proven that Marconi's transmissions could be intercepted and read. He wrote, "The plain question is, can Mr. Marconi so tune his Poldhu station that, working every day and all day, it does not affect the station at Porthcurno? Up to September 12th, on which date my personal supervision of the experiments at Porthcurno ceased, he had only succeeded in proving that he cannot do so."

Cuthbert Hall, Marconi's managing director, countered with a letter to *The Electrician* in which he wrote that "the evidence furnished of interception of our messages . . . is not conclusive." He argued that anyone could take the messages published in Solari's article and use a Morse inker to produce counterfeit tapes. "To have any value whatever as evidence, Mr. Maskelyne's article should have been published before, not after, Lieut. Solari's report."

Hall's argument must have struck Maskelyne as ironic, given Marconi's penchant for describing his own triumphs through trust-me testimonials that could not be counterchecked for validity.

In the next issue Maskelyne responded: "Clearly Mr. Hall is between the horns of a dilemma. He must either say I am a liar and a forger, or he must accept the situation as set forth in my article. . . . If it be the former, I shall know how to deal with him. If it be the latter, the airy fabric of over-sanguine and visionary expectation, which we have so long been called upon to accept as a structure of solid fact, must fall to the ground."

At Glace Bay silence prevailed. Nothing explained the persistent failure to receive signals from Poldhu. In Newfoundland, with kites bobbing in the air, he had received signals, but here at this elaborate new station with its 210-foot towers and miles of wire, he received nothing. He and Vyvyan decided to try something they so far had not attempted—reversing the direction of transmission, this time trying to send from Nova Scotia to England. They had no particular reason for doing so, other than that nothing else had worked.

They made their first attempt on the night of November 19, 1902, but the operators at Poldhu received no signals.

Marconi and Vyvyan made countless adjustments to the apparatus. Vyvyan wrote, "We did not even have means or instruments for measuring wavelengths, in fact we did not know accurately what wavelength we were using."

They tried for nine more nights, with no success. On the tenth night, November 28, they received a cable stating that the operators at Poldhu had received vague signals, but that they could not be read. This buoyed Marconi, though only briefly, for the next night Poldhu reported that once again nothing had come through. The silence continued for seven more nights.

On the night of Friday, December 5, Marconi doubled the length of the spark. Later that night he received word back, via cable, that Poldhu at last had achieved reception:

> *Weak readable signals for first half-hour, nothing doing during the next three-quarters, last three-quarters readable and recordable on tape.*

The next night Marconi tried exactly the same configuration.

Nothing.

The following night, silence again.

Marconi had borne these weeks of failure with little outward sign of frustration, but now he cursed out loud and slammed his fists against a table.

But he kept trying. Failure now, even rumor of failure, would be ruinous. Not surprisingly, word had begun to leak that he might be in trouble. On Tuesday, December 9, 1902, a headline in the *Sydney Daily Post* asked, "WHAT'S WRONG AT TABLE HEAD?" The accompanying article said, "Something strange seems to have happened at Table Head, but that something doesn't look very encouraging to the promoters of the scheme."

That night every attempt to reach Poldhu failed. Failure dogged him for the next four nights. On the fifth night, Sunday, December 14, after hours of pounding messages into the sky, a cable arrived from Poldhu: "Readable signals through the two hours programme."

Given all they had experienced since Marconi's Halloween arrival, this was cause for celebration. The men tore from the operator's room into the frozen night and danced in the snow until they could no longer stand the cold.

It seemed, for the moment, that by sheer chance Marconi had struck exactly the right combination of variables. Rather than wait to confirm this, as prudence might have dictated, Marconi now proceeded to the next step of his plan, to send the first-ever *public* message across the ocean by wireless. He made the decision to try it, according to Vyvyan, "owing to financial pressure and to quiet the adverse press criticism that was making itself noticeable."

This time he recognized that his testimony alone would not be enough to persuade a skeptical world of his achievements. He invited a reporter, George Parkin, Ottawa correspondent for the London *Times,* to write this historic message and to serve as a witness to the process. First, however, Marconi swore Parkin to secrecy until accurate reception of the message by the Poldhu station could be confirmed.

Marconi made the first attempt to send the message early on Monday, December 15, less than twenty-four hours after the cable from Poldhu that had caused so much celebration. He asked Parkin to make a change in the wording of his message just before transmission, to neuter any potential claim that Marconi's men in Britain had somehow acquired an advance copy. At one o'clock in the morning, Marconi grasped the heavy key and began levering out the message. "All put cotton wool in their ears to lessen the force of the electric concussion," Parkin wrote. He likened the clatter to "the successive explosions of a Maxim gun."

The message failed to reach Poldhu. At two o'clock Marconi tried again. This attempt also failed.

Marconi repeated the attempt that evening, first at six o'clock, then at seven, without success. Later that night, between ten and midnight, Parkin's message did at last reach Poldhu. It read:

> *Times London. Being present at transmission in Marconi's Canadian Station have honour send through Times inventor's first wireless transatlantic message of greeting to England and Italy. Parkin.*

Marconi arranged a celebration later that morning, during which the flags of Britain and Italy were raised with great ceremony.

A sudden gale promptly destroyed both.

PARKIN'S MESSAGE WAS NOT immediately relayed to *The Times*. Marconi's sense of protocol and showmanship required that first two other messages of greeting had to be transmitted, one to King Edward, the other to King Victor Emmanuel in Rome. Marconi had instructed Poldhu not to relay Parkin's message, and Parkin to hold back his story, until the two royal messages could be transmitted and their contents confirmed by return cable. This process required six days.

Parkin crafted an account that glowed with praise, including his "feeling of awe" at the fact that impulses sent from Glace Bay would reach Poldhu in one-thirtieth of a second. He neglected to mention the six-day delay.

Vyvyan, in his memoir, was more candid. "Although these three messages were transmitted across the Atlantic and received in England it cannot be said that the wireless circuit was at all satisfactory. There was a great element of uncertainty as to whether any message would reach its destination or not, and so far the cause of this unreliability had not been ascertained. All conditions remaining the same at the two stations, the signals would vary from good readable signals to absolutely nothing and often vary through wide degrees of strength in two or three minutes."

Further evidence of this unreliability struck close to home for Vyvyan. On January 3 his wife gave birth to a healthy daughter. There was celebration and of course a transatlantic message by wireless to *The Times* of London. But an atmospheric distortion converted a reference to *Jan* as in January, to the name *Jane*. The telegram as received in Poldhu had a Bluebeardesque cast:

> *Times London by transatlantic wireless Please insert in birth column Jane 3rd wife of R. N. Vyvyan Chief Engineer Marconi's Canadian Station of a daughter. Marconi.*

EMBOLDENED BY HIS NEW VICTORY, Marconi now prepared to capitalize on it with a final achievement that he hoped would at last empty the sea of doubt. On January 10, 1903, he left for Cape Cod, intent on sending the first all-wireless message from the United States to England. He carried in his pocket a greeting from President Theodore Roosevelt to be sent to King Edward. He believed he could not send the message directly from Cape Cod because the station did not have the necessary power, and planned instead to send it by wireless from South Wellfleet to Glace Bay in Nova Scotia, for relay across the ocean.

Roosevelt's message trudged from Cape Cod to Nova Scotia in fits and starts, as if Glace Bay were at the far side of the planet, not just six hundred miles to the northeast. Meanwhile, to everyone's astonishment, the message also traveled direct to Poldhu, where it arrived long before the relayed message came struggling in from Glace Bay.

For once the system had performed far better than expected. But now came a miscalculation, and a costly one.

The operators at Poldhu sent a return greeting from King Edward, for Roosevelt. They sent it, however, by conventional undersea cable.

Marconi had seen no other choice: His hard experience at Glace Bay had shown that for whatever reason the Poldhu station could not transmit messages to Nova Scotia. In reporting the event, however, countless newspapers reprinted the two messages one atop the other, a juxtaposition that suggested a fluid back-and-forth exchange entirely by wireless.

When it became clear that Edward's message had traveled the conventional route, Marconi's critics seized on the episode as evidence of the continuing troubles of wireless and accused Marconi of creating the false impression that two-way wireless communication across the sea had been achieved. In London managing director Cuthbert Hall claimed the decision to send the return message by cable was due solely to a need to be courteous to King Edward. He explained that the royal reply had been handed in on a Sunday, when the telegraph office nearest to Poldhu was closed. The telegram would not have been delivered to the operators at Poldhu until Monday morning at the earliest, and only then could they have begun their attempt to send it by wireless. It was far more respectful, Hall argued, to get the king's message out immediately, even if that meant sending it by cable.

Marconi's critics sensed blood. The head of the Eastern Telegraph Co., Sir John Wolfe Barry, cited Marconi's use of cable as further evidence that wireless never would become a serious competitor.

The *Westminster Gazette* sent a reporter to ask Marconi about the incident.

"I was not concerned with the reply, nor how it came," Marconi said. "You know the local telegraphic office near Poldhu was closed, and all that. But what I wished to demonstrate was that a message could be sent across the Atlantic. It did not matter from a scientific point of view whether it came from east to west or west to east. No scientific man would say that it did." But Marconi said nothing about the previous winter's struggle at Glace Bay to receive anything at all from Poldhu, let alone a complete message from a king. Instead, he told the reporter, "If it could go one way, why not the other?"

Yet Marconi and his engineers were fully aware of the shortcomings of his transatlantic system. Vyvyan wrote, "It was clear that these stations were not nearly in a position to undertake a commercial service; either more power would have to be used or larger aerials, or both." On January 22, 1903, at great cost to his company and to the dismay of his board, Marconi shut down all three stations for three months to reappraise their design and operation. He sailed for home aboard Cunard's *Etruria*.

ON RETURNING TO LONDON he discovered that Maskelyne's attacks had begun to resonate with investors and the public alike. In the *Morning Advertiser* a writer adopting the name Vindex proposed that Marconi could easily resolve public doubt about his invention by subjecting it to a test whose every aspect would be open to public scrutiny. He proposed that Marconi send a transatlantic message to Poldhu at a predetermined time, with transmission and receipt observed by the editors of four American newspapers and four English.

Dubbed immediately the "Vindex Challenge," the proposal gained popular endorsement. The public had grown accustomed to verifiable displays of progress, such as races between transatlantic ocean liners. Now Marconi was promising the ultimate in speed. If he wanted the world to believe his fantastic claims that he could send messages across the Atlantic in an instant, he should provide evidence and reveal his methods.

One reader wrote to the *Morning Advertiser*, "If 'Vindex' does no more than secure the demonstration for which he asks, he will be doing a great service to the Marconi Company, and a greater service to the public in destroying the rumors which are current about the Transatlantic service, and, further, in establishing the claim of the Marconi Company to the assistance of the public in its fight with the vested interest of the cable companies. . . .

"If Mr. Marconi successfully passes his test I am sure he will have the whole-hearted support not only of your paper but of every honest Englishman in his fight against capital and political influence."

He signed his letter, "A BELIEVER IN FAIR PLAY."

The *Westminster Gazette* put the question directly to Marconi: Why *not* give a demonstration for the press?

"Well, we have got beyond that," Marconi said. "It would be casting doubt upon what is clearly proved. What is there to demonstrate? It might have done some time ago, I admit; but not now, I think. But I should not mind showing to anyone of standing and position who does not start off from a sceptical point of view. I will not demonstrate to any man who throws doubt upon the system."

THE TIMING OF THIS CONTROVERSY was especially awkward. Even as it flared, Marconi and Fleming were preparing a series of tests meant to quash the equally prevalent skepticism about Marconi's ability to send tuned messages, and to address a new concern raised by critics as to whether a transmitter big enough to send signals across the Atlantic would disrupt communication with other stations. Marconi asked Fleming to devise an experiment to prove that high-power stations would not, as Fleming put it, "drown the feebler radiation" involved in communication between ships and between ships and shore.

Instead of trying to incorporate transmissions from actual ships into his experiment, Fleming installed a small marine set in a hut about one hundred yards away from the giant Poldhu aerial and connected it to a simple one-mast antenna. He planned to send messages from the big and small transmitters simultaneously, each on a different wavelength, to Marconi's station at the Lizard. He attached two receivers to the Lizard's antenna, one tuned to capture the high-power messages, the other to receive messages from the simulated ship.

Fleming created sixteen messages, eight to be sent from the high-power transmitter, eight from the low. He put each into an envelope, "no person except myself knowing the contents," and wrote on each the time at which the enclosed message was to be sent. Four messages were in code. Each high-power message was to be transmitted at the same time as a low-power message and repeated as many as three times.

On the day of the experiment, Fleming gave all the envelopes to an assistant "unconnected with the Marconi Company, in whose integrity and obedience I had confidence" and instructed him to deliver the envelopes to the operators at the times selected. The assistant signed an affidavit confirming that Fleming's instructions had been "precisely obeyed."

But as any of Fleming's peers in academic science instantly could see, Fleming's precautions—his sealed envelopes, the coded messages, the unknowing assistant—created only an illusion of scientific rigor. They reflected the tension between science and enterprise, openness and secrecy, that continued to shape the behavior of Marconi and his company and that in turn had the perverse effect of helping sustain the suspicions of his most steadfast critics.

By Fleming's account, all the messages arrived at the Lizard on schedule and were recorded on tape by two Morse inkers. Fleming collected the ink rolls and turned them over to Marconi for translation from Morse to English. "In every case he gave the absolutely correct message which was sent," Fleming reported.

Well, not *absolutely*. In the next sentence of his report Fleming dimmed the glow of his own testimonial. The first set of messages had been distorted. "Only in one case was there some little difficulty in reading two or three words, and that was in the messages sent at 2 p.m." Marconi's explanation, according to Fleming, was that the messages "had been slightly blurred in the attempt of two ships somewhere in the Channel to communicate with each other."

Though Fleming dismissed this as "some little difficulty," in fact the distortions were significant and gave further testimony to the problematic nature of wireless telegraphy. The garbling of "two or three words" was no small thing. The two o'clock message from the high-power station was in code and consisted of five words, "Quiney Cuartegas Cuatropean Cubantibus Respond." If only two words came through fractured, the distorted portion would amount to 40 percent of the message; if three words, 60 percent. The coding made the distortion even more problematic since the coded messages looked like gibberish anyway and the receiving operator would be unlikely to recognize that errors had occurred.

Nonetheless, Fleming and Marconi promoted the experiment as nothing less than a total success. In a much-publicized lecture on March 23, 1903, Fleming crowed that it proved beyond doubt that Marconi's tuning technology prevented interference. A week later Marconi applauded the experiment in a speech to shareholders at his company's annual meeting. Four days later Fleming wrote a letter to *The Times* in which he again extolled Marconi's tuning prowess.

At the Egyptian Hall, Nevil Maskelyne read Fleming's accounts and was struck by how much the sealed envelopes and other trappings of false rigor reminded him of techniques used by spirit mediums to convince audiences of their powers. He sensed fraud and longed for a way to reveal it.

A friend, Dr. Horace Manders, came to him with an idea: If Marconi would not willingly subject his system to public challenge, why not attempt to do so *without* his cooperation? Dr. Manders believed he knew of just such an opportunity.

Though somewhat wicked, the idea delighted Maskelyne, who later wrote that he "at once grasped the fact that the opportunity was too good to be missed." As for the wicked part, he argued that carrying out his plan was "something more than a right; it was a duty."

Soon, thanks to Maskelyne, Fleming would experience a vivid demonstration of the true vulnerability of wireless, one that would erode his status within the Marconi company, wound his friendship with the inventor, and shake the reputations of both.

In Nova Scotia, when winter and spring collide, an event called a silver thaw can occur. As rain falls, it freezes and sheathes everything it touches with ice until tree limbs begin to break and telegraph wires to fall. Marconi's men at Glace Bay had never experienced a silver thaw before, and they were unprepared for the phenomenon.

On April 6, 1903, the rain came. Ice accumulated on the station's four hundred wires until each wore a coat about one inch thick. It was lovely, ethereal. A giant crystal pyramid hung in the sky.

The weight of so much ice on so many miles of wire became too great. The whole array pulled free and crackled to the ground.

Blue Serge

For two of Belle's friends, John Nash and Lil Hawthorne, the news of Belle's death came as an especially harsh surprise. On March 23, 1910, the day before Crippen telegraphed the news, Nash and his wife had set sail for America, after a doctor recommended a sea voyage to ease Hawthorne's nerves. No one thought to send them word by wireless. After their arrival in New York, they paid a visit to Mrs. Isabel Ginnette, president of the guild, who also was in New York. To their shock, the Nashes now heard that Belle had died.

Nash promised Mrs. Ginnette that on his return to England he would go and talk to Crippen. Once safely back in London, the Nashes got together with their friends in the guild and discovered that no one believed Crippen's account of what had happened. Nash was appalled that his friends had done little to learn the truth. "I came over here and found that no one had had the courage and pluck to take up this matter," Nash said. "I therefore felt it my duty to take action myself."

Nash and his wife stopped by Crippen's office. "It was the first time we had seen him since his wife's death," Nash said. "He seemed much cut up—in fact, sobbed; he seemed very nervous, and was twitching a piece of paper about the whole time."

Crippen told him that Belle had died in Los Angeles, but then corrected himself and said it had happened in "some little town" around San Francisco. Nash knew San Francisco and pressed Crippen for a more precise location. Exasperated, Nash said, "Peter, do you mean to say that you don't know where your wife has died?"

Crippen said he could not remember but thought the place was called "Allemaio."

Nash changed direction. "I hear you have received her ashes."

Crippen confirmed it and said he had them in his safe. Nash did not ask to see them. Instead, he asked for the name of the crematorium and whether Crippen had received a death certificate.

"You know there are about four Crematoria there," Crippen said. "I think it is one of those."

"Surely you received a certificate."

Crippen became visibly nervous.

Nash said later, "I began to feel there was something wrong, as his answers were not satisfactory when a man cannot tell where his wife died or where her ashes came from."

Two days later, June 30, Nash and his wife set out to visit a friend who worked at New Scotland Yard. No mere functionary, this friend was Superintendent Frank C. Froest, head of the Yard's Murder Squad, established three years earlier as a special unit of its Criminal Investigation Department or CID.

ANYONE APPROACHING THE HEADQUARTERS of the Metropolitan Police from the north along the Victoria Embankment saw a building of five stories topped by a giant mansard roof, with Westminster Hall and Big Ben visible two blocks south. Huge rectangular chimneys marched along the top of the roof. Turrets formed the corners of the building and imparted the look of a medieval castle, giving their occupants—one of whom was the police commissioner—unparalleled views of the Thames. The lower floors were sheathed in granite quarried by the residents of Dartmoor Prison; the rest up to the roofline was brick.

The Nashes were anxious, but being creatures of the theater, they were also excited by the prospect of their interview with Froest. The building and its setting conveyed melodrama, and Froest himself was a man of some fame, for his pursuit in the mid-1890s of a crooked financier, Jabez Balfour, whom he captured in Argentina.

Froest listened with care as Nash and Lil Hawthorne told their story, then summoned a detective from the Murder Squad, one of its best. When the man entered, Froest introduced the Nashes and explained they had come because a friend of theirs seemed to have disappeared. Her name, he said, was Mrs. Cora Crippen, though she also used the stage name Belle Elmore and was a member of the Music Hall Ladies' Guild. Her husband, Froest said, was a physician "out Holloway way," named Hawley Harvey Crippen.

"Mr. and Mrs. Nash are not satisfied with the story the husband has told," Froest said. "Perhaps you had better listen to the full story."

The detective took a chair.

His name was Walter Dew; his rank, chief inspector. He was a tall man, built solid, with blue eyes and a large cow-catcher mustache, neatly groomed. He had joined the force at nineteen; he now was forty-seven. He had received his detective's badge in 1887 and shortly thereafter won the nickname "Blue Serge" for always wearing his best suit on duty. He took part in the Yard's investigation of the Ripper killings in 1888 and had the good luck, or bad, to be one of the detectives who discovered the remains of Jack's last and most horribly mutilated victim, Mary Kelly. "I saw a sight which I shall never forget to my dying day," Dew wrote in a memoir. "The whole horror of that room will only be known to those of us whose duty it was to enter it." What stayed with him most keenly, he wrote, was the look in the victim's eyes. "They were wide open, and seemed to be staring straight at me with a look of terror."

Now Nash told Dew his story:

"When we got back from America a few days ago, we were told that Belle was dead. Our friends said she had gone suddenly to America without a word of good-bye to any of them, and five months ago a notice was out in a theatrical paper announcing her death from pneumonia in California. Naturally, we were upset. I went to see Dr. Crippen. He told me the same story, but there was something about him I didn't like. Very soon after his wife's death Dr. Crippen was openly going about with his typist, a girl called Ethel Le Neve. Some time ago they went to a dance together and the girl was actually wearing Belle's furs and jewelry."

He told Dew, "I do wish you could make some inquiries and find out just when and where Belle did die. We can't get details from Dr. Crippen."

Froest and Dew asked a few more questions, then Froest said, "Well, Mr. Dew, that's the story. What do you make of it?"

Under ordinary circumstances, Dew would have been inclined to reject the inquiry and turn it over to the uniformed branch for handling as a routine missing-person case. Dew did not suspect foul play and sensed that the Nashes also did not. "What was really in the minds of Mr. and Mrs. Nash, what prompted them to seek the assistance of Superintendent Froest, I cannot say, but it is quite certain that neither of them dreamt for a moment that there was anything very sinister behind the affair," Dew wrote. "It is probably that they were actuated more than anything else by Crippen's lack of all decency in placing another woman so soon and so completely in the shoes of his dead wife."

But the Nashes were personal friends of Superintendent Froest, and Lil Hawthorne was a well-known music hall performer. It seemed important to demonstrate that Scotland Yard was taking their concerns seriously. Also, his own experience on the force had taught him, as he put it, "that it is better to be sure than sorry."

Dew said, "I think it would be just as well if I made a few inquiries into this personally."

Rats

THE FADING CREDIBILITY OF HIS COMPANY prompted Marconi to deploy Ambrose Fleming yet again, this time for a lecture on tuning and long-range wireless at the Royal Institution, on June 4, 1903. Fleming arranged to have Marconi send a wireless message from Poldhu to a receiver installed at the institution for the lecture, as a means of providing a vivid demonstration of long-range wireless. The recipient was to be James Dewar, director of the Royal Institution's Davy-Faraday Research Laboratory. Dewar was best known for his ability to chill things, in particular his successful effort in 1898 to liquefy hydrogen, which in turn had led a German technician to invent the Thermos bottle.

The timing was tricky. The message was to arrive in the closing moments of Fleming's lecture. It was pure theater, a variety show complete with a turn of magic, and if everything had gone according to plan, it might indeed have done much to help Marconi regain some of his lost credibility.

FLEMING BEGAN HIS LECTURE promptly at five o'clock. As always, every seat was taken. He spoke with confidence and flair, and the audience responded with murmurs of approval. One of his assistants, P. J. Woodward, took up position by the receiver and prepared to switch on the Morse inker to record Marconi's expected message to Dewar. But as the moment approached, something peculiar happened.

Another assistant, Arthur Blok, heard an odd ticking in the arc lamp inside the large brass "projection lantern" in the hall. As he listened, he realized the ticking was not simply a random distortion. The electric arc within the projector was acting as a crude receiver and had begun picking up what seemed to be deliberate transmissions. At first he assumed that Marconi's men in Chelmsford "were doing some last-minute tuning-up."

Fleming did not notice. During his tenure with the Marconi company he had become increasingly hard of hearing. The audience too seemed unaware.

Blok was experienced in sending and receiving Morse code. The clicking in the lantern was indeed spelling out a word—a word that no one in Chelmsford would dare have sent, even as a test. Yet there it was, a single word:

"*Rats.*"

As soon as Blok recognized this, "the matter took on a new aspect. And when this word was repeated, suspicion gave place to fear."

A few years earlier the expression "rats" had acquired a new and non-zoological meaning. During one action of the Boer War, British troops had used light signals sent by heliograph to ask the opposing Boer forces what they thought of the British artillery shells then raining down on their position. The Boers answered, "Rats," and the word promptly entered common usage in Britain as a term that connoted "hubris" or "arrogance."

As the time set for Marconi's final message neared, the assistant at the receiver turned on the Morse inker, and immediately pale blue dots and dashes began appearing on the ribbon of paper that unspooled from the inker. A new word came across:

There

Blok alternately kept an eye on the clock and on the tape jolting from the inker.

was

a

young

fellow

from

Italy

Blok was stunned. "There was but a short time to go," he said, "and the 'rats' on the coiling paper tape unbelievably gave place to a fantastic doggerel":

There was a young fellow from Italy
Who diddled the public quite prettily

The clicking paused, then resumed. Helpless, Blok and his colleague could only listen and watch.

Now entertain conjecture of a time
When creeping murmur and the poring dark
Fills the wide vessel of the Universe.
From camp to camp through the foul womb of night
The hum of either army stilly sounds,
That the fix'd sentinels almost receive
The secret whispers of each other's watch

Any well-schooled adult of the time, meaning almost everyone in the audience, would have recognized these lines as coming from Shakespeare's *Henry V.*

It was like being at a séance, that mix of dread and fascination as the planchette pointed to letters on the rim of an Ouija board. "Evidently something had gone wrong," Blok wrote. "Was it practical joke or were they drunk at Chelmsford? Or was it even scientific sabotage? Fleming's deafness kept him in merciful oblivion and he calmly lectured on and on. And the hands of the clock, with equal detachment, also moved on, while I, with a furiously divided attention, glanced around the audience to see if anybody else had noticed these astonishing messages."

Another pause, then came an excerpt from *The Merchant of Venice.* Shylock:

I have possessed your grace of what I purpose
And by our holy Sabbath have I sworn
To have the due and forfeit of my bond.

Still oblivious, Fleming talked on. Marconi's message from Poldhu, via Chelmsford, would arrive at any moment. Blok erased all anxiety from his own expression and scanned the audience for indications that the interloping signals had been detected. At first, to his great relief, he found none. The audience had entered the oblivion of complete engagement—"a testimony to the spell of Fleming's lecture." But then his gaze came to rest on "a face of supernatural innocence." He knew the man, Dr. Horace Manders; he knew him also to be a close associate of Nevil Maskelyne. In that instant, Blok understood what had happened but allowed no change in his own expression.

"By a margin of seconds before the appointed Chelmsford moment, the vagrant signals ceased and with such *sang froid* as I could muster I tore off the tape with this preposterous dots and dashes, rolled it up, and with a pretence of throwing it away, I put it in my pocket."

But the receiver clicked back to life. Was this more Shakespeare, more doggerel, or something worse? The tape unspooled. With scientific detachment, Blok and his colleague read the first blue marks.

The first letters across were *PD,* the call sign for Poldhu. Marconi's message was coming through. Dewar in his message to Poldhu earlier in the day had asked Marconi about the status of transatlantic communications. Now, as expected, Marconi was providing his answer.

> *To Prof. Dewar. To President Royal Society and yourself*
> *Thanks for kind message. Communication from Canada was*
> *re-established May 23.—Marconi.*

Fleming ended his lecture. The audience erupted with what Blok described as "unsuspecting applause." Fleming beamed. Dewar shook his hand. Other members did likewise and congratulated him on another fine performance, while marveling at how well he orchestrated the demonstration. To the audience, it seemed a testament to the increasing reliability and sophistication of Marconi's technology. Blok knew otherwise: In the end the lecture's success had hinged on something far outside the control of Marconi and his supposed new ability to avoid interference and interception. Had the pirate signals continued, Marconi's message would have come through grossly distorted or not at all, at great cost to the reputations of both Marconi and Fleming. Smug mockery would have filled the pages of *The Electrician.*

After the handshaking and congratulations had subsided, someone, perhaps Blok or Woodward, told Fleming what had occurred and about the presence in the audience of Maskelyne's associate, Dr. Manders. Fleming was outraged. To attempt to disrupt a lecture at the Royal Institution was tantamount to thrusting a shovel into the grave of Faraday. But the affair also inflicted a more personal wound. A man of brittle and inflated dignity, Fleming was embarrassed on his own behalf, even though no one in the audience other than his assistants and Dr. Manders had appeared to notice the intrusion.

All night Fleming steamed.

MASKELYNE WAITED, DISAPPOINTED.

He was indeed the pirate behind the wireless raid on Fleming's lecture; in fact, he had hoped his intrusion would cause an immediate uproar of satisfying proportions. As he confessed later, "The interference was purposely arranged so as to draw Professor Fleming into some admission that our messages had reached the room."

But Marconi's men had been too cool and quick; also, Maskelyne had not appreciated the extent of Fleming's loss of hearing. But he guessed that Fleming's assistants eventually would tell him of the intrusion. He understood well the inner character of his prey, his need for approval and respect. Fleming could not help but respond.

The trap was well set. An immediate outcry would have been far more satisfying, but Maskelyne believed he would not have to wait long for Fleming himself to make the phantom signals public, at which point Maskelyne intended to make both Marconi and Fleming squirm.

And this would be satisfying indeed.

THE MORNING AFTER THE LECTURE, Fleming wrote a letter about it to Marconi. "Everything went off well," he began, but then added: "There was however a dastardly attempt to jamb us; though where it came from I cannot say. I was told that Maskelyne's assistant was at the lecture and sat near the receiver."

In a second letter soon afterward Fleming told Marconi that Dewar "thinks I ought to expose it. As it was a purely scientific experiment for the benefit of the R.I. it was a ruffianly act to attempt to upset it, and quite outside the 'rules of the game.' If the enemy will try that on at the R.I. they will stick at nothing and it might be well to let them know."

Marconi's responses to these letters are lost to history, but if he or anyone else counseled Fleming as to the benefits of letting dogs sleep, the advice went unheeded.

On June 11, 1903, in a letter published by *The Times,* Fleming first reminded readers of his lecture at the Royal Institution and his demonstration, then wrote: "I should like to mention that a deliberate attempt was made by some person outside to wreck the exhibition of this remarkable feat. I need not go into details; but I have evidence that it was the work of a skilled telegraphist and of some one acquainted with the working of wireless telegraphy, whilst at the same time animated by ill-feelings towards the distinguished inventor whose name is always popularly and rightly connected with this invention.

"I feel certain that, if the audience present at my lecture had known that in addition to the ordinary chances of failure in difficult lecture experiments the display was carried through in the teeth of a cowardly and concealed attempt to spoil the demonstration, there would have been a strong feeling of indignation."

Fleming allowed that tapping Marconi's wireless communications might indeed constitute fair play, but disrupting a lecture to the Royal Institution was out of bounds. "I should have thought," he sniffed, "that the theatre which has been the site of the most brilliant lecture demonstrations for a century past would have been sacred from the attacks of a scientific hooliganism of this kind."

He wrote that he did not yet know who had attempted this sacrilege and urged any reader who might "happen to obtain a clue" to pass the information to him. "There may not be any legal remedy against monkeyish pranks of this description; but I feel sure that, if the perpetrators had been caught red-handed, public opinion would have condoned an attempt to make these persons themselves the subject of a 'striking experiment.' "

From Fleming's perspective, the letter was perfect, a jewel of subtle threat. He could not prove beyond doubt that Maskelyne was the pirate and therefore could not accuse him openly, but he had crafted his letter so as to transmit to the magician a warning that such behavior would not be tolerated. It is easy to imagine his satisfaction at opening *The Times* that Thursday morning and seeing those few inches of black type, knowing full well that not just Maskelyne but all of Britain's scientists, statesmen, barristers, thinkers, and writers, perhaps even the king, would read them, and that Maskelyne's teacup would by then be chattering against its saucer as the chill of impending danger crept down his spine.

THE LETTER *WAS* PERFECT—EXACTLY what Maskelyne had hoped for. Better, actually, given the charmingly veiled threat that Fleming might stoop to inflicting physical harm. If his teacup chattered, it was from delight at the prospect of composing his reply. He posted his own letter on Friday, June 12, from the Egyptian Hall. *The Times* published it the next day.

"Sir," Maskelyne wrote, "The matter referred to in your columns, yesterday, by Professor Fleming has a public importance far greater than he appears to imagine. It is a case in which members of the public are driven to take extreme measures in order to obtain information to which they are justly entitled."

He wrote, "The Professor complains that, during his lecture on the 4th inst., the Marconi instruments were disturbed by outside interference, and desires to know the names of those who perpetrated the 'outrage.' His suggestions of public opprobrium, legal proceedings, and personal violence may, of course, be dismissed as mere crackling of thorns beneath the pot. Personally, I have no hesitation in admitting my complicity as an accessory before the fact, the original suggestion having been made by Dr. Horace Manders."

He countered Fleming's charge of hooliganism by asking, "If this be described as 'scientific hooliganism' and the like, what epithets must we apply to the action of those who, having publicly made certain specific claims, resent being taken at their word?

"We have been led to believe that Marconi messages are proof against interference. The recent Marconi 'triumphs' have all been in that direction. Professor Fleming himself has vouched for the reliability and efficacy of the Marconi syntony. The object of the lecture was to demonstrate this." He wrote that he and Manders merely had put Fleming's claims to the test. "If all we had heard were true, he would never have known what was going on. Efforts at interference would have been effort wasted. But when we come to actual fact, we find that a simple untuned radiator upsets the 'tuned' Marconi receivers—"

Here he twisted the knife.

"—and Professor Fleming's letter proves it."

Immediately the press leaped into the fray. If Fleming simply had kept the cap on his inkwell, the whole matter likely would have remained dormant. As the *Morning Leader* of June 15, 1903, noted, "Nothing would have been heard of it had not Professor Fleming sent his indignant letter to the 'Times' denouncing the 'scientific hooligans' who upset him. That was just what Mr. Maskelyne hoped for; and now he is chuckling at his success in 'drawing the badger.' " In an interview in the Saturday, June 13, edition of the *St. James Gazette*, Maskelyne noted that he himself had composed the "diddling" verse. Now he added another, deeper dimension to his attack. "The Professor called up the name of Faraday in condemning us for what we did. Supposing Faraday had been alive, whom would he have accused of disgracing the Royal Institution—those who were endeavouring to ascertain the truth or those who were using it for trade purposes?"

He accused Fleming of giving two lectures that afternoon. "The first was by Professor Fleming the scientist, and was everything that a scientific lecture ought to be; the second was by Professor Fleming, the expert adviser to the Marconi Company."

That December Marconi declined to renew Fleming's contract.

AH

CHIEF INSPECTOR DEW BEGAN HIS INQUIRY by paying a visit to the Music Hall Ladies' Guild at Albion House, accompanied by an assistant, Det. Sgt. Arthur Mitchell. They were careful to keep their presence from being discovered by Crippen, whose office—"curiously enough"—was in the same building.

Over the next six days the detectives interviewed Melinda May, the Burroughses, and the Martinettis and talked again with John Nash and his wife, Lil. Dew heard about the rising sun brooch, and examined the correspondence that had taken place between Crippen and various members of the guild in the months since Belle's alleged disappearance. He learned that Belle had been "a great favorite with all whom she came in contact with." He collected details about her relationship with Crippen. Maud Burroughs described Belle "as always having her own way with her husband and going about just as she liked, which he apparently was content to submit to."

Dew wrote a sixteen-page report on his findings and turned it in to Froest on July 6, 1910. Dew had doubts about whether further inquiry would turn up anything criminal. On the first page he wrote, "The story told by Mr. and Mrs. Nash and others is a somewhat singular one, although having regard to the Bohemian character of the persons concerned, is capable of explanation."

Still, the story did contain contradictions that Dew considered "most extraordinary." His recommendation: "without adopting the suggestion made by her friends as to foul play, I do think that the time has now arrived when 'Doctor' Crippen should be seen by us, and asked to give an explanation as to when, and how, Mrs. Crippen left this country, and the circumstances under which she died. . . . This course, I venture to think, may result in him giving such explanation as would clear up the whole matter and avoid elaborate enquiries being made in the United States."

Superintendent Froest agreed.

On Friday morning, July 8, at ten o'clock, Chief Inspector Dew and Sergeant Mitchell walked up the front steps to No. 39 Hilldrop Crescent. The knocker on the door was new; the house seemed prosperous and well kept.

A girl in her late teens answered the door. Dew asked, "Is Dr. Crippen at home?"

The girl was French and spoke little English but managed to invite Dew and Mitchell into the front hall. A few moments later a woman appeared whom Dew judged to be between twenty-five and thirty years old. "She was not pretty," Dew recalled, "but there was something quite attractive about her, and she was neatly and quietly dressed."

He noticed that she was wearing a diamond brooch and knew at once it must be the rising sun brooch he had heard so much about.

"Is Dr. Crippen in?" Dew asked again.

He was not, the woman said. She explained that he had gone to his office at Albion House, in New Oxford Street.

"Who are you?" Dew asked.

"I am the housekeeper."

Dew said, "You are Miss Le Neve, are you not?"

Her cheeks turned a faint rose, he noticed. "Yes, that's right."

"Unfortunate the doctor is out," he said. "I want to see him rather urgently. I am Chief Inspector Dew of Scotland Yard. Would it be asking too much for you to take us down to Albion House? I am anxious not to lose any time."

He of course knew exactly where Albion House was but did not want to give Le Neve an opportunity to telephone Crippen and warn him that two detectives were on the way. Le Neve went upstairs and returned with her coat. Dew noticed she had removed the brooch.

A few moments later they were aboard the electric tram on Camden Road. They rode it to Hampstead Road, where they caught a cab for the remainder of the journey through Bloomsbury to Albion House.

ETHEL'S RECOLLECTION OF THIS encounter differed from Dew's. Hers made no mention of the brooch or her initial claim to be the housekeeper but added a plume of detail that illuminated the moment and the personalities involved.

She was helping straighten up the house, "making beds and so on," when she heard the knocker on the front door. It surprised her because tradesmen always used the side door. She listened at the top of the stairs as the French maid opened the door and a man asked, "Is Dr. Crippen at home?"

The maid did not understand the question. "Yes," she said.

"What a stupid creature that is!" Ethel whispered to herself, then came down the stairs and saw that two men stood at the door. "I had not the faintest idea who they were or what they wanted," she wrote.

"He is not at home," she told the men, "and will not return until after six o'clock this evening."

One of them looked at her "in a curious way," she recalled. He said, "I beg your pardon, but I am informed that Dr. Crippen is still here, and I wish to see him on important business."

"Well, you have been wrongly informed," Ethel said. She told him that the doctor had left at his usual hour, just after eight o'clock.

"I am sorry to doubt your word," the man said, "but I am given to understand that Dr. Crippen does not go to his office until after eleven. I feel quite sure he is in the house, and I may as well tell you at once I shall not go until I have seen him. Perhaps if I tell you who I am you will find Dr. Crippen for me."

He then identified himself as Chief Inspector Walter Dew of Scotland Yard, and his partner as Sgt. Arthur Mitchell.

"All the same," she said, "I cannot find Dr. Crippen for you. He is out." She was angry now. "You will have to stay a long time if you want to see him here," she said. "He will not be home until after six this evening. As you decline to believe me when I say he is not at home, you had better come inside and look for him."

She led them into the sitting room.

Dew repeated that his visit "was most important" and that he would not leave until he had spoken with Crippen. He told Ethel to be "a sensible little lady" and get him.

Ethel laughed at Dew. She repeated that Crippen was not home. She offered to telephone him at his office.

Dew asked her instead to accompany them to the office.

"All right," she said, "but you must give me time to dress properly."

She went to the bedroom to change. She wrote, "I had no compunction in making them wait a good long time while I arranged my hair, put on a blouse, and generally made myself look presentable."

Their visit puzzled her. "Yet I can honestly say that I was not much alarmed," she wrote, "—only a little bewildered and more than a little annoyed."

When she went back downstairs, she found Dew to be a changed man, suddenly affable and friendly and "inclined for conversation." He asked her to sit down. "I would very much like to ask you a few questions," he said.

He asked when she had come to live at Hilldrop and about Mrs. Crippen's departure. Sergeant Mitchell took notes. Ethel told them what she knew and mentioned Crippen having received cables telling him of his wife's illness and, later, her death.

"Did you see the cables?" Dew asked.

"No. Why should I? I do not doubt Dr. Crippen's word."

"Ah," Dew said.

Ethel wrote, "He was always saying that little word, 'ah,' as though he knew so much more than I did."

Again Dew asked her to accompany them to Crippen's office. Now she resisted. She told him she was a woman "of methodical habits" and did not like having the day disrupted.

"No; I quite understand that," Dew said, "but you see this is a matter of very special importance to Dr. Crippen. It is for his sake, you see."

She assented.

AT ALBION HOUSE Ethel went to the upstairs workroom to get Crippen. She found him sitting at a table working on dental fittings, alongside his partner, Rylance. She touched him to get his attention and whispered, "Come out, I want to speak to you."

Crippen asked why.

"There are two men from Scotland Yard," Ethel said. "They want to see you on important business. For heaven's sake, come and talk to them. They have been worrying me for about two hours."

"From Scotland Yard?" Crippen said. "That's very odd. What do they want?"

He was utterly calm, she wrote. She accompanied him down the stairs. The time was now, by Ethel's recollection, about eleven-thirty A.M.

DEW WAITED. A FEW moments later Le Neve reappeared "with an insignificant little man at her side." For Dew it was a revelatory moment. So this was the doctor he had heard so much about. Crippen was a small man, balding, with a sandy mustache. His most notable feature was his eyes, which were blue and protruded slightly, an effect amplified by his spectacles, which had thick lenses and thin wire rims. If Crippen was at all troubled by a visit from two detectives, he gave no sign of it whatsoever. He smiled and shook hands.

Dew kept it formal: "I am Chief-Inspector Dew, of Scotland Yard. This is a colleague of mine, Sergeant Mitchell. We have called to have a word with you about the death of your wife. Some of your wife's friends have been to us concerning the stories you have told them about her death, with which they are not satisfied. I have made exhaustive inquiries and I am not satisfied, so I have come to see you to ask if you care to offer any explanation."

Crippen said, "I suppose I had better tell the truth."

"Yes," Dew said, "I think that will be best."

Crippen said, "The stories I have told about her death are untrue. As far as I know she is still alive."

AH.

The Girl on the Dock

For Marconi the first half of 1904 proved a time of disillusionment and sorrow. His father, Giuseppe, died on March 29, but Marconi was so consumed by the difficulties of his company that he did not travel to Italy for the funeral. In May he set off on a voyage aboard the Cunard Line's *Campania* to conduct more long-range tests but found that by day his maximum distance was 1,200 miles, by night 1,700, not much better than the results he had gotten in February 1902 during similar tests on the *Philadelphia*. Two full years had passed with no significant improvement. His stations, he decided, had to be even bigger and more powerful, though enlarging them would further strain his company's financial health. On Nova Scotia he faced a choice—invest more money in the Glace Bay station, or abandon it and find a site far larger for a wholly new station. He chose the latter path. He envisioned the installation of an antenna three thousand feet in diameter.

The new station would impose great strain on his company's increasingly fragile financial health, to say nothing of taxing his board's willingness to support his transatlantic quest—especially now, in the face of the grave threat posed by Kaiser Wilhelm II and his international conference on wireless. The conference had taken place the previous August in Berlin, and the nations in attendance had agreed in principle that every station or ship should be able to communicate with every other, regardless of whose company manufactured the equipment involved. They agreed also that companies must exchange the technical specifications necessary to make such communication possible. For the moment the agreement had no effect, but eventual ratification seemed certain.

Back in London Marconi confronted skepticism and suspicion that seemed to have deepened. He found it hard to comprehend. Wireless worked. He had demonstrated its power time and again. Lloyd's had endorsed the system. More and more ships carried his apparatus and operators. News reports testified to the value of wireless. The previous December, for example, the Red Star Line's *Kroonland* had lost her steering, but thanks to wireless all her passengers had been able to notify family that they were safe. Even Kaiser Wilhelm's conference testified, albeit perversely, to the quality and dominance of Marconi's system.

Yet here it was, 1904, and the author of a newly published book on wireless still felt compelled to write: "Notwithstanding the great mass of positive evidence, there are many conservative people who doubt that wireless telegraphy is or will be an art commercially practicable. Public exhibitions have so often proved disappointing that a great deal of disparaging testimony has circulated."

Marconi turned thirty on April 25. The context was bittersweet. "At thirty," his daughter, Degna, wrote, "his nerves were dangerously frayed, he was disheartened, and near the end of his endurance."

He told his friend Luigi Solari, "A man cannot live on glory alone."

MARCONI, HOWEVER, WAS NOT exactly leading a life of misery. In London, when he wasn't immersed in business matters, he dined in elegant restaurants, certainly the Criterion and Trocadero, and was coveted as a guest at dinner parties in Mayfair and at the country homes of the titled rich. Marconi loved the company of beautiful women and was pursued by many, albeit in the sotto voce fashion of the day. The fact that he was Italian put him a rank or two below the kind of suitor that British parents considered ideal, but still, as Degna put it, he "was internationally considered a brilliant second-best."

While conducting experiments from his base at the Haven Hotel in Poole, he often would sail to nearby Brownsea Island for lunch with his friends Charles and Florence Van Raalte, who owned the island and lived there in a castle. In the summer of 1904 the Van Raaltes had houseguests, a young woman named Beatrice O'Brien and her mother, Lady Inchiquin. Beatrice was nineteen years old and one of fourteen children of the fourteenth Baron Inchiquin, Edward Donagh O'Brien, who had died four years earlier, possibly from parental exhaustion. Beatrice and her siblings were accustomed to castle life, having grown up in a large one on the family estate, Dromoland, in County Clare, Ireland.

On a day when Marconi was expected, Mrs. Van Raalte dispatched Bea to the dock to meet him. She put on her favorite dress, a satin number meant for evening wear that she had sewn herself. To her, it seemed lovely; to everyone else, merely peculiar. While walking on the dock, she broke a heel off one of her shoes. She stood waiting, a bit off kilter, as Marconi's boat arrived.

Two observations struck him: first, as he said later, that "the dress she had on was *awful,*" and second, that she was utterly beautiful. He was thirty, eleven years her senior, but in those moments on the dock he fell in love.

Suddenly his wireless troubles did not seem so overwhelming. He came to Brownsea more often, not just for lunch but also for dinner and high tea. When Beatrice left Brownsea for the family's mansion in London, Marconi dropped his experiments and followed.

In London one evening Marconi went to the Albert Hall to attend a charity banquet organized by Beatrice's mother. He had little interest in the charity. He found Beatrice at the top of a long flight of iron stairs. He asked her to marry him.

To this point, marriage had not entered her thoughts. She did not love him, at least not in the way that marriage might require. She asked for more time. He bombarded her with letters, using the post office's express mail service, which dispatched messenger boys to carry important letters directly to their destinations.

At last Beatrice invited Marconi to tea. She told him, gently, that she would not become his wife.

He fled for the Balkans, behaving, Degna said, "like the jilted suitor in a romantic Victorian novel." He contracted malaria, which would plague him with intervals of fever and delirium for the rest of his life.

BEATRICE WAS SURPRISED at how sad Marconi's departure from her life had made her. Stricken with the grief of failed romance, she returned to Brownsea Island for another long stay. Mrs. Van Raalte promised her "solemnly," according to Degna, that Marconi would not learn of her presence. But Mrs. Van Raalte liked Marconi and believed he and Beatrice constituted an ideal match.

Without telling Beatrice, Mrs. Van Raalte wrote to Marconi, still sulking in the Balkans, to tell him of Beatrice's heartbreak. In the great conspiratorial tradition of Englishwomen of title, she invited Marconi to the island as well, this time as a houseguest.

Marconi accepted at once and returned to England as quickly as possible. Beatrice was stunned to see him but charmed by the fact that his ardor had not diminished. They took walks, and sailed, and fell into what Degna described as an "easy comradeship." On December 19, 1904, as they walked through the heather on a headland overlooking the sea, Marconi again asked her to marry him. This time she said yes—on condition that her sister Lilah approved.

This meant another delay, for Lilah was in Dresden. Beatrice was unsure how to tell her sister the news and needed two days to compose her letter. "It's so serious I don't know how to break it to you," she wrote. "I'm not crazy; it's only this, I've settled the most serious thing in my life. Can you guess it—I am engaged to be married to Marconi. . . . I don't love him. I've told him so over and over again, he says he wants me anyhow and will make me love him. I do like him so much and enough to marry him." She added, "And to think I never meant to marry! I had always arranged to be an old maid."

She did not yet reveal the engagement to her mother. First, she told one of her brothers, Barney, who approved and urged her to go to London to tell her mother and, more important, her eldest brother Lucius, who after the death of their father had become the ranking male, the fifteenth Baron Inchiquin. Nothing could happen without his consent.

Beatrice and Marconi set out for London. Soon after their arrival

Marconi bought her a ring, which Degna described as "tremendous," then paid a visit to the O'Brien family's London mansion to ask Beatrice's mother, Lady Inchiquin, for her daughter's hand.

Nothing was easy. Lady Inchiquin was a bulwark of propriety. She said no. Lucius, the fifteenth Baron Inchiquin, seconded the refusal. While Marconi was indeed famous and was believed to be rich, he was still foreign. Neither the baron nor Lady Inchiquin knew anything of his heritage.

Predictably, their opposition had a perverse effect on Beatrice. Now, in the grand tradition of the daughters of titled Englishwomen, Beatrice steeled her resolve. She would marry Marconi, no matter what.

MARCONI WAS CRUSHED AND probably furious, fully aware of the underlying reasons for the O'Briens' refusal. He had lived among the wealthy of England long enough to know that their welcome had boundaries. He fled again, this time for Rome.

Troubling news drifted back. A German governess in the O'Brien family happened to read in a European newspaper that Marconi had been spotted often in the company of a Princess Giacinta Ruspoli. The governess told Lady Inchiquin. Worse news followed the next day. Another item reported that Marconi and the princess were engaged.

In the O'Brien mansion there were tears and rage and underneath it all a certain smug sense that the inevitable had occurred. Marconi was, after all, an Italian. Lady Inchiquin and the fifteenth baron accepted the reports as hard fact and saw them as affirming the righteousness of their decision. Beatrice wept. She insisted the news was false.

The situation called for lofty counsel. Lady Inchiquin took Beatrice to the home of an ancient and imposing aunt, Lady Metcalfe, "whose opinions," according to Degna, "were often invoked in times of crisis." Over tea, as Beatrice sat silent, the Ladies Inchiquin and Metcalfe railed and condemned, oblivious to her presence—"as though Beatrice, obviously reduced to the status of a naughty child, were not there."

Marconi was not the sole object of Lady Metcalfe's scorn. At one point she turned to Lady Inchiquin and asked, "What can you be thinking of, to let this child become engaged to a *foreigner*?"

IN ROME, MARCONI too read the reports of his alleged engagement and, realizing their likely impact in the O'Brien household, he immediately left for London. The reports were false, he assured Lady Inchiquin, and then he began a campaign to charm her into permitting his marriage to Beatrice. Surprisingly, he succeeded. He and Lady Inchiquin became friends. She took to calling him "Marky."

The engagement moved forward. Beatrice and Marconi scheduled their wedding for the following March.

From the first there were warning signs. "She was a born flirt," Degna wrote, and was "incapable of suppressing her adorable, flashing smile at every male who came near her."

At such times Marconi flared with jealous rage.

IN HIS LABORATORY at University College, London, Fleming directed the hurt of his rejection into work aimed at winning his way back into Marconi's favor. He invented a device that in time would revolutionize wireless, the thermionic valve. He wrote to Marconi, "I have not mentioned this to anyone yet as it may become very useful." Soon afterward he invented another device, the cymometer, that at last provided a means for the accurate measurement of wavelength. He told Marconi about this as well.

Oliver Lodge, meanwhile, became a bona-fide competitor to Marconi. In Birmingham, in moments when he wasn't busy managing his university or teaching or conducting research or investigating strange occurrences, Lodge helped his friend Alexander Muirhead attempt to find buyers for their wireless system. In 1904, while seeking a contract from the Indian government to provide a wireless link to the Andaman Islands in the Bay of Bengal, they wound up in direct competition with Marconi.

And won.

Hook

The new century raced forward. Motorized taxis and buses clogged Piccadilly. The fastest ocean liners cut the time for an Atlantic crossing down to five and a half days. In Germany the imperial war fleet rapidly expanded, and British anxiety rose in step. The government began talks with the French, and in 1903 Erskine Childers published his one and only novel, *The Riddle of the Sands,* in which two young Britons stumble across preparations for a German invasion of England. Prophetically, the German villain captains a ship named *Blitz.* In Germany the authorities ordered the book confiscated. In Britain it became an immediate best-seller and served as a rallying cry. The last sentence in the book asked, "Is it not becoming patent that the time has come for training all Englishmen systematically either for the sea or for the rifle?"

But this question raised a corollary: Were the men of England up to the challenge? Ever since the turn of the century, concern had risen that forces at work in England had caused a decline in masculinity and the fitness of men for war. This fear intensified when a general revealed the shocking fact that 60 percent of England's men could not meet the physical requirements of military service. As it happened, the general was wrong, but the figure 60 percent became branded onto the British psyche.

Blame fell upon the usual suspects. A royal commission found that from 1881 to 1901 the number of foreigners in Britain had risen from 135,000 to 286,000. The influx had not merely diminished the population; it had caused, according to Scotland Yard, an upsurge in crime.

Most blame was attributed to the fact that Britain's population had increasingly forsaken the countryside for the city. The government investigated the crisis and found that the percentage of people living in cities had indeed risen markedly from the mid-nineteenth century but had *not* caused the decay of British manhood, though this happy conclusion tended to be overlooked, for many people never got past the chilling name of the investigative body that produced it, the Inter-Departmental Committee on Physical Deterioration. A month later the government launched another investigation with an equally disheartening name, the Royal Commission on Care and Control of the Feeble-Minded, and discovered that between 1891 and 1901 the number of mentally defective Britons had increased by 21.44 percent, compared to the previous decade's increase of just over 3 percent. There was no escaping it: Insane, weak and impoverished, the British Empire was in decline, and the Germans knew it, and any day now they would attempt to seize England for their own.

In London on the night of December 27, 1904, at the Duke of York's Theater, a new play opened and immediately found resonance with that part of the British soul that ached for a past that seemed warmer and more secure. The action opened in the nursery of a house in what the playwright, James M. Barrie, described as a "rather depressed street in Bloomsbury," and it involved children led off to adventure by a mysterious flying boy named Peter. The *Daily Telegraph* would call the play "so true, so natural, so touching that it brought the audience to the writer's feet and held them captive there."

There were pirates and Indians, and danger. At the end of Act IV the audience gasped as Peter's fairy companion, Tinker Bell, drank poisoned medicine meant for him.

Peter turned to the audience. "Her light is growing faint, and if it goes out, that means she is dead! Her voice is so low I can scarcely tell what she is saying. She says—she says she thinks she could get well again if children believed in fairies!"

He turned and spread his arms. "Do you believe in fairies? Say quick that you believe! If you believe, clap your hands!"

And oh yes, on this cold night in London in December 1904, they did believe.

THEN CAME HOOK, the pirate captain, and the audience chilled at the intimation of coming evil.

"How still the night is," said Hook. "Nothing sounds alive."

Part V

The Finest Time

Beatrice O'Brien.

The Truth About Belle

Crippen led the detectives into his office—"quite a pleasant little office," Dew said. It was now about noon. A clamor of hooves and engines rose from the street outside, and the increasingly prevalent scent of gasoline tinged the air. Sergeant Mitchell sat at a small table, pencil and paper at hand. Dew began asking questions, and Crippen answered each without hesitation. "From his manner," Dew wrote, "one could only have assumed that he was a much maligned man eager only to clear the matter up by telling the whole truth."

The interview had barely begun when all realized it was time for lunch. Dew and Mitchell invited Crippen to join them, and the three left Albion House for a nearby Italian restaurant. Le Neve watched them go, chafing at Dew's order to remain in the office and at his lack of courtesy in failing to notice that she might wish to have lunch as well. "Meanwhile," she wrote, "I was absolutely fainting with hunger."

Over lunch the men talked. Crippen ordered a steak "and ate it with the relish of a man who hadn't a care in the world," Dew wrote. He found himself liking Crippen. The doctor was gentle and courteous and spoke with what appeared to be candor. Nothing in his manner suggested deception or anxiety.

Once back in Crippen's office, Dew continued his interview. He asked a question one way, then later asked it again in a different form to test the consistency of Crippen's story.

"I realized that she had gone," Crippen said, "and I sat down to think it over as to how to cover up her absence without any scandal." He wrote to the guild that she had gone away. "I afterwards realized that this would not be a sufficient explanation for her not coming back, and later on I told people that she was ill with bronchitis and pneumonia, and afterwards I told them she was dead from this ailment."

To "prevent people asking me a lot of questions," he said, he placed a death notice in the show-business journal *The Era.*

He said, "So far as I know she did not die, but is still alive."

Dew watched Crippen closely. "I was impressed by the man's demeanor," he wrote. "It was impossible to be otherwise. Much can sometimes be learned by an experienced police officer during the making of a statement. From Dr. Hawley Harvey Crippen's manner on this our first meeting, I learned nothing at all."

The detectives then reduced Crippen's story to a written statement. Crippen initialed each page and signed the last.

IT WAS ABOUT FIVE o'clock. Six hours had passed since the detectives had first come to Hilldrop Crescent. Ethel was hungry and annoyed but also fearful. With each hour the detectives had spent closeted with Crippen, her concern had deepened.

Now it was her turn, as Dew put it.

Ethel told the detectives about Belle's sudden departure, her illness, and her death. Mitchell took careful notes. "The girl showed some signs of embarrassment when she came to the admissions about her relations with Crippen," Dew wrote later. "But making due allowance for this, there was nothing in Miss Le Neve's manner which gave rise to anything in the nature of suspicions."

As had been the case with Crippen, nothing about the way she spoke suggested an attempt at deception. She seemed to be telling the truth, or at least the truth as she knew it, but Dew wanted to make sure.

He turned to her abruptly. "He told you a lie," he said. "He has just admitted to us that, as far as he knows, his wife [is] still alive, and that the story of her death in America was all an invention."

Any last doubt about her candor now disappeared.

"I WAS STUNNED," Ethel wrote. "I could not believe it. It seemed impossible to me that Belle Elmore might still be alive." Crippen would never have lied, she believed, and yet here was Dew confirming that he had done so. "Stricken with grief, with anger, with bewilderment, I answered all the questions put to me about my relations with the doctor, my love for him, and my life. But all the time I was thinking of the way I had been deceived if this story about Mrs. Crippen were true."

She signed her statement, but the ordeal, she now learned, was not yet concluded.

TO BE THOROUGH, DEW wanted to search Crippen's house. He knew, however, that no judge would give him the legal authority to do so. "There was not enough evidence against the man—indeed any evidence at all—on which I could have asked a magistrate for a search warrant." He asked Crippen's permission, and Crippen readily assented. Shortly after six that evening all four climbed into a growler and rode to Hilldrop Crescent, Le Neve and Crippen at one end of the cab, the detectives at the other. It was a long, silent ride. "I seemed to be living in a nightmare," Ethel wrote. "I felt rather faint and sick."

The detectives began their search, but there was nothing in particular that Dew hoped to find. "I certainly had no suspicion of murder," he wrote.

First the detectives walked around the garden, then entered the house and went through each room, searching wardrobes, cupboards, dressing tables. They found signs that Crippen and Le Neve had been packing for a move, including filled boxes and rolled carpets. They found nothing that shed light on Belle's current whereabouts, but they did find "plenty of evidence that Belle Elmore had a passion for clothes," as Dew put it. "In the bedroom I found the most extraordinary assortment of women's clothing, and enough ostrich feathers to stock a milliner's shop. The whole would have filled a large van."

No search, of course, would have been complete without a visit to the coal cellar. "I had no special motive in looking there on this occasion," Dew wrote. "It was just that I wanted to make certain that I had covered the whole of the house."

Crippen led the detectives down a short passage that ran from the kitchen to the cellar door.

Ethel waited upstairs, struggling to absorb the news of Crippen's deception. She sat in the sitting room, "quite stupefied and dazed," she recalled. "What were these men doing? Would they never go? It grew dark, and I sat there in the gloom. My head was aching furiously."

Crippen watched from the cellar doorway.

The cellar was cramped—nine feet long and six feet, three inches wide. "The place was completely dark," Dew wrote, "I had to strike matches to see what it contained and what sort of a place it was. I discovered nothing unusual. There was a small quantity of coal and some wood which looked as though it had been cut from the garden trees." The floor was brick, coated with a fine layer of dust.

The detectives and Crippen next went to the breakfast room off the kitchen and took seats at the table. There Dew asked his last questions and examined the jewelry Belle had left behind, including the rising sun brooch. He told Crippen, "Of course I shall have to find Mrs. Crippen to clear this matter up."

Crippen, helpful as always, agreed and promised to do everything he could to assist. "Can you suggest anything?" Crippen asked. "Would an advertisement be any good?"

Dew liked that idea, and together he and Crippen composed a brief advertisement for placement in newspapers in America. Dew left that task to Crippen.

It was after eight when the detectives said good night and exited the house. They had found Crippen's story, especially his fear of scandal, entirely plausible, though the fact that Crippen clearly had lied was troubling. In a deposition Dew said, "I did not absolutely think that any crime had been committed."

He told at least one observer that for all intents and purposes the case was closed.

ETHEL FELT GREAT RELIEF that she and Crippen at last were alone, "but I am bound to say," she wrote, "that I was angry and hurt, and that I felt in no mood for conversation. One thought only was in my head. It was, that the doctor had told me a lie. He had been untruthful to me for the first time in ten years—to me, of all people in the world, who was certainly the one to know the truth and all the truth. I had been faithful to him. I loved him. I had given up all things for him, and it hurt me frightfully that he should have deceived me."

Crippen attempted to cheer her up. He made supper and coaxed her down to the breakfast room. She could not eat and said nothing. At ten she went up to the bedroom and sat in a chair fully dressed, too tired to get ready for bed. Soon Crippen came up.

"For mercy's sake," she said, "tell me whether you know where Belle Elmore is. I have a right to know."

"I tell you truthfully that I don't know where she is."

Crippen told her that he had concocted the story of Belle's disappearance and death to avoid scandal, but now in the wake of the detectives' visit, everyone would know the truth and his and Ethel's reputations would be destroyed. It would be impossible to face the ladies of the guild. The scandal, Crippen feared, would be far more damaging to Ethel. He would do anything, he said, to spare her the inevitable humiliation.

It seemed to Ethel that Crippen had a plan in mind. She asked him what he intended.

"My dear," he said, "there seems to me only one thing possible to do."

The Prisoner of Glace Bay

Beatrice and Marconi married on March 16, 1905. They expected a relatively private ceremony but arrived at St. George's Church to find Hanover Square filled with what one newspaper called "a vast crowd of onlookers." That morning Alfred Harmsworth's latest creation, the *Daily Mirror,* the first British newspaper to make regular and lavish use of photographs, had filled its front page with half-tone images of Marconi and Beatrice, a display technique the newspaper had pioneered a year earlier when it published a full page of photographs of the king and his children. The police stood guard, not as protection from the crowd but because two days earlier the O'Briens had received a letter that warned that Marconi would be killed as he approached the church. The ceremony came off peacefully. Marconi gave Beatrice a diamond coronet, which she suspected had been her mother's idea. He also gave her a bicycle. "That," she said, "was really his own idea."

They retired for their honeymoon to Beatrice's ancestral home, Dromoland, in Ireland. When she was growing up, the castle had been full of clamor, generated by her thirteen siblings and their friends, but now it struck her as gloomy and lonely. They were assigned rooms in the "visitors" part of the castle, apparently for privacy, but this only amplified the alien feel of the place.

Alone now with her husband (apart, that is, from a small battalion of servants), Beatrice quickly discovered that Marconi was not always the gentleman of charm and good cheer she had come to know on Brownsea Island. He revealed himself to be moody and volatile. They fought, and afterward he would storm from the castle and walk off his rage in the woods, alone. They ended their honeymoon early, after only a week, ostensibly because Marconi had to get back to London on business.

In London they first checked into a small hotel near Marconi's office, but Marconi realized it was hardly the place for his new bride. They moved to something far grander, the Carlton Hotel at Haymarket and Pall Mall, which the *Baedeker's Guide* called "huge and handsome." For Beatrice, despite her wealthy upbringing, the experience of the Carlton was novel and wonderful.

She found the hotel's location irresistible and one day decided to take a walk, alone, to explore the surrounding streets. The National Gallery and Trafalgar Square, with Nelson's column, were two blocks east, St. James's Park just south. Piccadilly was an easy walk northwest, but as a destination was not, for the time being, a terribly appealing one. The city had resolved that because of increased traffic the street had to be made far wider. Demolition was under way and soon would bring the destruction of many treasured places, among them Nevil Maskelyne's Egyptian Hall.

"When she got back," Degna wrote, "her husband met her at the door of their room, storming that she was henceforth to tell him before she left exactly how long she would be gone, and street by street where she planned to be."

ONCE AGAIN HE PLUNGED into work. He was compelled to acknowledge that his company now confronted a difficult choice. His ship-to-shore business had grown slowly, but it had indeed grown, until by the end of 1904 his company had equipped 124 ships and 69 land stations in Britain, America, Canada, and elsewhere. The Italian Navy had selected his equipment for its warships, and the Italian government had contracted for a giant station in Coltano, now under construction. Moreover, Parliament at last had enacted a law that eased the strictures of the British Post Office monopoly over telegraphy by allowing customers for the first time to turn in messages at their local telegraph offices for delivery to ships at sea. Marconi had also agreed to take

Ambrose Fleming back into the company, not because of some new-found adoration for the man but because he recognized that Fleming's two new inventions, his cymometer and his thermionic valve, had the potential to greatly improve wireless transmission. One clause of the agreement—and no doubt the most important—gave Marconi the rights to use the inventions while allowing Fleming to retain ownership of the underlying patents.

But this empire had become complex and costly, and it promised to become more so. In Canada, the company opened nine new shore stations along the routes ships took when approaching the St. Lawrence River. Such stations tended to be remote and required that their operators and managers live on the grounds, a reality that imposed its own set of costs. The new station at Whittle Rocks, for example, required six kitchen chairs, for a total of $2.88; one armchair, for $1.75; two kitchen tables, for $5; two dresser stands, $22; one rocking chair, $3; and one reading chair, $4.25. Every station got a clock, for $2.35, and at least one bed. And telegraphers had to eat. In April 1905 the new station at Belle Isle paid $42.08 for salt pork, Fame Point spent $43.78 on bacon, and Cape Ray spent $42.37 on lard. And of course every station needed a rubber stamp. Each got one, for 34 cents.

Costs accumulated quickly. In August 1905 Marconi's bookkeepers found that the cash expenditures of his Canadian operations for just that one month totaled $46,215, more than triple the amount in the previous August. Onetime expenses accounted for much of this increase, but once the new stations began operating and breaking down and freezing and losing aerials to windstorms; once they began hiring employees and cleaners and stationers and freight haulers; once they started buying acid for batteries and paying for postage, telephone service, and land-line telegrams—once all these expenses became routine, they too began to increase, like yeast in a warm oven. Especially wages and salaries. In 1904 Glace Bay alone paid wages and salaries totaling $8,419. This figure would more than double by the end of 1907. Living and general operating expenses at the station increased at an even faster rate, as they did throughout Marconi's empire.

To Marconi, all this was just business. It didn't interest him. As always his true passion lay in transatlantic communication, and here things still were not going well. His Glace Bay station had been disassembled, and the timber and other components had been moved to a new inland location nicknamed Marconi Towers. He had recognized that his Poldhu station also had become obsolete and would have to be replaced by one even larger and more powerful.

For the first time he began to wonder whether he should jettison his transatlantic dream and settle for something more quotidian, perhaps focus his company on ship-to-shore communication. There was little doubt about the direction his directors would choose if left to decide on their own.

To better evaluate his future course, Marconi decided to visit the new station in Nova Scotia. In spring 1905 he booked passage aboard the *Campania* for himself and Beatrice. Despite the deepening financial crisis facing his company, they traveled first class, reflecting yet again a trait that Degna Marconi considered elemental to his character. As she put it, "All he asked of life was the best of everything."

On this voyage Beatrice would come to feel herself more prisoner than passenger and would learn that her keeper was quirkier than she had imagined.

THEY SETTLED INTO THEIR STATEROOM, which had more in common with a Mayfair flat than a ship-borne cabin. Beatrice was startled when Marconi began pulling numerous clocks from his trunk and placing them at various points in the cabin. She knew of his fixation on time. He gave her many wristwatches, but she consigned them all to her jewelry box where, according to Degna, "she left them all unwound to be spared their multiple ticking." Now she found herself surrounded by ticking clocks, which Marconi set to display the time in Singapore, Chicago, Rangoon, Tokyo, Lima, and Johannesburg.

Once the voyage was under way, he disappeared into the ship's wireless cabin, to conduct further experiments and to manage the receipt of news for the onboard newspaper, the *Cunard Bulletin*.

Left to herself, Beatrice sought to enjoy the ship, one of the most luxurious in Cunard's fleet. The dining saloon that served first-class passengers had Corinthian columns and ten-foot ceilings. A central shaft rose thirty feet through the decks above to a dome of stained glass at the top of the ship. The crew of 415 included more than one hundred stewards and stewardesses, who sought to satisfy every legal need. The ship served four meals a day, prepared by forty-five chefs and bakers and their helpers. As Beatrice walked the thrumming decks, she delighted in the celebrity accrued to her by the fact of her marriage to Marconi.

The attention given her once again caused Marconi to react with jealous anger. Degna wrote, "When her husband did emerge from the wireless cabin and found her talking to the other passengers, he led her stonily to their stateroom and lectured her about flirting." Marconi taught her Morse code, though she had little interest in learning. Degna suspected he did so in part to prevent Beatrice from wandering the decks and returning the smiles of the *Campania*'s other male passengers.

One day Beatrice entered their stateroom to find Marconi consigning his dirty socks to the sea through a porthole. Stunned, she asked him why.

His explanation: It was more efficient to get new ones than wait for them to be laundered.

THEY STOPPED BRIEFLY in New York and traveled to Oyster Bay on Long Island to have lunch with Theodore Roosevelt. They met his daughter Alice, who later reported that they were a handsome couple and seemed very happy with each other. They sailed to Nova Scotia, where snow still lay on the ground and the four towers of the newly completed station stood over the landscape like sentries. They moved into the nearby house, which they were to share with Richard and Jane Vyvyan and their daughter. The child was nearly one and a half years old, not the easiest age to manage, especially in close quarters. And these quarters were close. Beatrice had grown up in a castle with rooms seemingly beyond number. This house had a living room, a dining room, two bedrooms, and a single small bathroom. Marconi left Beatrice with Jane and her daughter and immediately joined Vyvyan at Marconi Towers, where they began adjusting and tuning the apparatus.

The new station encompassed two square miles. The four towers stood at its center. Next came a ring of twenty-four masts, each 180 feet tall, and beyond them another ring, consisting of forty-eight poles, each fifty feet tall. Over it all was draped an umbrella of wire with a diameter of 2,900 feet, comprising fifty-four miles of wire. Another fifty-four lay in ditches below.

Every day Marconi walked down the "corduroy" road of felled trees to the station compound and remained there for most of each day, while back at the house Beatrice confronted a situation wholly new to her experience. She possessed only limited domestic skills but nonetheless tried to help around the house, only to have Mrs. Vyvyan refuse her offers of assistance in a manner as cold as the weather outside. At first Beatrice kept her unhappiness from Marconi, but after days of enduring such behavior, she broke down and, weeping, told Marconi about all that had happened.

The news made Marconi furious. He was ready to charge out to the living room to confront the Vyvyans, but Beatrice stopped him. She knew how much Marconi depended on Vyvyan. She resolved to confront Mrs. Vyvyan herself.

Now it was Jane Vyvyan who burst into tears. She confessed that she had feared that Beatrice, as the daughter of a lord, would act superior and dominate the house or, worse, treat her as if she were a servant. Jane had hoped to assert her own superiority from the start.

Their talk cleansed the atmosphere. Almost immediately they became friends—and just in time.

WHILE WORKING AT THE NEW station under its great umbrella of wire, Marconi became convinced once more that transatlantic communication could succeed. He made arrangements to return to London, again aboard the *Campania,* for a summit with his board and to use the *Campania*'s wireless to test the reach of the new station.

Inexplicably, given how prone he was to jealousy, Marconi left Beatrice behind. She found little to occupy herself. Nova Scotia was a male realm, full of male pursuits, like ice hockey, hunting, and fishing. She found it dull.

In contrast, Richard Vyvyan gauged life in Nova Scotia as "on the whole quite pleasant." Especially the fishing, which he described as "superlatively good." Winter, he conceded, could be "trying at times," but even then the landscape took on a frigid beauty. "The stillness of winter in the country in Canada is extraordinary, when there is no wind. All the birds have left, except a few crows, and although the tracks of countless rabbits are to be seen they themselves are invisible. Not a sound can be heard but one's own breathing, beyond the occasional sharp crack of frost in a tree. The winter air is intensely exhilarating and the climate is wonderfully healthy."

Beatrice did not agree. There was no place to walk, save for the barbed-wire grounds of the station, and there she felt imprisoned. She would have loved to bicycle, but there were no roads in the vicinity of sufficient quality to make bicycling possible. She was sad and lonely and became ill with jaundice, possibly the result of contracting a form of hepatitis. And, her daughter wrote, always there was that silence, "so intense it made Bea's ears ring."

Marconi did not return for three months.

DURING THE VOYAGE Marconi was the toast of the vessel. Though he spent most of his time in the *Campania*'s wireless cabin, he always emerged for meals, especially dinner, where he sat among the richest and loveliest passengers, in a milieu of unsurpassed elegance.

During the first half of the voyage transmissions from the new Glace Bay station reached the ship strong and clear. Daylight reception reached a maximum of eighteen hundred miles—a good result, though he had hoped the range would be far greater, given the station's size and power.

In England he persuaded his directors to continue investing in his transatlantic quest. He volunteered his own fortune to the effort and sought new capital from investors in England and Italy.

At Poldhu he inaugurated a new series of experiments.

First he concentrated merely on trying to achieve communication between Poldhu and Nova Scotia. He tuned and adjusted the Poldhu receiver and via cable directed Richard Vyvyan to make other changes at

Marconi Towers. At last, at nine o'clock one morning in June, the Poldhu station received readable messages—a major breakthrough for the simple fact that this transmission occurred when both stations were in daylight.

Resorting as always to trial and error, Marconi next tested different antenna configurations. He shut down segments of each to gauge the effect on reception. Again, endless variables came into play. He adjusted power and tried different wavelengths. He believed, as always, that the longer the wavelength, the farther waves would travel, though why this should be the case remained a mystery to him.

He began to see a pattern. An antenna consisting of a single wire stretched *horizontally* and close to the ground seemed to provide better reception and transmission than its vertical equivalent. He found too that direction mattered. A wire stretched along an east-west axis could send signals most effectively to a receiving wire erected along the same axis. These discoveries freed Marconi from the need to build taller and taller aerials and more complex umbrella arrays. In theory, a single wire or series of parallel wires stretched over a long distance would produce wavelengths longer than anything he had so far achieved.

He instructed Vyvyan at Nova Scotia to simulate that kind of directional antenna by disconnecting portions of the umbrella array, then learned that his hunch was correct. Transmission and reception improved.

He realized now that Poldhu was not merely obsolete—the site would have to be abandoned entirely and another location found that had enough land to allow him to stretch a horizontal antenna up to one mile long. The new Nova Scotia station too would have to be replaced and its power-generation equipment enlarged to produce ten times more power.

The expense would be staggering, but Marconi saw no other path.

HE RETURNED TO NOVA SCOTIA, and to Beatrice. He was appalled at her condition. Her jaundice was jarringly apparent. He promised to take her back to England.

Beatrice assumed this would mean a return to London, and friends and family, and city life. It had been nearly half a year since she had seen a hansom cab or felt the rumble of a subterranean locomotive racing through the darkness under her feet.

But Marconi, her keeper, had a different plan in mind.

Liberation

On Saturday morning, July 9, 1910, Crippen left Hilldrop Crescent at his usual hour and went to his office at Albion House. At around ten he approached his assistant, William Long, and asked him to go to a nearby men's shop, Charles Baker, and buy a few articles of clothing. Crippen gave him a list of things to acquire that included a brown suit cut for a boy, two collars, a tie, two shirts, a pair of suspenders, and a brown felt hat. He was to buy a pair of boots as well, from a shop on Tottenham Court Road. Crippen gave him the necessary money.

Ethel meanwhile took a taximeter cab to the home of her sister Nina and arrived there at about eleven. She asked the driver to wait.

Nina came to the door and exclaimed with delight at this surprise visit from her sister, but her joy quickly changed to concern. Ethel looked "rather troubled," Nina said, and asked hurriedly if anyone else was at home. She was pale, agitated. When Nina stepped close to put her arms around her sister, she found that she was trembling.

Ethel said, "I had two detectives call to see me yesterday morning about quarter past eight, soon after Harvey had gone."

(*Harvey*—not Peter. It raised the possibility that Peter was a name appended at Belle's whim; that she had not only dressed Crippen but named him as well.)

Ethel said, "Belle Elmore's friends don't seem to think she is dead." Her voice wavered. "Who am I?" she cried. "Everyone will think I am a bad woman of the streets." She broke down. Nina tightened her embrace.

After a few moments Ethel calmed. "I can't stop long with you," she said, "but I could not go without coming and saying goodbye."

This startled Nina. She asked where Ethel was going.

"I don't know," Ethel said. Crippen hadn't told her. She promised that once settled she would send Nina her address.

But Nina could not understand *why* Ethel had to leave.

"What good is it for me to stop without means, and my character gone?" Ethel said.

And there was another reason, she said. Crippen had told her he wanted to find the person who had sent the cable about Belle's death and, in so doing, perhaps locate his wife. Only by finding her, Ethel said, could he end this scrutiny by Scotland Yard. "For all I know she may not have gone to America at all," Ethel told Nina, "she may still be in London and have got somebody across the water to send a bogus telegram informing of her death." Ethel feared a conspiracy by Belle—that out of pure malice she might simply be hiding somewhere, waiting until Ethel and Crippen got married, and then, as Ethel put it, "confront us with bigamy."

Ethel and Nina hugged again. Ethel said good-bye and stepped back into the taxi. She told the driver to head for Bloomsbury, to Albion House.

That morning at New Scotland Yard Chief Inspector Dew considered what to do next regarding the disappearance of Belle Elmore. It was tempting to do nothing, but he had been in the police department long enough to know that doing nothing could be ruinous to a man's career. He did not suspect foul play but recognized the case could not be closed with confidence until Belle was found. The doctor's advertisement would help, but something more was needed, if only to prove to Superintendent Froest that he had done all he could for the Nashes.

Dew composed a circular in which he described Belle Elmore and classified her as a missing person. He arranged to have it sent to every police division in London. It was a routine step, unlikely to bear fruit, but necessary all the same.

AROUND NOON CRIPPEN and Ethel met in the work room of Yale Tooth, on the fourth floor of Albion House. Ethel's spirits had improved. Her anger of the night before was gone, and having completed the sorrowful task of saying good-bye to her sister, she now found herself caught up in the daring of the moment.

Crippen showed her the suit of clothes that William Long had bought earlier that morning. "You will look a perfect boy in that," Crippen said. He grinned. "Especially when you have cut off your hair."

"Have I got to cut my hair?" she cried.

His delight increased. "Why, of course," he said. "That is absolutely necessary."

She wrote, "Honestly, I was more amused than anything. It seemed to me an adventure."

She removed her clothes.

ETHEL'S BROTHER, SIDNEY, planned to visit Hilldrop Crescent that same day. Ethel had made the invitation a few days earlier, before everything changed, but was unable to reach him to cancel his visit.

Now he walked up the ten steps to the front door at No. 39 and knocked. The French maid gave him a note from Ethel.

"Dear Sid," it said, "Am sorry to disappoint you to day; have been called away. Will write you later. My love dear to you and all and kisses. From your loving Sis, Ethel."

AT ALBION HOUSE Ethel stood before Crippen in a white shirt, suspenders, tie, vest, brown jacket and pants, and a new pair of boots. In trying on the pants she had split the seat, but she reconnected the seam with safety pins. "It was not a good fit," she wrote. "It was ludicrous." To complete the outfit she put on the brown felt hat.

She laughed "at the absurdity" of dressing up as a boy. "Dr. Crippen was just as gay as I was at this transformation. It seemed a merry joke to him."

Crippen picked up a pair of scissors.

"Now for the hair," he said.

He began to cut. Hair flurried around her. "I did not think twice about this loss of my locks," she wrote. "It was all part of the adventure." She put the hat back on and walked back and forth across the room, trying to get used to the alien feel of the clothing. "I was like a child," she wrote, "and strutted up and down, and very soon felt quite at ease, although for a time I missed the habit of holding my skirt."

Crippen watched and smiled. "You will do famously," he said. "No one will recognize you. You are a perfect boy."

She feared she would not be able to muster the courage to wear her disguise on the street. It felt so odd. The nape of her neck was cold. The collar chafed. The boots hurt. The sensations reported to her brain from all quarters were strange. It was hard to imagine men wearing these things day in, day out, and not going mad from constriction and abrasion.

Crippen reassured her that she looked exactly like a sixteen-year-old boy. He instructed her to leave first, by the stairs, and to meet him at the Tube station at Chancery Lane, a dozen blocks east on High Holborn—the street along which, in the distant past, condemned men traveled on their way to be executed at Tyburn, at the northeast corner of Hyde Park. To enhance her costume, Ethel placed a cigarette in her mouth and lit it, "another novelty for me which I did not much appreciate."

She set off for the stairs and soon was outside. "I was terribly self-conscious," she wrote, "but the crowds surged past, and my disguise did not cause one man to turn his head. I suppose I must have had a certain amount of pluck. I was highly strung with excitement, and the adventure was amusing to me." She waited at the entrance to the Chancery Lane station.

Soon Crippen arrived but without his mustache. He smiled and asked happily, "Do you recognize me?"

They made their way by subterranean railway to the Liverpool Street station, where eighteen platforms served a thousand trains a day. Crippen planned to catch a train to Harwich and there to book passage aboard one of the steamships that regularly sailed to Holland. They arrived at the station just after a Harwich train departed and now faced a three-hour wait for the next one, scheduled to leave at five o'clock.

Crippen suggested a bus ride, just for fun, and Ethel agreed. "Strange as it may seem," she wrote, "I was now quite cheerful, and, indeed, rather exhilarated in spirits. It seemed to me that I had given the slip, in fine style, to all those people who had been prying upon my movements"—meaning the ladies of the guild. "I had gone in disguise past their very door in Albion House, and no longer would they be able to scan me up and down with their inquisitive eyes. That made me feel glad, and I had no thought whatever of any reason for escape except this flight from scandal."

That evening, in Harwich, they boarded the night boat to Hoek van Holland, which sailed at nine o'clock. They reached Holland at five the next morning, Sunday, and had breakfast, then caught a seven o'clock train to Rotterdam, where they spent a few hours walking and seeing sights. At one point they took seats in an outdoor café, where Ethel realized how good her disguise really was. Two Dutch girls began flirting from afar, one remarking, "Oh, the pretty English boy!"

Soon afterward they boarded a train for Brussels. That afternoon they checked into a small inn, the Hotel des Ardennes, at 65 Rue de Brabant. Crippen identified himself in the hotel's register as "John Robinson," age fifty-five, and listed his occupation as "merchant." At entry number 5, *"De Naissance,"* or place of birth, he wrote "Quebec," and beside *"De Domicile"* wrote "Vienna." He identified Ethel as "John Robinson, Junior," and explained to the innkeeper's wife, Louisa Delisse, that the boy was ill and that his mother had died two months earlier. They were traveling for pleasure, he said, and planned to visit Antwerp, The Hague, and Amsterdam.

The innkeepers noticed that the Robinsons carried only a single suitcase, measuring about twelve inches by twenty-four. They observed too that the boy spoke only in whispers.

Later that Sunday Chief Inspector Dew went over Crippen's statement and realized that in the cause of thoroughness he ought to meet with the doctor one more time. He planned a visit to Albion House the next day, Monday, July 11.

A Loss in Mayfair

MARCONI DID NOT TAKE BEATRICE back to London. He brought her instead to the Poldhu Hotel, adjacent to his wireless compound. She was pregnant and felt ill nearly every day.

Marconi was oblivious, distracted by his experiments and by his company's financial troubles. The expenses of his transatlantic venture were mounting rapidly, as was pressure from his board and investors. Even so he began looking for a location to replace Poldhu and found one near Clifden in County Galway, Ireland. He envisioned a station that would produce 300,000 watts of power, four times that of his original Glace Bay station, with a horizontal antenna more than half a mile long stretched across the tops of eight two-hundred-foot masts. To fuel the boilers needed to power the station's generators, he planned to use peat from a bog about two miles away and to build a small railway to deliver it to the station. Once erected, the condenser building would house eighteen hundred plates of galvanized iron, each five times the height of a man and suspended from the ceiling.

By this point he had invested his personal fortune in his quest. Another failure now would ruin not just his company but himself as well. He kept the situation a secret from Beatrice. She said, years later, "I was almost too young to realize the strain he was under during the first year of our marriage. In view of my condition he kept his increasingly pressing financial difficulties from me. He was dreadfully overworked yet he couldn't allow himself to neglect his experiments."

Beatrice grew weary of her new isolation and resolved to move to London. Thinking still that Marconi was rich, her mother, Lady

Inchiquin, leased for her an expensive house in Mayfair. After moving in, Beatrice saw little of Marconi. The journey from Poldhu to London took eleven hours; the round trip consumed the better part of two workdays. It was time Marconi did not want to lose.

Some trips were unavoidable. In February Beatrice bore a daughter, Lucia. Marconi immediately headed for London to meet this newest member of his family. After a brief stay he left again for Poldhu.

The family doctor pronounced Lucia "a more than usually healthy baby," but after several weeks the baby fell ill. Her body grew hot and she seemed to suffer abdominal pain. Her condition worsened rapidly. Beatrice, still weak from the ordeal of childbirth, was terrified. One night Lucia had convulsions, a consequence possibly of meningitis. Shortly after eight o'clock the next morning, a Friday, the baby died. There had not even been time to have her baptized.

Marconi came back to London to find Beatrice bedridden from grief and illness. He wrote to his own mother, "Our darling little baby was taken away from us suddenly on Friday morning." Beatrice, he wrote, had received "a most awful shock and she is now very weak."

He sought to arrange Lucia's burial but found that cemeteries refused to accept her because she had not been baptized. Now he endured what Degna Marconi called "the ghastly experience of driving around London in a taxi, trying to find a cemetery that would bury his baby." Eventually he found one, in west London.

Beatrice's sister, Lilah, came to the house to tend to Beatrice, and Marconi again left for Poldhu.

Marconi's financial troubles worsened, and he at last revealed the true state of his financial affairs to Beatrice. She was startled but vowed henceforth to conserve money whenever she could.

Now Marconi fell ill. His malaria flared again and compelled him to return to London, to the Mayfair house, where he collapsed into bed and remained for three months. On April 3, 1906, an employee wrote to Fleming that Marconi's "condition is unchanged, and the Doctor has now given strict instructions that Mr. Marconi must not be disturbed."

During this time Beatrice learned something else about her husband—that he was a morbid, difficult patient.

He insisted on knowing the contents of every medicine and was impatient with the overly tactful manner of English doctors and nurses. At intervals he exploded, "They take me for an *idiot*!"

He clipped funeral advertisements from newspapers and displayed them on a bedside table. Beatrice, grieving her lost daughter and anxious about her husband's health, did not think this funny.

At one point she stepped out for a walk and to bring a new prescription to a nearby chemist's shop. She returned to find Marconi standing on his head in the bedroom. She was convinced he had gone mad.

Once he was upright again, he explained that he had bitten his thermometer and broken it and swallowed some of the mercury. Standing on his head had seemed the most efficient means of getting the mercury out of his body.

HIS ILLNESS LINGERED through much of the summer and cost him time, during which his critics and competitors remained active. Nevil Maskelyne, his magic shows now lodged in a new location farther up Regent Street from Piccadilly, acquired the rights to new wireless technology from America and formed a company, Amalgamated Radio-Telegraph, to develop it into a competing wireless system. He recruited Marconi's opponents to join him and claimed that his new apparatus allowed him to transmit messages 530 miles.

Meanwhile the secretary of Lloyd's of London, Henry Hozier, grew disenchanted with Marconi and his company. In a letter to Oliver Lodge, marked, "Private and Confidential," dated May 11, 1906, Hozier wrote, "We find that the administration of the Marconi Company is so unsatisfactory, and so difficult to deal with, that we must take precautions to have some other system available for Lloyd's business as soon as our present agreement with the Marconi Company comes to an end, and I should be very glad to have an opportunity of discussing this matter with either yourself, or Dr. Muirhead, or possibly your business manager."

Muirhead arranged to have a test station constructed on a field owned by his brother.

But Lodge's focus wavered. Mrs. Piper, the medium, returned to England with her daughters and stayed at his house, where he conducted a series of sittings. Impressed anew, he wrote a 153-page report on the experience for the *Proceedings of the Society for Psychical Research.* Once again Lodge found himself convinced of her gift and deeply distracted.

Germany's hostility to Marconi continued unabated, as British fears of German invasion deepened. In 1906, in response to Germany's growing naval power, Britain launched the most powerful battleship ever built, the HMS *Dreadnought.* That year a widely read novel, *The Invasion of 1910* by William Le Quex, fanned British anxiety and planted the fear that Germany might already have secreted spies throughout England. Commissioned by Alfred Harmsworth, the novel appeared first in serial form in his *Daily Mail* and described a future invasion in which German forces crushed all resistance and occupied London—until a heroic counterattack expelled them. Harmsworth sent men dressed as German soldiers into the streets wearing sandwich boards to promote each new installment. One witness described a line of men "in spiked helmets and Prussian-blue uniforms parading moodily down Oxford Street."

The book immediately became a bestseller in Britain, but German readers loved it too. The publisher of the German-language edition had chosen to omit the counterattack.

On September 11, 1908, Marconi was in America when he received word that Beatrice had given birth to another baby girl. Immediately he booked passage for England. During the voyage he happened to read a history of Venice, in which he spotted a name that he found appealing. The child became Degna.

The birth did little to bridge the growing distance between Marconi and his wife. They fought with increasing frequency.

An Inspector Returns

At one o'clock Monday afternoon, just as the sun emerged for the first time in a week, Chief Inspector Dew and Sergeant Mitchell set out for Albion House to have a second conversation with Dr. Crippen. Upon their arrival they learned disturbing news. Crippen's associate, William Long, told them he had last seen the doctor on Saturday leaving the office with a suitcase. He showed the detectives a letter he had received from Crippen that day, in which the doctor had written, "Will you do me the very great favour of winding up as best you can my household affairs." Crippen had enclosed enough money to cover the previous quarter's rent for the house on Hilldrop Crescent. Long chose not to mention Crippen's curious order of a boy's suit.

Dew and Mitchell secured a taximeter cab and sped to Hilldrop Crescent, through streets suffused with sunshine. The entrance to the crescent appeared as a blue-black tunnel of shade, pierced here and there by shards of golden light. The detectives were greeted by the French maid, Lecocq, who told them in a mix of French and English that Crippen and Le Neve had left and she did not expect them to return.

Dew asked if he might come in and look around the house. Lecocq understood little of what he asked but led him inside all the same. Once in the house, the two men discovered William Long's wife, Flora, hard at work packing up Belle's clothing, of which mountains remained.

The detectives searched again, this time more attentively. As before, they entered every room, paying special attention to the cellar. They found nothing to indicate the whereabouts of Belle Elmore, but Dew did find a five-chambered revolver, fully loaded. Mitchell found a box of cartridges and several targets made of cardboard.

The detectives made arrangements to send Lecocq home the next day and returned to New Scotland Yard. That evening Dew sent a request to officers throughout London to interview cab drivers and movers as to whether any had removed boxes or packages from No. 39 Hilldrop Crescent since January 31. He composed detailed descriptions of Crippen and Le Neve and arranged to have flyers distributed to police at ports in England and abroad, asking them to keep an eye out for the couple but not to attempt an arrest.

Though the case was growing more mysterious by the hour, Dew still was not convinced a crime had been committed.

IN BRUSSELS THE "ROBINSONS" delighted in their new freedom. The hotel's owner noticed that they left each day at about nine-thirty, returned by one o'clock, and remained in their room until four, at which point they left again. They returned by nine each evening, had dinner, then retired for the night.

Ethel loved touring the city. They walked all over, "north and south, east and west, and in the country parts beyond," she wrote. "Dr. Crippen showed no sign of nervousness or any desire to keep me indoors at the Hotel des Ardennes, where we put up. Never did he express a wish for me to avoid public places. Never did he evince a trace of anxiety about himself."

They visited palaces, museums, and galleries and spent hours in the Bois de la Cambre, where they walked and listened to a band and to the songs of birds. Ethel wrote, "It all seemed very beautiful, very peaceful, and they were happy days."

ON TUESDAY DEW ordered that a photograph of Crippen be circulated as well. He and Mitchell returned to Hilldrop Crescent and again searched the house—their third search thus far—and again found nothing. They made other inquiries in the surrounding neighborhood until well into the evening. Late that night as Dew tried to sleep, his thoughts kept returning to the house, in particular to the coal cellar. It "stuck in my mind," he wrote. "Even in bed, what little I got of it during those hectic days, I couldn't keep my mind from wandering back to the cellar."

The next morning, Wednesday, July 13, another brilliant but cool day, Dew and Mitchell returned to Albion House, and there Dew confronted Crippen's assistant, William Long. By now Dew had spoken to him twice, but each time had gotten the sense that Long was holding something back. Now Dew warned him to speak up or else.

At last Long disclosed his shopping trip of Saturday morning.

Dew returned to Scotland Yard and composed a new circular that included the possibility that Ethel Le Neve might be dressed in boy's clothing.

Next, acting on instinct, and for want of fresh leads to pursue, Dew proposed to Sergeant Mitchell that they search Crippen's house yet again, their fourth visit, and this time really scour the cellar.

Amid flickering candlelight, the detectives got down on their hands and knees and examined the floor brick by brick. The unusually cool temperatures outside made the chamber feel especially cold and dank. They saw nothing unusual. Dew found a small poker and used it to tap the bricks and probe the earthen gaps between them. He and Mitchell worked in silence, "too tired to say a word," Dew wrote. The light shimmied; the poker clanged against brick.

At one gap the poker drove downward with little resistance. One of the bricks moved. Dew pulled it up. Its neighbors now loosened. He pulled up several more.

Mitchell went to the garden for a spade.

The Mermaid

Now and then the marriage between Beatrice and Marconi flared back to life, and in the fall of 1909 Beatrice discovered that once again she was pregnant. At the time she and little Degna were living in a house in Clifden, as remote and austere a place as Glace Bay and Poole. Perhaps she felt a need to escape, or simply wanted for once to see the look on her husband's face when he first heard the news rather than wait for a reply by telegram, but now she plotted a surprise. She knew when his ship was due and traveled to Cork, where she managed to talk her way onto a tugboat scheduled to rendezvous with the ship. She planned to surprise Marconi with her presence and her exciting news.

Marconi, meanwhile, was reveling in the voyage and the luxuries of the ship, and in the attention lavished upon him by his fellow first-class passengers, in particular Enrico Caruso, destined to become a friend. In future years, when circumstances allowed, Marconi would stand with Caruso offstage to ease the anxiety the great tenor felt before each performance. Marconi was particularly enthralled with the young women traveling with Caruso, a group of alluring and flirtatious actresses.

Suddenly Beatrice appeared.

She had expected him to be delighted by her surprise visit, but instead, according to Degna, his welcome "was like a pail of icy water poured over her head. Returning to his bachelor habits, he was having a gay time with the ship's passengers. . . . The last thing he expected or wanted to see, popping out of the sea like a mermaid, was his wife's face."

Beatrice fled to Marconi's cabin, where she spent the night in tears.

The next morning Marconi apologized and urged her to come join the group. Beatrice refused. She felt awful, and believed she looked awful, and did not feel up to competing for her husband's attention among such a glamorous crowd. She stayed in the cabin until the ship reached Liverpool.

The Mystery Deepens

Where the bricks had lain, Dew found a flat surface of clay. He broke into it with the spade and found that the soil underneath seemed to be loose, or at least looser than it would have been had it lain there undisturbed for a period of years. He thrust the spade in deeper. The unmistakable odor of putrefaction struck him full in the face and sent him reeling. "The stench was unbearable," he wrote, "driving us both into the garden for fresh air."

Outside in the brilliant cool green, Dew and Sergeant Mitchell steeled themselves. With one last full breath they reentered the cellar, where the odor now suffused the entire chamber. Dew removed two more spadefuls of earth and found what appeared to be a mass of decomposing tissue. Once again he and Mitchell were driven from the house. They gulped the fresh cool air, found brandy, and took long draughts of it before entering the cellar a third time. They uncovered more tissue and viscera, enough to convince them that the remains were those of a human being.

At five-thirty Dew called his immediate boss, Superintendent Froest, head of the Murder Squad, and told him of the discovery. Froest notified Assistant Commissioner Sir Melville Macnaghten, in charge of the entire Criminal Investigation Department. As Macnaghten left his office, he grabbed a handful of cigars, with the idea that Dew and Mitchell might need them to counter the awful stench. He and Froest set out immediately in a department motorcar. Traveling first along the Embankment, they moved through air gilded with sun-suffused haze, the Thames a lovely cobalt edged with black shadow.

DR. THOMAS MARSHALL, divisional surgeon for Scotland Yard's "Y" Division, which encompassed Hilldrop Crescent and the surrounding district, walked over from his practice on nearby Caversham Road. His task would be to lead the postmortem examination once the remains were removed from the house.

He and Dew watched the constables dig. Lanterns had replaced candles, and the close work had begun, the constables on their knees pushing dirt away with their hands as macabre shadows played on the surrounding walls. The men concentrated on an opening at the center of the floor about four feet long by two feet wide.

What Dew saw before him evoked recollection of his discovery of Jack the Ripper's last victim and begged comparison: This was worse. The remains bore no resemblance to a human body, and the distortion had nothing to do with decomposition. In fact, much was well preserved, surprisingly so, though why this should be the case was itself a mystery. As Dew would note in a report entitled "Particulars of Human Remains," the largest mass consisted of one long connected train of organs that included liver, stomach, lungs, and heart. All the skin—"practically the whole of the soft covering of a body"—had been removed and lay in a pile, like a coat dropped to the floor.

Most notable, however, was all that was absent. There was nothing to confirm sex. No sign of hands or feet. No teeth. The head and scalp were missing. And there were no bones whatsoever. None. Dew wrote, "Someone had simply carved the flesh off the bones and laid it there."

The scope of the challenge ahead immediately became clear. It was one thing to infer from the circumstances of the case that the remains had once been Belle Elmore; it was quite another to prove it beyond doubt. The first step was to confirm that the remains were human. That proved simple: The organs were in such good condition that Dr. Marshall on first viewing was able to confirm their provenance.

It was equally obvious, however, that nothing else would be so easy. The next challenge was to identify the sex of the victim, yet no reproductive organs, pelvic bones, or other physical markers of gender could be found, save for one lump of tissue that seemed, at first, as though it might have been a portion of a female breast. Once the sex was confirmed—if ever—then Dew would have to prove that the woman was Belle Elmore. Next he would have to find the cause of death, to determine beyond doubt whether she had been murdered or had died from illness or accident. Finally he would have to determine who killed her.

What lay before him in the cellar was an affront to his working hypothesis that the killer was Dr. Crippen. It defied physics and common sense. Crippen was five feet four inches tall and of slight build. Everyone Dew had interviewed described him as kind, gentle, and affectionate. How could he kill a woman so much larger and more robust than he and then marshal the physical and mental stamina to bring her to the basement, strip the flesh from her body, remove her head, denude every bone, somehow dispose of head, bones, teeth, and sexual organs, and then bury the remainder in his cellar, all without showing signs of physical or emotional duress?

According to witnesses, on the day after Belle had last been seen alive Crippen was his usual calm and peaceful self, cheerful and ready with a smile. That day he had stopped by the Martinettis to check on Paul, and Mrs. Martinetti had noticed nothing unusual about his demeanor.

But three facts were beyond challenge:

—a mass of human remains lay in Crippen's cellar;

—Belle had disappeared; and

—Crippen and his typist, Miss Le Neve, seemed to have fled.

Macnaghten and Froest arrived, bearing cigars. Dew showed the men the cellar and walked them through the rest of the house. What most struck Macnaghten was how near the burial site was to Crippen's kitchen and breakfast area. "From the doctor's chair at the head of the dining-room table to the cellar where the remains had been found was a distance of only some fifteen or twenty feet," Macnaghten wrote. It would have taken a character of cool temperament indeed to have continued cooking and dining while aware of what lay buried beyond the next door.

After seeing the remains thus far exposed, Macnaghten telephoned a friend, Dr. Augustus Pepper, at St. Mary's Hospital. Pepper was a surgeon and one of the foremost practitioners of the emerging field of forensic pathology, "the beastly science," and as such had helped investigate many of England's ugliest murders. Recognizing that the hour was late and that much work had yet to be done to expose fully the remains, Macnaghten asked Dr. Pepper to come to the house first thing the next morning.

Macnaghten authorized Froest and Dew to spare no effort in solving the case. Dew prepared another circular, this for distribution to police throughout the world. He added photographs of Crippen and Le Neve and samples of their handwriting. He described each suspect in detail, including Crippen's habit of throwing his feet out as he walked and his "slight Yankee accent," and Le Neve's penchant for appearing to listen "intently when in conversation." He titled the circular "MURDER AND MUTILATION."

The hunt for Crippen and Le Neve began in earnest, and suddenly Dew found himself at the center of a storm of effort and press scrutiny surpassed in his recollection only by the days of the Ripper.

THAT AFTERNOON IN LONDON two detectives from Scotland Yard's Thames Division, Francis Barclay and Thomas Arle, began visiting ships moored at Millwall Docks to alert crews to the manhunt underway. Among the vessels they boarded was a single-screw steamship, the SS *Montrose,* owned by the Canadian Pacific Railway's shipping division. After a ship's officer informed them that the *Montrose* was not going to pick up any passengers in London, the detectives disembarked and continued on their way, but soon afterward they learned from another source along the wharf that while the *Montrose* would not accept passengers in London, it would do so in Antwerp, its next destination. The detectives returned to the ship and there met one of its junior officers.

The detectives told him about the recent discovery at Hilldrop Crescent and suggested that he might want "to take a few particulars." The officer had a taste for mystery and invited Barclay and Arle into his cabin, where they conversed for about an hour. The detectives suggested that the fugitives might attempt to join the ship in Antwerp and described several likely ruses that Crippen and Le Neve might deploy. Crippen, they said, might be masquerading as a clergyman, and Miss Le Neve might try "to disguise herself in youth's clothing."

The ship's officer said he would keep an alert watch and would pass the information to his captain, Henry George Kendall. The detectives departed and continued their canvass of the wharves.

IN BRUSSELS ETHEL began to feel that she was falling out of touch with the outside world. She could not read French, though Crippen could and bought copies of *L'Etoile Belge*. He spoke little about what he found in its pages.

"I asked him several times to try and get an English paper," Ethel wrote, "but he never did."

THE DYNAMITE PRIZE

SLOWLY, THROUGH GREAT EFFORT and endless experimentation, Marconi forced his transatlantic service into operation, despite foul weather and frequent malfunctions and in the face of competition that seemed to grow more effective and aggressive by the day. Germany's Telefunken, marketing the Slaby-Arco-Braun equipment, was particularly energetic. It seemed that every time Marconi's men approached a new customer abroad, they discovered that Telefunken's salesmen already had been there. They described the German company's omnipresence as "The Telefunken Wall." To make matters worse, in 1908 the provisions of Kaiser Wilhelm's international wireless conference at last took effect. Marconi ordered his men to continue shunning other systems, especially Telefunken, except in case of emergency; Telefunken engineers likewise refused to accept communication from Marconi-equipped ships. Later, Germany banished all foreign wireless systems from its vessels.

Marconi's new transatlantic service was slow and fraught with problems. A company memorandum dated August 4, 1908, showed that from October 20, 1907, through June 27, 1908, the total traffic between Clifden and Glace Bay was 225,010 words—an average of only 896 words a day. Another company report revealed that in March, the best month, the average time needed to complete transmission of a message was 44 minutes; the maximum was 2 hours and 4 minutes. The next month, however, the average climbed to more than 4 hours; the maximum was 24 hours and 5 minutes, an entire day to send one message.

But the system worked. Marconi had achieved the impossible. These were not merely three-dot messages but full-length dispatches, many of which were sent by correspondents based in America for publication in *The Times* of London, and Marconi knew, with his usual certainty, that improvements in speed and reliability would come.

In 1909 he received at last the kind of recognition that had eluded him for so many years, amid the sniping of Oliver Lodge, Nevil Maskelyne, and others. In December the overseers of the eight-year-old Nobel prizes awarded the prize for physics to Marconi, for wireless, and to Karl Ferdinand Braun, for inventing the cathode ray tube, which years later would make television possible. This was the same Braun who had joined with Slaby and Arco to produce the wireless system that Telefunken was so aggressively selling throughout the world.

To Marconi, the prize was an immense honor and utterly unexpected, for he had never considered himself a physicist. In the opening moments of his Nobel lecture in Stockholm, Marconi conceded that he was not even a scientist. "I might mention," he said, "that I never studied Physics or electrotechnics in the regular manner, although as a boy I was deeply interested in those subjects." And he frankly admitted that he still did not fully understand why he was able to transmit across the Atlantic, only that he could. As he put it, "Many facts connected with the transmission of electric waves over great distances still await a satisfactory explanation."

He acknowledged that other mysteries remained as well. "It often occurs that a ship fails to communicate with a nearby station, but can correspond with perfect ease with a distant one," he told the audience. He did not know why this was the case. Nor had he found, yet, a persuasive explanation for why sunlight so distorted communication, though he was "inclined to believe" in a theory recently put forth by physicist J. J. Thomson, that "the portion of the earth's atmosphere which is facing the sun will contain more ions or electrons than that portion which is in darkness" and therefore absorb energy from the waves being transmitted. He had found too that sunrise and sunset were times of especially acute distortion. "It would almost appear as if electric waves in passing from dark space to illuminated space, and vice versa, were reflected or refracted in such a manner as to be deviated from their normal path."

But a few moments later, with particular satisfaction, Marconi said, "Whatever may be its present shortcomings and defects there can be no doubt that Wireless Telegraphy even over great distances has come to stay, and will not only stay, but continue to advance."

HE HAD COME FAR. Though his company was struggling financially, he believed its troubles soon would ease. Ships now routinely hailed each other at midocean. Shipboard newspapers were becoming common. The term *Marconigram* had entered the lexicon of travel. Despite the competition rising everywhere, especially in Germany and America, his company had clearly achieved dominance in the realm of wireless, and in large part this was a consequence of his transatlantic gamble and the knowledge it had yielded. In Stockholm, receiving the prize, it seemed as though success had crept up unawares and had overtaken him only there at the podium, as men in black and women in gowns rose and applauded.

The biggest hurdle that remained was the skepticism that still confronted long-range wireless. For reasons he could not understand, the world continued to see it as an invention of limited use, and nothing he did seemed capable of draining once and for all that vast and persistent reservoir of doubt.

Five Jars

On Thursday, July 14, 1910, two men from the Islington Mortuary Chapel of Ease on Holloway Road came to Hilldrop Crescent to collect the remains and bring them back to the mortuary for a formal postmortem examination, to be conducted the next morning by Drs. Pepper and Marshall. The mortuary's men brought a coffin. Two constables placed the remains inside, using only their bare hands.

Dew and the two doctors watched this process closely and from time to time selected items to be placed on a tray beside the excavation. They found a Hinde's curler with hair still crimped to its vulcanite core; two pieces of what appeared to be a woman's "undervest," or camisole, with six buttons and lace around the neck; and a large man's handkerchief, white, with a reef knot connecting two corners, the portion opposite torn through. Affixed to the handkerchief were a number of strands of fair hair.

Dew also retrieved a length of "coarse string" fifteen inches long, and a second piece eleven inches long, and theorized that these, along with the knotted handkerchief, "might well have been used for strangulation, or for dragging portions of the body along."

The mortuary's men sealed the coffin and loaded it into an undertaker's van. As appalled neighbors looked on, the men drove slowly from the crescent onto Camden Road.

The next morning Pepper, Marshall, and Dew gathered at the Islington Mortuary for the formal postmortem. Pepper long ago had ceased to be squeamish about work such as this and saw the examination not as a horrific task but as the first step in resolving an engrossing puzzle, far more compelling, certainly, than conducting a routine examination of a victim who had died of a gunshot wound or been bludgeoned with a drainpipe.

First, with delicacy, he probed the mass of tissue and teased out all organs, muscles, and tendons that he was able to recognize. "There was one large mass which comprised the liver, stomach, gullet, lower 2½ inches of the windpipe, 2 lungs, the heart in its bag intact, the diaphragm or septus between the chest and abdomen, the kidneys, the pancreas, spleen, all the small intestines and greater part of the large"—all of this in one continuous chain. (In fact, as Pepper later realized, one kidney was missing.)

The connectedness was noteworthy. "It would not be a difficult thing to remove all this mass in one part from the body, but it would be a difficult thing to do it as it was done," Pepper said. "There was no cut or tear in any of the viscera, except where it was necessary for removal. There is a cut at the upper part where the gullet and windpipe were severed and at the large intestine and lower part. This showed that the person who removed the viscera was possessed of considerable dexterity: this must have been done by someone with either a considerable anatomical knowledge or someone who had been accustomed to the evisceration of animals (including human beings)."

Amid the discarded skin he found a few individual pieces that seemed worthy of extra attention. One measured seven by six inches. It had a gray-yellow hue that deepened in places to blackish gray and carried an odd mark on its surface. Pepper set it aside for closer study. He also examined the strands of hair caught in the Hinde's curler that Dew had found in the cellar. The longest strand was eight inches, the shortest, two and a half. That the hair had not come from a wig was obvious, for each strand was cut only at one end. "False hair," as Pepper put it, inevitably was cut at both ends. Where the hair was trapped around the core of the curler, its color ranged from yellow to light brown, clear evidence that the hair had been bleached.

As Pepper probed, he found additional man-made articles, including the sleeve of a pajama jacket made of white cotton with broad green stripes, and the "right posterior portion" of what appeared to be the same jacket, in which he found a label: "Shirtmakers, Jones Brothers, Holloway, Limited." This portion was stained with blood.

Pepper's initial examination suggested the victim was a woman, though the evidence was only circumstantial and was in part rebutted by the presence in the remains of a man's handkerchief and pajama top. The bleached hair, however, gave Pepper and Chief Inspector Dew confidence that the remains were indeed female and thus increased the likelihood that the victim was Belle Elmore. According to her friends in the Music Hall Ladies' Guild, she had bleached her hair blond.

Dr. Pepper placed certain organs and the reserved man-made articles into five large jars, for safekeeping. The pajama arm went into jar number four by itself; the rear portion with collar went into jar number five. The jars were stoppered, covered with white paper, tied with tape, then secured with the seal of the coroner's office.

Dew found the pajamas particularly interesting. He and Sergeant Mitchell returned to Hilldrop Crescent for another search, this time with a specific goal in mind.

ETHEL GREW WEARY of Brussels. "I had exhausted all the shop windows, which I had gazed into at first with such delight, and now I wanted to move on somewhere else."

She told Crippen of her ennui.

"Tired of Brussels already?" he said. "Very well, we will push on. How about Paris?"

"No," she said, "not Paris. Somewhere else."

Crippen suggested America.

On Friday, July 15, as Dew and the doctors probed the remains from Hilldrop Crescent, Crippen and Ethel stopped in at a ticket office and learned that one ship, the SS *Montrose,* was to depart Antwerp for Quebec the following Wednesday, July 20. They learned too that the ship carried only two classes of passengers, second and steerage. Crippen bought a cabin in second class. For purposes of the passenger manifest, he identified himself as John Philo Robinson, a fifty-five-year-old merchant from Detroit, and Ethel as his son, John George Robinson, age sixteen, a student. No one asked to see identification.

They planned to leave Brussels on July 19, spend that night in Antwerp, and board the ship first thing in the morning.

AT HILLDROP CRESCENT Chief Inspector Dew and Sergeant Mitchell concentrated on searching boxes and wardrobes and anything else in which clothing was stored. They found dresses and furs and shoes in quantities they still found staggering.

In a bag in Crippen's bedroom Dew discovered two complete suits of green-striped pajamas that seemed similar to the fragments found with the remains, except that these were new and apparently never worn. He checked their collars for labels and found "Shirtmakers, Jones Brothers, Holloway, Limited."

His search also turned up a single pair of pajama bottoms, white with green stripes, that showed signs of having been "very much worn." He could not locate a matching jacket.

THE LONDON *TIMES* GAVE the mystery a name, "The North London Cellar Murder." The *Daily Mirror* published photographs of the house and of the fugitive couple. The case seized the imagination of editors abroad, and soon news of the remains found at No. 39 Hilldrop Crescent was the stuff of breakfast conversation for readers from New York to Istanbul. "There has never been a hue and cry like that which went up throughout the country for Crippen and Miss Le Neve," Dew wrote.

The case dominated conversation everywhere, from the City to the Metropolitan Cattle Market, among the guards and prisoners at Holloway and Pentonville prisons, and at the Long Bar at the Criterion, and in the great clubs, the Bachelor's, Union, Carlton, and Reform. "It was the one big topic of conversation," Dew wrote. "On the trains and buses one heard members of the public speculating and theorizing as to where they were likely to be."

Suddenly reports of sightings of Crippen and Le Neve began to arrive at New Scotland Yard. They came by telephone and telegram and by that latest miracle, the Marconigram. The urgency and number of these tips became amplified when the home secretary, Winston Churchill, authorized a reward of £250—$25,000 today—for information leading to the fugitives' capture. "Not a day passed without Crippen and Miss Le Neve being reported to have been seen in some part of the country," Dew wrote. "Sometimes they were alleged to have been in a dozen places at the same time." Nearly every lead had to be examined. "One couldn't afford to ignore even the slenderest chance," he wrote, "and all such reports were carefully inquired into."

One man who resembled Crippen found himself arrested twice and released twice. "On the first occasion he took the experience in good part," Dew wrote, "but when the same thing happened a second time he was highly indignant, and said it was getting a habit."

On this score the police were especially wary, for Scotland Yard was still smarting from the infamous example of Adolph Beck, a Norwegian engineer who over the preceding decade and a half had been erroneously imprisoned for fraud, not once but twice, on the basis of eyewitness testimony, while the look-alike who actually had done the crimes remained free. The most important lesson of this "lamentable business," wrote Sir Melville Macnaghten, "was unquestionably the extreme unreliability of personal identification."

Dew met with the Crippen duplicate and found no particular likeness. "I did what I could to pour oil on troubled waters, offering the man my profound apologies; and after a while I was able to make him see that the police officer who had made the mistake was really only doing his duty."

On Friday, July 15, Dew and Mitchell visited Emily Jackson for the first time and heard her tell of Le Neve's miscarriage and the period in late January 1910 when she had seemed so depressed and perturbed. They revisited Clara Martinetti, this time at her bungalow on the Thames, and collected details of the dinner at the Crippens' house when she had last seen Belle alive. They interviewed Marion Louisa Curnow, a manager at Munyon's. She reported that on the day he disappeared she had cashed a check for him in the amount of £37, more than $3,700 today. She paid him in gold.

At every stop Dew and Mitchell and the detectives working with them heard anew how kind and good-natured Crippen was. Witness after witness portrayed him as too gentle to cause harm to anyone. A former neighbor, Emily Cowderoy, told one detective how she had never heard Crippen speak crossly to his wife. "They were on exceedingly good terms with each other," she said. The phrase that police heard most often in describing Crippen was "kind-hearted."

Yet there in Crippen's house at No. 39 Hilldrop Crescent, Dew had seen the eviscerated remnants of a human being who in all likelihood had once been Crippen's wife. What kind of strength, both psychic and physical, did one need to fillet one's helpmate?

It stretched plausibility to envision Crippen conducting the many different acts of dissection necessary to reduce so robust a woman to the mass unearthed in the cellar. How had he done it? Where did he begin? At the head? Perhaps a quick decapitation with a butcher's knife, maybe the same knife he had used to carve the "joint" of beef during that last dinner with the Martinettis on January 31. Or did he start with the feet, working his way up from the easy portions and coping with each new challenge as he went along? No bones remained, not even the tiny bones of the hands and feet. No doubt he simply had disposed of these extremities, but as he moved upward, then what? What tools did he use to strip muscle and tendon from the rib cage? By what means did he dislocate and detach the upper arms from the shoulders? As he advanced, did he experience elation, or was each step a source of sorrow and bittersweet recollection?

And what of the janitorial aspects? How did he cleanse the house of blood and viscera so well as to leave no apparent trace? On that score

Crippen's bull terrier had perhaps proved an able assistant. The missing portions—the head, pelvis, and outer extremities—clearly had been disposed of elsewhere.

At Dew's direction, police searched the garden. They probed with spades and in places dug deep but found none of the missing components. They searched neighboring yards and mused about likely repositories—perhaps the rendering pits and waste basins and hog sloughs of the Metropolitan Cattle Market, or the nearby channel of the Regent's Canal, which ran through North London toward Regent's Park. The canal passed under Camden Road three-quarters of a mile south of Hilldrop Crescent, an easy walk for a man with a satchel; an even easier journey if one dared carry such macabre cargo on the electric tram.

Could Crippen have done all this and, further, could he have done it without help? If so, how had he steeled himself, and how had he then managed to erase the knowledge of the act from his eyes and visage?

By Wednesday, July 20, the challenge confronting Chief Inspector Dew had become far more daunting. Somehow Crippen and Le Neve had evaded detection despite a manhunt of an intensity that Sir Melville Macnaghten believed had been surpassed only once in the history of Scotland Yard: the hunt for Jack the Ripper. Eleven days had elapsed since Crippen and Le Neve left Albion House and disappeared. The fastest ocean liners could cross the Atlantic in less than a week. The fugitives quite literally could be anywhere.

And indeed, sightings now poured in from around the globe. One caller swore she saw Crippen and Le Neve strolling along the Seine arm in arm. Another spotted them on a ship in the Bosporus. They were in Spain—and Switzerland.

Mrs. Isabel Ginnette, the president of the Music Hall Ladies' Guild, happened to be in New York City and volunteered her services to the police. Accompanied by detectives, she visited the wharves as liners arrived and watched closely for any sign of Crippen and the typist. Mrs. Ginnette and the police boarded one of the newest and most celebrated ships, Cunard's *Lusitania,* the first of the great liners to cross the Atlantic in under five days, but she recognized no one. Over the next few days she and her police escorts monitored the arrivals of the *Lorraine* from Le Havre, the *St. Paul* from Southampton—the ship Marconi had made famous—and the *Cedric* from Liverpool. In a letter to the guild's secretary, Melinda May, Mrs. Ginnette wrote, "Up till today we have met, and searched every passenger of five boats from England and France." She added, "May we soon catch him!"

On July 20 New York police arrested a passenger who had arrived aboard the *Kroonland* of the Red Star Line, believing him to be Crippen. He was, in fact, the Rev. William Laird, rector of an Episcopal church in Delaware. Mrs. Ginnette expressed dismay that the police had not taken her on that inspection as well. She told a reporter, "The reverend gentleman looked about as much like Crippen as I do."

The lack of forward motion in the investigation was discouraging and a source of mounting anxiety for Dew. There had been one recent bit of progress, however. It had come two days earlier, by chance, just after the close of the first coroner's inquest on the remains.

The proceeding itself had buoyed Dew's spirits, for the coroner in his opening remarks had praised the chief inspector. "Many a man might have gone into that cellar and made no discovery. It remained for a detective with a genius for his work to go a step further."

Afterward, in the hall outside, Dew happened to be standing near a group of women, one of them Clara Martinetti, and overheard her say something about Belle having once had a serious operation.

He took her aside and asked if he had heard correctly.

"Oh, yes," Mrs. Martinetti said. "Belle had an operation years ago in America. She had quite a big scar on the lower part of her body. I have seen it."

Dew recognized that this could be a vital clue. If evidence of that operation could be found among the remains now stored at the Islington Mortuary Chapel of Ease, it would greatly support Dew's presumption that the victim was Belle Elmore. He relayed the information to Dr. Pepper.

Nonetheless, as of Wednesday, July 20, Dew was keenly aware that his investigation, the biggest and most scrutinized of the new century, had stalled. He knew also that not everyone shared the coroner's appreciation of his investigative genius. At least one newspaper, the *Daily Mail,* asked why Scotland Yard had not kept Crippen under surveillance during its initial inquiry into the disappearance of Belle Elmore. A member of Parliament asked Home Secretary Churchill if he would be so kind as to state for the record "who is responsible for allowing Dr. Crippen to get out of their hands." Churchill declined to answer.

TESTAMENT

IN THE SPRING OF 1910, Marconi was again at sea when Beatrice gave birth to a son, Giulio. By this point Marconi had traveled so much and so far that Bea had no idea what ship he was aboard, only that he was somewhere in the Atlantic. That he would sail so near the time when his wife was expected to give birth was not surprising, given his obsession with work and his social blindness; that he would depart without leaving behind the name of his ship was something else entirely, a reflection of the decline of their marriage.

Beatrice sent him the news anyway, addressing the message only "Marconi-Atlantic."

He got it. The message was transmitted from station to station, ship to ship, until it reached him in the middle of the ocean.

It would be hard to imagine a better testament to his achievement of eliminating the isolation of the deep sea, yet a better and more public proof—one that would galvanize the world and rupture the reservoir of doubt once and for all—was soon to occur.

With the technology at last in place, the stage was set.

AT EIGHT-THIRTY WEDNESDAY MORNING Hawley Harvey Crippen and Ethel Clara Le Neve, disguised as the Robinsons, father and son, stepped onto a gangplank at the Canadian Pacific wharf in Antwerp and walked aboard their ship, the SS *Montrose*. No one gave them a second glance, despite the fact that in this age of steamer trunks and bulky coats and dressing for dinner, all they carried was a single small suitcase.

"It was without the slightest sensation of nervousness that I stepped on board the big steamer in my boy's clothes," Ethel wrote. "The change of scene seemed to me a delightful thing to look forward to."

She felt the same sense of adventure that she had felt on the night she and Crippen had sailed from England for Holland. This was escape of the purest kind. She was leaving behind a life corseted by class and disapproval, and doing it, moreover, in the guise of a male. She had shed not only her past but her sex as well.

She wrote, "I was quite easy and free from care when I followed Dr. Crippen on to the deck of the *Montrose*."

Part VI

Pursuit by Thunder

Crippen and Ethel Le Neve
aboard the *Montrose*.

The Robinsons

Ethel and Crippen settled into cabin number five, which Ethel found to be "quite cozy." The air, the sea, the throb of the engines, the miraculous crackle of the liner's wireless, all of it thrilled her. "The whole ship was wonderful."

By now her disguise was as natural to her as dresses once had been. "I felt so sure of myself," she wrote. At one point she and an adolescent boy became "rather chummy," as she put it. She could tell that he believed she too was a boy. To her amazement, she soon found herself chatting with him about football. Crippen observed the encounter. Later he told her, with a laugh, "How nicely you are getting on!"

She and Crippen spent hours on the deck, sitting and walking, "but, naturally, I kept rather aloof from the other passengers, and did not speak very much," she wrote. "On the other hand, when any of the officers spoke to me I did not hesitate to reply, and did not feel in the least embarrassed."

She marveled at the fact that even the captain gave her a good deal of attention. He was as gracious and accommodating as a steward. "I found plenty to amuse me," Ethel recalled, "for Captain Kendall supplied me with plenty of literature in the shape of novels and magazines, not forgetting some detective stories."

The captain also produced books for Crippen, who took a particular interest in Dickens's *Pickwick Papers* and two novels of the age, *Nebo the Nailer* by Sabine Baring-Gould and *A Name to Conjure With* by John Strange Winter, the mercifully truncated pen name for Henrietta Eliza Vaughan Palmer Stannard. Like many passengers, Crippen often checked the ship's track chart, updated regularly, to see where the ship was and to gauge how many days remained of the eleven the *Montrose* typically required to reach Quebec. The ship's open-sea velocity was thirteen knots.

As the weather grew colder, Ethel found that walking the deck with Crippen became less and less pleasant. The thin material of her boy's suit offered little protection from the wind, and she had nothing else to wear. "So with a rug wrapped round me I used to tuck into a corner of the lounge with a novel before me, and read quite fanciful adventures," she recalled. "I was as happy as I could expect to be."

DURING LUNCH THAT FIRST DAY, as the Robinsons and their fellow passengers dined in the second-class saloon, Kendall slipped into their cabin and conducted a brief search. He found their hats and examined them. The inside of the older man's had been stamped *"Jackson, blvd du Nord, Bruxelles."* There was no label in the brown felt hat the boy wore, but Kendall saw that the inner rim had been packed with paper—a means, he presumed, of improving the fit.

The morning of the second day at sea Kendall told his first officer, Alfred Sargent, of his suspicions. He asked Sargent to take a discreet look and see what he thought. Sargent reported back that Kendall's appraisal might be correct.

Kendall still did not feel certain enough to alert police by wireless, though he knew that after the ship exited the English Channel and entered the open Atlantic, his ability to send such a message would become limited. The shipboard transmitter had a range of about 150 miles, though its receiver could pick up signals at as great a distance as 600 miles. There was always the possibility of relaying a message via another ship closer to land, but to be absolutely certain of contact, he would have to send a message soon.

Kendall ordered Sargent to collect every English newspaper aboard and to say nothing of their suspicions to anyone else.

"I warned him," Kendall wrote, "that it must be kept absolutely quiet, as it was too good a thing to lose, so we made a lot of them, and kept them smiling."

SUICIDE

THE WORLD SEEMED GALVANIZED.

In Chicago police arrested a man named Albert Rickward, despite the fact that he was English and only twenty-nine years old, two decades younger than Crippen. They searched him and found English notes with a value in American currency of about $2,000. This increased their suspicions. They held him for hours as they interrogated him and searched his luggage, which he had left at the train station. Rickward was furious. Eventually the police let him go, without apology.

In Marseille a shipping agent notified police that he had spotted Crippen and Le Neve boarding a steamer bound for Antwerp. French detectives and the British consul raced to the wharf but found the ship had just departed.

In Halifax, Nova Scotia, police intercepted a steamer, the *Uranium,* just as it arrived. They kept everyone on board while they searched it from bow to stern. They found no one of interest.

From Brussels came the report that the owner of a café outside the city had noticed two customers who exactly fit the description of the fugitives. One of them, the café keeper reported, was a woman dressed as a man. He was sure of it.

In fact, this last report was likely correct, but it was impossible to know which reports to take seriously, which to discard. As the *New York Times* noted, "Many meek looking men with glasses have been looked on with suspicion, and the number of people who have been shadowed by amateur detectives anxious to gain the police reward of $1,250 is beyond count."

And then came this, from the French city of Bourges:

On the night of Wednesday, July 13, a lovely young woman registered at the Hotel France. She wore an elegant dress and carried herself in a refined manner. She was about twenty-five years old, brunette, and of slight build. Overall she had "a prepossessing appearance." The name she gave was Jeanne Maze. She claimed to be French, though no one at the hotel believed it.

Upon receiving her key, she went directly to her room.

One hour later hotel staff heard three gunshots. They hunted for the source and eventually came to the woman's room, which they found locked. Using a spare key, they entered. The woman was sprawled across the bed. A note lay on a nearby table.

"I request that my identity be not sought. The cause of my suicide is known to me alone. I ask to be allowed to rest tranquilly in my tomb.

"I am a foreigner. I leave 100 francs to defray my funeral expenses.

"Life to me, alas! appeared unsmiling."

The local police investigated but learned nothing of the woman's identity and let the matter rest. Clearly she was a victim of failed romance. Only when they received Dew's circular did they realize the young woman could be—had to be—the fugitive typist, Ethel Le Neve.

They found the resemblance uncanny.

A Message from the Sea

On board the *Montrose* Captain Kendall continued his investigation to test his theory about the secret identities of the Robinsons, and he did so with enthusiasm and subtlety. He read over the descriptions conveyed to the ship by Scotland Yard when it had been moored in London, and he examined the photographs of Le Neve and Crippen published in the continental *Daily Mail.* The man in the photograph had a mustache and spectacles—Robinson did not. Using chalk, Kendall erased both the mustache and the rims around Crippen's lenses, and found the likeness a close one. In talking with Robinson on deck Kendall noticed marks on either side of his nose, where the frames of spectacles would have rested. He noticed too that Robinson's supposed son filled out his clothing in a decidedly feminine way. Once, a gust of wind raised the back of the boy's jacket, and Kendall saw that the rear seam of the pants was held together with large safety pins.

Kendall invited the Robinsons to join him at his table for dinner and found that the boy's table manners "were most lady-like." The boy plucked fruit from his plate in a dainty fashion, using only two fingers instead of the full-fisted approach that a lot of men deployed. His father cracked nuts for him, gave him half his salad, and generally attended to him with the kind of solicitude men reserved for women.

During dinner Kendall told stories meant to make Robinson laugh out loud, to gauge whether in fact he had false teeth as mentioned in the police circular. "This ruse was successful," Kendall noted.

The next morning, Thursday, the second day of the voyage, Kendall engaged Robinson in a conversation about seasickness. He remarked that neither Robinson nor his son seemed to exhibit any symptoms at all. Kendall hoped through this conversation to determine whether Robinson possessed a knowledge of medicine, and indeed found that Robinson immediately began deploying medical terminology to describe certain remedies. "I was then fully convinced that he was a medical man," Kendall wrote.

Other fragments of damning evidence accumulated. Kendall overheard Robinson speaking French to a pair of other passengers. According to the police circular, Dr. Crippen spoke French. One afternoon Kendall spotted the Robinsons strolling ahead of him and called out, "Mr. Robinson!" But the man took no notice. Kendall tried again, and again Robinson was oblivious, until his son nudged him to get his attention. The father turned with a smile and apologized for not hearing, explaining that the cold weather had made him deaf. (In fact, Crippen by now had developed a hearing deficit and was known at times to use a hearing aid in the shape of a tiny funnel, of brass, which resides today in a display case at the Museum of London.)

Early in the morning of the third day, Friday, July 22, the *Montrose* left the English Channel and passed the giant Marconi station at Poldhu. Kendall knew that if he was going to alert the police, he would have to do it by evening or pass beyond the transmission range of the ship's Marconi apparatus.

Kendall composed a message for his superiors at the Canadian Pacific office at Liverpool and sent for his wireless operator, Llewellyn Jones of the Marconi company. At three that afternoon, Greenwich mean time, when the ship was about 130 miles west of the Lizard, Jones began tapping a sequence of dots and dashes destined to become one of the most famous messages in the history of marine wireless.

> *Have strong suspicions that Crippen London Cellar Murderer and accomplice are amongst saloon passengers. Moustache taken off growing beard. Accomplice dressed as boy voice manner and build undoubtedly a girl. Both traveling as Mr. and Master Robinson. Kendall.*

Kendall received nothing in reply; he had no idea whether his message reached Liverpool or not. He kept the Robinsons under close observation.

"Mr. Dewhurst"

Kendall's message tore through the atmosphere at the speed of light. Its train of waves struck the giant receiving antenna at Poldhu, and every other wireless antenna within range, and was received by Marconi's new magnetic detector, a device operators nicknamed the "maggie." The detector in turn activated a secondary circuit connected to a Morse inker, and immediately a tape bearing pale blue dots and dashes began to emerge. The operators relayed the message by landline to Canadian Pacific's office in Liverpool, where officials summoned police. Liverpool detectives, in turn, sent a message to Scotland Yard, in which they repeated the contents of Kendall's Marconigram. A messenger carried it to the office occupied by the CID's Murder Squad.

"It was eight o'clock in the evening," Dew said. "Almost completely worn out with the strain of work, I was chatting with a confrere in my office at the Yard when a telegram was handed to me."

As he read it, his fatigue "instantly vanished."

There had been thousands of leads, from all over the world. At that moment detectives in Spain and Switzerland were exploring two seemingly solid reports. Countless other supposedly good leads had dissipated like smoke. This new message, however, bore a level of authority hitherto absent. It had come from the captain of a ship at sea, owned by a large and respected company. It had been read by company officials, who presumably would not have forwarded it to police if they had harbored doubts about the captain's credibility. One portion of the message carried a particular resonance: "Accomplice dressed as boy voice manner and build undoubtedly a girl."

Dew read it over again. He checked a shipping schedule and made a series of telephone calls, the last to Sir Melville Macnaghten, the Criminal Investigation Department chief, at his home. Macnaghten was in the midst of dressing for dinner.

"Read it to me," Macnaghten said. When Dew was finished, Macnaghten was quiet a moment, then said, "Better come over for a chat."

Dew dashed down to the lobby and out to the Victoria Embankment, where he caught a cab to Macnaghten's house. Dew instructed the driver to wait. Inside, Dew showed Kendall's message to Sir Melville, who was now fully adorned in formal black and white. According to Dew, Macnaghten read the telegram with raised eyebrows.

Now Macnaghten looked at Dew. "What do you think?"

"I feel confident it's them."

"So do I. What do you suggest?"

Dew said, "I want to go after them in a fast steamer." He told Macnaghten that a White Star liner, the *Laurentic,* was set to depart Liverpool the next day for Quebec. "I believe it is possible for her to overtake the *Montrose* and reach Canada first." He proposed to book passage and intercept Crippen before he disembarked at Quebec.

Macnaghten smiled at the boldness of the idea but took a few moments to consider its implications. "It was a serious step to take to send off the Chief Inspector," Macnaghten wrote. Dew was the leader of the investigation and as such was the only man in Scotland Yard who understood every element of the case and every lead thus far explored. Moreover, the Murder Squad now found itself taxed with two additional killings to investigate, one in Slough, the other a gunshot murder in Battersea. Macnaghten worried that Dew's voyage "might well turn out to be a wildgoose chase." If so, the loss of Dew for the seven days of the crossing would prove a costly error indeed and a significant embarrassment to the department.

A decision had to be made quickly. Macnaghten walked to his desk and began to write. "Here is your authority, Dew," he said, "and I wish you all the luck in the world."

They shook hands.

"That night could not fail to be one of anxiety," Macnaghten wrote; "but the die was cast, the Rubicon was crossed. If the coup happened to come off, well and good, but, if otherwise, why, then, the case would have been hopelessly messed up, and I didn't care to dwell on the eventualities of its future."

Dew returned to the waiting taxi and rode it back to Scotland Yard. He sent a telegram to the Liverpool police, requesting that they buy him a ticket for the *Laurentic* under a false name. He went home to pack and sleep. He kept the true nature of his mission even from his wife, telling her only that he had been called abroad "on a matter of great urgency." The next day he took a cab to Euston station and caught the 1:40 P.M. "special" to Liverpool, scheduled expressly for passengers intending to sail aboard the *Laurentic.* An officer with the Liverpool police booked him passage under the name Dewhurst. Only the ship's captain and wireless operator and a couple of officers knew his true identity. To further protect the mission's secrecy, Dew gave it a code-name, Handcuffs. He was met at the Liverpool station by an inspector wearing a red rose in his coat.

The *Laurentic* departed at 6:30 P.M., on schedule. Dew knew the race would be a close one. The *Montrose* required eleven days to reach Quebec, the *Laurentic* only seven, but by now the *Montrose* had been under way for three days. If all went well, that is, perfectly, Dew's ship would beat Crippen's by a day. Given the vicissitudes of long-distance sea travel—fog, storms, mechanical failure—a single day was almost no margin at all.

Dew spent hours in the *Laurentic*'s wireless cabin as the ship's Marconi operator sent off message after message to Kendall. He heard nothing to indicate receipt. "It was hopeless," he wrote. "The answering signals simply would not come."

Macnaghten arrived at Scotland Yard early Saturday morning. "I assumed an air of nonchalance which I was very far from feeling," he wrote. He met with Superintendent Froest, Dew's immediate boss, and asked him for his candid appraisal of the night's decision to send Dew across the sea in pursuit. Froest thought it foolhardy, as did the other inspectors in the Murder Squad. They all had talked it over, Macnaghten wrote, "and had come to the conclusion that the probabilities were all against the very sanguine view that I had taken as to the correctness of the news conveyed in the marconigram."

Macnaghten's anxiety increased when a telegram arrived from Antwerp describing the father and son who had booked passage on the *Montrose.* These descriptions, Macnaghten wrote, "in no wise corresponded with those of Dr. Crippen and Miss Le Neve."

Detectives continued to explore fresh leads. New York police boarded more ships. A French rail guard swore he had seen the couple on a train. A traveler on an English train was convinced he had shared a compartment with Crippen.

In Brussels a Scotland Yard detective named Guy Workman went to the Hotel des Ardennes and photographed the registration entry of two travelers identified in the book as father and son. He learned that the innkeepers had not been fooled by the boy's clothes and had concocted a romantic explanation for why an older man would travel with a lovely young woman in disguise. The innkeeper's wife dubbed the girl "Titine" and nicknamed the man "Old Quebec" because he often spoke of the city. To her it was clear the girl had fallen in love with a teacher, and now the two were on the run.

Such passion, such adventure. It was impossible not to wish the couple well.

AT SEA THE *LAURENTIC* CLOSED ON THE *MONTROSE* at a rate of about four nautical miles each hour.

Despite the frustration of being unable to reach Kendall by wireless, Dew began to enjoy his voyage. When he needed to relax, he could walk the deck. The captain treated him with generosity and respect, and the ship was lovely and comfortable.

He believed his identity and purpose remained a complete secret.

An Intercepted Signal

For twenty-four hours Kendall heard nothing to indicate that Scotland Yard had received his message. He and his officers maintained their watch on the Robinsons and became more confident than ever that the two were indeed the fugitives Crippen and Le Neve—though none of them could quite imagine Crippen doing what the police claimed. He was polite and gentle and always solicitous of the needs of his companion.

Kendall did all he could to make sure the couple stayed relaxed and happy and unaware that their true identities had been discovered.

For Crippen and Ethel, the hours passed sweetly. Compared to life before the departure of Belle Elmore, this was heaven. No one stared, and there were no furtive meetings in secret rooms. They felt free to love each other at last.

"The doctor was as calm as ever, and spent as much time in reading as myself," Ethel wrote. "He was very friendly with Captain Kendall, and at meal times many amusing stories were told over the table, which kept us in a good humor. All the officers were very courteous to us, and used often to ask me how I was getting on."

She imagined the letter she would write to her sister Nina once settled in America. "Oh! Such a letter! I had been saving up all my little adventures in Rotterdam and Brussels. How she would laugh at my boyish escapade. How she would marvel at my impudence!"

ON SUNDAY NIGHT, July 24, the *Montrose*'s Marconi operator, Jones, intercepted a message from a London newspaper meant for someone aboard another ship, the White Star *Laurentic*. The contents were intriguing enough that Jones passed the message along to Captain Kendall.

It asked: "What is Inspector Dew doing? Are passengers excited over chase? Rush reply."

Only then did Kendall realize that his own message had gotten through and—far more amazing—that Scotland Yard was pursuing his ship across the Atlantic.

He understood, too, that the story was now public knowledge.

Cage of Glass

ABOARD THE *LAURENTIC* FELLOW PASSENGERS knew Chief Inspector Dew only as Mr. Dewhurst, and on the *Montrose* Crippen and Le Neve remained the Robinsons, but suddenly millions of readers around the world now knew, or at least suspected, their real identities. On Sunday Scotland Yard released a brief statement:

"It is believed that 'Dr' Crippen and Miss Le Neve are now on board a vessel bound for Canada. Chief Inspector Dew has left Liverpool for Canada, and hopes to overtake the fugitives and arrest them on arrival."

It took only a bit more effort by reporters to learn the names of the ships involved and the contents of Kendall's initial message. Marconi operators on other ships had intercepted it and passed it on. Aboard inbound liners the news moved from passenger to passenger. Some ships may even have reported the telegram in their onboard newspapers. Foreign correspondents based in London passed the news by cable to their editors in New York, Berlin, Stockholm, and New Delhi, and soon the front pages of newspapers around the world bloomed with maps of the Atlantic showing the supposed relative positions of the *Montrose* and the *Laurentic.*

The story consumed the editors of the *Daily Mail,* who offered a reward of £100—about $10,000 today—for information about Crippen and Le Neve. On Tuesday the paper reported, "At noon to-day the *Laurentic* will be only 253 knots (285 miles) behind the *Montrose.*" The article predicted that Dew would try to intercept the ship at Father Point in the St. Lawrence River, where pilots boarded large ships to guide them to Quebec. One article speculated that Crippen would realize, eventually, that he had been discovered—"that he will not before long have gauged the fact that the cracking, snapping, in the 'wireless' cabin means that messages about him are flying to and fro across the hundreds of miles of sea. All on board will assume that nothing is amiss, and even those who know most will pretend an ignorance of the fact that the air is quivering with wireless messages transmitted, perhaps, by intervening ships. It will be a voyage which no one aboard will be likely to forget."

For editors around the world, one point seemed obvious: Wireless had made the sea less safe for criminals on the run. "Mysterious voices nowadays whisper across it," a writer for the *Daily Mirror* observed: "invisible hands stretch out upon it; viewless fingers draw near and clutch and hold there." A French newspaper, *Liberté,* proclaimed that wireless "has demonstrated that from one side of the Atlantic to the other a criminal lives in a cage of glass, where he is much more exposed to the eyes of the public than if he remained on land."

The thing that most entranced readers was that Crippen and Le Neve knew nothing of the pursuit by Dew. To be able to watch the chase as it happened, from afar, was unprecedented, almost miraculous. J. B. Priestley wrote, "The people, who have a sure instinct in these matters, knew they had seats in a gallery five hundred miles long for a new, exciting, entirely original drama: *Trapped by Wireless*! There were Crippen and his mistress, arriving with a smile at the captain's table, holding hands on the boat deck, entirely unaware of the fact that Inspector Dew . . . was on his way to arrest them. While they were looking at the menu, several million readers were seeing their names again in the largest type."

Crippen had made a serious error, Priestley wrote: "he had forgotten, if he ever knew, what Marconi had done for the world, which was now rapidly shrinking. So we see two hunted creatures, say a fox and a hare, with millions of hounds baying and slavering after them."

IN LONDON, SCOTLAND Yard and the forensic scientists of the Home Office continued to puzzle over what had killed the victim found in the cellar at No. 39 Hilldrop Crescent. Yes, the body had been mutilated, but paradoxically the state of the remains told nothing about the proximal cause of death. For all anyone knew, the victim could have died by accident or from an illness and been eviscerated after the fact.

At St. Mary's Hospital in London, William Henry Willcox, a famed forensic chemist and senior scientific analyst for the Home Office, took delivery of the five jars of remains held at the Islington Mortuary and began a detailed examination of their contents. He was an expert on poisons and testified so often that reporters gave him a nickname, "The King's Poisoner." He took the first steps toward determining whether poison might have been the cause of death, a painstaking process that he expected would take another two or three weeks to complete.

In the meantime he asked the police surgeon, Dr. Marshall, to return to the mortuary and probe the mass of remains for additional organs. He most wanted the second kidney, the remainder of the liver, and more intestine. This was gruesome, taxing work. "The remains," Dr. Marshall noted, "had greatly changed." He succeeded in locating a portion of liver weighing 16½ ounces and a length of small intestine at 13½ ounces, but he could not locate the other kidney. He placed his finds in a sixth jar, along with a fresh discovery, another Hinde's curler with hair attached. He delivered the jar to Willcox.

But poison was only one possibility. Crippen had possessed a revolver—perhaps he had shot his wife. Or bludgeoned her and removed her head to eliminate the evidence. Or stabbed her to death and then simply kept on carving.

Superintendent Froest assigned Sgt. C. Crutchett to revisit Hilldrop Crescent and talk to the occupants of neighboring houses about anything suspicious that they might have seen or heard. He began canvassing on Wednesday, July 27, and immediately heard stories that seemed to merit further investigation.

At No. 46 Brecknock Road, which overlooked the Crippens' back garden, Crutchett interviewed a Mrs. Lena Lyons, who told him that one night at the end of January or beginning of February, while lying awake in bed, she heard "distinctly" the sound of a gunshot. It was dark at the time, though she placed the hour at about seven in the morning. Moments later a lodger, an elderly woman with the improbable name Mrs. May Pole, burst into her bedroom and said, "Did you hear a shot, Mrs. Lyons?" Mrs. Pole occupied the upstairs-rear bedroom of the house, from which she had a clear view of the Crippens' garden. She was terrified and sat on the end of Mrs. Lyons's bed. An instant later there was another gunshot. Mrs. Pole stayed at the end of the bed until daylight.

Another neighbor, Franziska Hachenberger, told Crutchett that early one morning at the end of January or in early February, at about one-thirty, she heard a scream. She lived with her father, a musician, in a house on a street adjacent to Hilldrop Crescent, and through her window she had an unobstructed view across the back gardens of the Crippens and their neighbors. "I only heard the one long scream and it ceased suddenly," she told Crutchett. She did not hear a gunshot, but she immediately suspected foul play. "As soon as I heard the scream I thought of murder," she said. "It gave me a nasty turn." For the next week she checked newspapers and the placards of newspaper hawkers for word of a murder in the neighborhood but found nothing.

The most detailed report came from a man who lived at No. 54 Brecknock and whose garden was only several yards from the rear of the Crippens' house. A metalworker and auditor of two private clubs, Frederick Evans had been out with friends at a public house, the Orange Tree, and returned home at about 1:18 A.M. He knew the time because each night when he rounded the corner of Brecknock Road, he made it a practice to stop in front of a jewelers' shop and adjust his watch. "I had been indoors a few minutes and taken my boots off when I heard a terrible screech which terminated with a long dragging whine," he said. As best he could tell, whoever had screamed had been outdoors or in a room with the windows wide open. "It startled me and I at once thought of the Ripper murders, and knowing the locality and that Parmetes Row, a turning out of Hilldrop Crescent, is frequented by prostitutes, I thought it was one of these poor creatures in trouble."

He put his boots back on, checked on his wife, and went outside and quickly walked through the neighborhood, along Brecknock and Hilldrop Crescent and Camden Road, but saw nothing suspicious. Like Miss Hachenberger he checked the newspapers to see if a crime had been discovered.

He also told Crutchett that a few days later, while he was in his garden, he scented what he called a "strong burning smell." At first he thought it was the odor of burning leaves, but it was midwinter and there were no leaves left to burn. He concluded that "perhaps somebody had stripped a room and was burning the old wallpaper." The fires continued over the next several mornings. At one point he looked over his wall and saw smoke rising from the Crippens' garden. He had never known the Crippens to burn refuse before. He also told Crutchett that another neighbor had seen Crippen carrying a "burning substance" in a white enamel pail, which Crippen then emptied into a dustbin and stirred.

Evans said he missed Belle. He and his wife had enjoyed hearing her sing as she worked in the garden.

Crutchett tracked down the "dustman" employed by the Islington Borough Council to collect refuse from Hilldrop Crescent. Every Wednesday for nine years William Curtis had come to No. 39 Hilldrop Crescent to empty the accumulated waste. He told Crutchett that in February 1910 he and another dustman, James Jackson, took from the Crippens' back garden four and a half baskets of partially burned material, in addition to the ordinary contents of the dustbin. "There was burnt stuff of all description, paper, clothing, womens petticoats, old skirts, blouses," Curtis said.

On the next two Wednesdays Curtis and Jackson took away additional baskets of burned refuse, though it had been reduced entirely to ash. In his years as a dustman Curtis had learned to tell one kind of ash from another. This ash, he told Crutchett, was not ordinary fireplace ash; nor was it the ash one would expect to find after incinerating paper. "It was very light stuff, white ash," Curtis said. He added, "I did not see any bones amongst it."

Just how much credence could be given to all these accounts was unclear. None of the witnesses had seen fit to report the shots and screams to police at the time they occurred. As any detective knew, one had to treat such belated reports with a good deal of skepticism, especially in the midst of an investigation as celebrated as this one. It was likely that stories had circulated through the neighborhood many times over, each time gaining detail and color. Still, the accounts were consistent and thus worthy of record.

Sergeant Crutchett had each witness sign a statement; he gave them to Superintendent Froest.

ON WEDNESDAY, JULY 27, at midnight, far out in the Atlantic, Dew's ship passed the *Montrose*. The vessels never came close enough for visual contact. Their courses, though parallel, were separated by a vast swath of deep ocean. But now, for the first time during the voyage, the ships came within wireless range of each other. At last Dew was able to contact Kendall directly. "Will board you at Father Point," Dew's message read. "Please keep any information till I arrive there strictly confidential."

Kendall replied, by wireless, "What the devil do you think I have been doing?"

CORRESPONDENTS FROM CITIES throughout North America began making their way to Quebec and from there to Father Point and Rimouski on the St. Lawrence. Provincial towns that had never seen a reporter now saw dozens trooping through with valises, stenographic notepads, and cameras.

Within Scotland Yard, however, a good deal of skepticism remained as to whether Crippen and Le Neve really were aboard the *Montrose*. Alternative leads continued to reach the Murder Squad, including a report that Crippen and Le Neve had escaped to Andorra, a small republic situated between France and Spain.

"Speaking for myself," Superintendent Froest told a reporter, "I am keeping a perfectly clear mind on the subject. We have so many of these houses built with cards which fall down when the last of the pack is placed on top, and for this reason we are pursuing every clue which comes to us, just as if the *Montrose* incident had never occurred."

Quivering Ether

For Captain Kendall, it was irresistible. Here they were, Crippen and Le Neve, aboard his ship, utterly unaware of the messages rocketing back and forth all around them. Relayed from ship to ship, at least fifty Marconigrams arrived at the *Montrose* wireless room from editors and reporters. The *Daily Mail* said, "Kindly wireless on business terms good description of how Crippen and Le Neve arrested." The *New York World* tried to reach Crippen himself and promised, "Will gladly print all you will say." Kendall withheld the message.

The captain loved the attention. Suddenly his modest ship was the most famous vessel afloat. It was indeed "too good a thing to lose." Aware that he had an audience of millions around the world, Kendall prepared an account for the *Daily Mail* of how Crippen and Le Neve spent their days. When the *Montrose* was about one hundred miles east of Belle Isle, an island just north of Newfoundland that marked a vessel's entry into the Gulf of St. Lawrence, Kendall instructed his Marconi man to send his story to the newspaper's correspondent in Montreal.

He knew, however, that his account would gain much wider distribution. Once reduced to the invisible confetti of Morse, his story would hurtle from ship to ship, station to station, until it suffused the atmosphere, available to any editor anywhere.

As the *Montrose* entered the Gulf of St. Lawrence, Ethel's excitement rose. She could not wait to disembark and proceed to America. Crippen, however, seemed to grow anxious. He came to their cabin looking "very serious" and handed her £15 in notes. "My dear," he told her, "I think you had better take charge of these."

"Why?" she asked. "I have nowhere to put them except in these pockets. You can keep them, can't you, until you get to Quebec?"

He paused. "I may have to leave you."

"Leave me!" His remark left her "astounded," she wrote. "It seemed to me incredible that I should have come all this way and then should be let alone."

Crippen said, "When you get to Quebec you had better go on to Toronto. It is a nice place and I know it fairly well. You have not forgotten your typewriting, have you, and you have got your millinery at your fingers' ends?"

She relaxed. She had misunderstood, she thought. Crippen was not in fact planning to abandon her but rather wished to scout opportunities in America first, on his own, and then would send for her, "that we might settle down in peace in some out-of-the-way spot."

She asked, "How about these clothes?"

Crippen smiled. "Are you tired of being a boy?"

They worked out a plan where as soon as they left the ship, they would check into a hotel. He would go out immediately and find a dress shop and buy all the clothes she needed. The prospect restored her optimism. She wrote, "I looked forward with keen delight to an adventurous life in Canada."

Crippen went back up on deck. Ethel returned to her reading. The deck had little appeal for her now. The weather was too cold, and she found the periodic fogs unbearable.

On Friday, July 29, the London *Daily Mail* published Kendall's dispatch, transmitted by wireless from the *Montrose,* snared by the wireless station at Belle Isle, and relayed to London by undersea cable—and doubtless touched up by the newspaper.

Kendall began by describing his own detective work, starting with his discovery of the Robinsons holding hands. "Le Neve squeezed Crippen's hand immoderately," Kendall wrote. "It seemed to me unnatural for two males, so I suspected them at once."

He described Le Neve as having "the manner and appearance of a very refined, modest girl. She does not speak much, but always wears a pleasant smile. She seems thoroughly under his thumb, and he will not leave her for a moment. Her suit is anything but a good fit." A wave of sympathy must have risen from women around the world. "Her trousers are very tight about the hips, and are split a bit down the back and secured with large safety-pins."

Crippen was growing a beard but continued to shave his upper lip to prevent the reappearance of his mustache, Kendall reported. The doctor still had marks on his nose from his glasses. "He sits about on deck reading, or pretending to read, and both seem to be thoroughly enjoying their meals." Crippen seemed knowledgeable about Toronto, Detroit, and California, Kendall wrote, "and says that when the ship arrives he will go to Detroit by boat if possible, as he prefers it."

Kendall listed some of the books Crippen had been reading and noted that at the moment he was engrossed in a thriller called *The Four Just Men,* a novel by Edgar Wallace in which anarchists assassinate Britain's prime minister despite every effort of Scotland Yard to protect him.

"At times both would sit and appear to be in deep thought," Kendall wrote. "Though Le Neve does not show signs of distress and is perhaps ignorant of the crime committed, she appears to be a girl with a very weak will. She has to follow him everywhere."

One evening about midway through the voyage there was a concert on board, which Crippen and Le Neve both seemed to enjoy. The next morning Crippen told Kendall "how one song, 'We All Walked Into the Shop,' had been drumming in his head all night, and how his 'boy' had enjoyed it and had laughed heartily when they retired to their room. In the course of one conversation he spoke about American drinks and said that Selfridge's was the only decent place in London to get them at."

Kendall wrote, "You will notice I did not arrest them: the course I am pursuing is the best as they have no suspicion, and with so many passengers it prevents any excitement."

To readers around the world, this report was magic. They knew what books the fugitives were reading. They knew about their contemplative moments, and how much they enjoyed the ship's concert. They saw Crippen laughing at the captain's jokes and Le Neve deploying her feminine manners to pluck fruit from a tray. The London *Times* said, "There was something intensely thrilling, almost weird, in the thought of these two passengers traveling across the Atlantic in the belief that their identity and their whereabouts were unknown while both were being flashed with certainty to all quarters of the civilized world." From the moment of their departure, the paper said, the two "have been encased in waves of wireless telegraphy as securely as if they had been within the four walls of a prison."

One newspaper invited W. W. Bradfield, one of Marconi's principal engineers, to write about the unfolding saga. Bradfield described a ship's Marconi room as resembling "a magician's cave" and said wireless had forever altered the prospects of criminals. "The suspect fugitive flying to another continent no longer finds immunity in mid-ocean. The very air around him may be quivering with accusatory messages which have apparently come up out of the void. The mystery of 'wireless,' the impossibility of escaping it, the certainty that it will come out to meet a fugitive as well as follow him in pursuit, will from henceforth weigh heavily on the person trying to escape from justice."

On one occasion Kendall found Crippen sitting on deck looking up at the wireless antenna and listening to the electric crackle coming from the Marconi cabin. Crippen turned to him and exclaimed, "What a wonderful invention it is!"

Kendall could only smile and agree.

The St. Mary's Cat

At St. Mary's Hospital, London, Dr. Willcox conducted an initial series of experiments to rule out certain easy-to-detect poisons, such as arsenic, antimony, and prussic acid. He found trace amounts of arsenic and carbonic acid but attributed them to a disinfectant that a police officer enthusiastically if unwisely had applied to the sides of the excavation in the Hilldrop cellar before the remains were removed. Willcox found the traces only in some organs, not in all, which reassured him that the arsenic was a contaminant, not the cause of death. Now he turned to the more complex and time-consuming task of determining whether the remains contained any poisons of the alkaloid variety, such as strychnine, cocaine, and atropine, a derivative of deadly nightshade. He estimated this phase would take about two weeks.

"It is necessary," Willcox said, "to weight the different parts of the remains where it is supposed that [an] alkaloid might possibly be. Those are mixed up quite fine, and then placed in rectified spirits of wine. The spirits of wine is drawn off after twenty-four hours, and then what is left of the mixed up flesh is placed in another lot of spirits, which again is drawn off after another twenty-four hours, and so on as long as the liquid which comes away is coloured—about five times. When the liquid ceases to get coloured we stop."

He found that an alkaloid of some sort was indeed present, then applied a well-known process, the Stas extraction method, to pull the alkaloid from the spirit solution in pure form. He weighed each amount. This was precise work. He found, for example, that his sample of intestines contained one-seventh of a grain of the alkaloid, the stomach only one-thirtieth.

Now came an important, yet startlingly simple, test that would if successful rule out a whole class of alkaloid poisons and greatly simplify Willcox's investigation. For this he needed a cat.

ABOARD THE *LAURENTIC* CHIEF Inspector Dew refined his plan. His ship was by now well ahead of the *Montrose,* as the world knew. Like all large ships, it would stop at Father Point in the Gulf of St. Lawrence, near the village of Rimouski, to pick up a pilot who would guide the ship along the St. Lawrence River to Quebec City, a route notorious for its sudden obliterating spells of fog.

He realized he would need clearance to disembark without first going through the quarantine station at Quebec, and now by wireless made the necessary arrangements.

Almost immediately each of the fifty reporters gathered at Father Point also knew the plan.

IN HIS LABORATORY at St. Mary's Hospital, Dr. Willcox mixed a bit of his alkaloid extract into a solution and, with the help of an assistant, placed a couple of droplets into the cat's eye. Moments later the cat's pupil expanded to many times its ordinary size. This was an important clue, for it meant the substance he had isolated was "mydriatic," that is, it had the power to dilate pupils. He knew of only four alkaloidal poisons with that power: cocaine, atropine, and two derivatives of henbane, hyoscyamine and hyoscine.

He shined a bright light directly into the cat's eyes and found that the pupil held its new diameter. This allowed him to rule out cocaine, because its mydriatic powers were less pronounced. When exposed to a powerful light, a pupil dilated by cocaine will still contract.

Willcox prepared for the next and most exacting series of tests with which he would narrow the identity to one of the three remaining possible alkaloids.

He dismissed the cat. His laboratory associates immediately named it Crippen. Adopted by a medical student, it would live for several years and bear a litter of kittens, before meeting its end in the jaws of a dog.

Whispers

On Friday, July 29, as the *Montrose* entered the vast Gulf of St. Lawrence, Captain Kendall sent a new message, stating that Crippen and Le Neve still had no idea that they were under surveillance.

At one point, Kendall reported, Crippen had spent about ten minutes at the door of the Marconi cabin listening as Llewellyn Jones transmitted a dispatch. Fascinated by the spark and thunder, Crippen asked who the recipient might be.

Jones proved himself an agile liar. Without expression he told Crippen it was a message to another liner, the *Royal George,* asking if her captain had spotted any ice in the vicinity of Belle Isle.

Crippen returned to his walk.

The Inspector Arrives

The *Laurentic* slowed to a stop off Father Point at about three o'clock on Friday afternoon, July 29. As Chief Inspector Dew emerged from a portal in the immense black hull and climbed gingerly down to the pilot boat, *Eureka,* he saw that its decks were crammed with reporters who shouted and waved. He was appalled and gauged it a display of unruly behavior unlike anything he had experienced in London, yet he confessed he also was relieved to see it because until this moment, despite assurances from the captain of the *Laurentic,* he had not quite believed that he truly had beaten the *Montrose* to Father Point. If the reporters were still here, he knew, the other ship had yet to arrive. In fact, he held a lead of about a day and a half.

Cameras were thrust in his face, questions shouted. "I was importuned to say something, but I need hardly say that I refused."

This did not sit well with the reporters, most of whom seemed to be Americans who clearly expected a higher level of police cooperation. They shouted and jostled, and when Dew refused to speak, they had the audacity actually to grow angry. Dew wrote, "I cannot refrain from saying that the whole affair was disgraceful and should and could have been avoided, and I was fearful lest this should in any way mar the success of my mission."

On shore Dew was met by two inspectors from the Quebec City police, who escorted him to a temporary lodging in one of the few structures—"shacks," Dew called them—near the Father Point lighthouse. Dew found Father Point to be a "lonely little place . . . with scarcely more than a dozen cottages and a Marconi station on it."

A fog had risen, adding to the desolation, but Dew himself was anything but lonely. The gentlemen of the press gathered in the other cottages and raised a clamor, shouting and joking and apparently singing, in short behaving as reporters throughout time have behaved when collected together in small places on the eve of an important event. Dew wrote, "The lighthouse foghorn combined with the vocal and musical efforts of my friends the reporters made sleep impossible."

The following evening, Saturday, one reporter gave Dew a tip that was profoundly unsettling. Reporters for one newspaper—an American paper, of course—were planning to construct a raft and sail it down the St. Lawrence posing as shipwrecked sailors, with the intent of being "rescued" by the *Montrose* and thus scooping everyone else. "Now I don't pretend to know whether there ever was any serious intention to carry out this ambitious scheme," Dew wrote, "but from what I had seen of the American newspaper men I did not put it beyond them."

He called all the reporters together and asked them to be patient. If indeed the passengers proved to be Crippen and Le Neve, he would ask Captain Kendall to blow the ship's whistle three times, at which point the reporters would be free to come out to the ship. He learned that most of the reporters, possibly all, had a legal right to board the ship—they had bought tickets for the twelve-hour voyage from Father Point to Quebec.

The reporters did not like being constrained but agreed all the same.

Dew still had doubts as to whether the passengers on the *Montrose* really were the fugitives. He spent a restless night wondering if under the gaze of the entire world he had just spent eleven days on a false hunt of historic dimension.

In London Superintendent Froest of the Murder Squad remained skeptical. Already there had been one initially persuasive report that Crippen and Le Neve had been spotted aboard a ship. For a time the world had been convinced that they were passengers on the *Sardinian*, the same ship that a decade earlier had brought Marconi to Newfoundland for his first transatlantic experiment. The *Sardinian*'s captain ordered his crew to conduct a search. Suspense mounted until at last the captain sent a wireless message to Scotland Yard stating that his men had found no one resembling Crippen or Le Neve.

Now Dew was off chasing a different ship across the Atlantic on the strength of another captain's suspicions. It too could prove a false trail—but if so, the consequences for the reputation of Scotland Yard would be grave. Every day the newspapers of London charted the positions of the two ships. Even the home secretary, Winston Churchill, had become caught up in the drama. His confidential clerk had called to notify Scotland Yard that Churchill wished to be informed immediately at his office of any new developments in the case.

So Froest kept the Murder Squad working at the same intensity as before Dew's departure. In Dew's absence he placed Sergeant Mitchell in direct charge. The squad hunted Crippen but also sought to fill in elements of the overall story and to better understand the characters involved.

They learned, for example, that Le Neve had been seen often at two public houses in Hampstead, the Stag and the Coach and Horses, accompanied by a young man whom at least one witness believed to be her "sweetheart." The CID's Sgt. William Hayman tracked him down and identified him as John William Stonehouse.

In a formal statement Stonehouse revealed that until the preceding October he too had been a roomer in Emily Jackson's house on Constantine Road and had come to be friends with Ethel Le Neve. *Just* friends, he was careful to note. Through Stonehouse, Sergeant Hayman discovered that after Ethel's first move from the boardinghouse she had taken a room in a building on Store Street. One day Stonehouse had walked her home. He said, "I accompanied her to the door and in conversation I understood that she was uncomfortable."

He added, "There was never any undue familiarity between us."

A room on Store Street—the same street where Crippen and his wife once had lived, and so very near Albion House. It did not take a detective to infer the use to which this nearby residence was put.

Churchill's confidential clerk telephoned again. The home secretary was now at his home, 33 Eccleston Square, and wished to have news of the Crippen case sent directly there.

Later the clerk called to say that Churchill was now at the Heath Golf Club, Walton. The news should go there.

A Boat in the Mist

On Saturday night fog settled over the St. Lawrence and forced Captain Kendall to slow the *Montrose.* The blue spark of the ship's wireless lit the suspended droplets and made the Marconi cabin seem as if it truly were a magician's cavern. Even with the door now shut against the weather, the crack of the spark generator was audible on the deck outside.

Fog during a voyage was never pleasant, but in so heavily traveled a channel as the St. Lawrence, it was especially unnerving. "The last night was dreary and anxious, the sound of our foghorn every few minutes adding to the monotony," Kendall wrote. "The hours dragged on as I paced the bridge; now and then I could see Mr. Robinson strolling about the deck."

Kendall told Robinson that he ought to consider getting up early so that he could be on deck in time to watch the pilots come aboard from Father Point. The captain suggested he might find the experience interesting.

At four-thirty the next morning, Sunday, Kendall blew the *Montrose*'s whistle to alert Father Point of the ship's imminent arrival.

Crippen followed Kendall's suggestion and rose early. He and Ethel had breakfast, then returned to their cabin, where Ethel snuggled up with her latest book, *Audrey's Recompense* by Mrs. Georgie Sheldon, the pen name of Sarah Elizabeth Forbush Downs. Crippen urged her to come up on deck. "I don't think I will," she told him. "It's very wretched up there, and I would rather stay down here and finish this book before lunch."

Crippen left "quietly," Ethel recalled, and went up alone. On deck he began to walk. Inside the lining of his vest he had sewn four diamond rings, a pin in the shape of a butterfly, and a gold brooch studded with diamonds that evoked a rising sun.

THE SHIP'S SURGEON, Dr. C. H. Stewart, also came up on deck early. He knew of the trap about to be sprung and wanted to see the whole thing unfold. At around eight o'clock he encountered Mr. Robinson, and the two began to chat. They stood together at the rail on the ship's port side. The fog had thinned to mist, and now rain began to fall.

Robinson seemed nervous. Stewart noticed, too, that Robinson had clipped off his new beard and had cut his upper lip, apparently while shaving. What most struck Stewart, however, was that Robinson looked nothing like the man in the photographs published in the *Daily Mail.*

A boat emerged from the pewter mist and gained definition.

"What a lot of men in that small boat," Robinson said. He turned to Dr. Stewart. "Why so many?"

Stewart shrugged. "There is only one pilot for the ship," he said. "Perhaps the others are his friends, who are going to take a little excursion as far as Quebec."

Robinson asked if the men might be medical officers. Dr. Stewart said he did not think that was the case.

They continued to watch.

KENDALL WENT TO HIS CABIN and found his revolver. As a precaution he placed it in his pocket. He returned to the bridge.

Treacherous Waters

Early that morning Dew and the reporters had gotten up well before dawn. At four-thirty amid the bleating of the foghorn, they heard a ship's whistle. The reporters raced to board the pilot boat, *Eureka,* and civilian spectators climbed into a flotilla of small boats. The police kept the crowd from shoving off.

Dew realized he would have to change his plan. He had intended to ride the *Eureka* to the ship, but now he saw that all those reporters jamming its decks would give the trap away long before he reached the *Montrose.* There was still a chance the Robinsons were not in fact Crippen and Le Neve. It would be best, he reasoned, if he could board the *Montrose* wearing some kind of disguise, so that he could get a look at Crippen without being detected. A disguise might also prevent the fugitive from panicking and doing something unexpected, like leaping into the river or drawing a gun. Dew and Mitchell had found one revolver at Hilldrop Crescent; Crippen might be carrying another.

Dew asked the chief pilot if he might borrow his uniform and cap. The chief agreed. Dew then arranged to go out to the ship in the company of the regular pilot but not aboard the *Eureka.* Instead they would take a large rowboat. The two Quebec inspectors would come along.

They launched the boat from a location well away from the reporters. Four sailors did the rowing. Soon the liner came into view, its long black hull barely visible in the mist and rain. Dew pulled the visor of his pilot's cap low over his face.

Crewmen on the deck high above threw over a ladder, which jolted to a rest just above the waterline. The real pilot climbed first. Dew followed, as did the Quebec detectives. All went directly to the bridge, where Dew introduced himself to Kendall. They shook hands and exchanged greetings. At that moment on the deck below a man of slight stature emerged from behind the ship's funnel. Dew watched him.

Kendall watched Dew. The captain looked for some sign that Dew recognized the passenger below. The inspector said nothing. Kendall led the party to his cabin and sent for Mr. Robinson. A few moments later the man appeared, looking unconcerned and cheerful.

Kendall stood. Discreetly he put his hand in his pocket and gripped the revolver. He said, "Let me introduce you."

Dew stepped up, still in his cap. The passenger smiled and held out his hand. Dew took it and with his free hand pulled off his cap. He said quietly, "Good morning, Dr. Crippen."

The expression on the passenger's face changed rapidly, Dew wrote. First came surprise, then puzzlement, then recognition. Finally, in a voice Dew described as being "calm and quiet," Crippen now said, "Good morning, Mr. Dew."

Once the details became public knowledge, all of Britain seemed to agree that the understated drama of this encounter had been equaled only once before, when Stanley caught up with Livingstone.

Now Dew told Crippen, "You will be arrested for the murder and mutilation of your wife, Cora Crippen, in London, on or about February last."

Dew's gamble had paid off. "During my long career as a detective I have experienced many big moments," he wrote, "but at no other time have I felt such a sense of triumph and achievement." But he also felt what he described as a "pang of pity" for the little doctor. Crippen, he wrote, "had been caught on the threshold of freedom. Only twelve hours more and he would have been safely at Quebec."

The Canadian police handcuffed Crippen and led him to a vacant cabin. Now Kendall led Dew to cabin number five, which Crippen and Le Neve had occupied during the voyage.

Dew tapped lightly on the door, then entered. To his great satisfaction, he saw that Ethel was indeed wearing a suit of boy's clothing.

She looked up from her book.

He said, "I am Chief Inspector Dew."

The introduction was unnecessary. Despite his pilot's uniform, she recognized him immediately. She gave a cry and stood, then fell unconscious as abruptly as if someone had struck the back of her head with a crowbar. Dew caught her in midswoon.

Epilogue

Into the Ether

Crippen and Le Neve during preliminary hearing at Bow Street Police Court.

The Table of Drops

Two days after the arrest detectives learned for the first time of Crippen's January purchase of five grains of hyoscine hydrobromide. Soon afterward Dr. Willcox, at St. Mary's Hospital, confirmed that the alkaloid he had isolated was indeed hyoscine. He was able to extract two-fifths of a grain from the available remains but knew that if he had been able to analyze all of the body, the amount would have been far greater. Just a quarter grain could have been lethal. "If the fatal dose were given," he said, "it would perhaps produce a little delirium and excitement at first; the pupils of the eyes would be paralyzed; the mouth and the throat would be dry, and then quickly the patient would become drowsy and unconscious and completely paralyzed, and death would result in a few hours."

By now Willcox and colleagues were confident the remains were those of a woman, though this conclusion was based entirely on circumstantial evidence, namely the curlers, the bleached hair, and the fragments of a woman's underclothing found in the excavation. The question of identity remained daunting until Dr. Pepper happened to reexamine the pieces of skin still held at the Islington Mortuary Chapel of Ease. One piece—the fragment measuring six by seven inches—had a mark on it about four inches long. Having learned from Chief Inspector Dew that Belle once underwent an abdominal operation, Pepper now took a closer look. It was possible, he decided, that the mark was a scar. He gave it to Willcox, who passed it on to the youngest member of the Home Office's elite forensic group, Dr. Bernard Spilsbury, an expert on scars.

Investigators made another important discovery. Upon close

examination, the torn pieces of pajama jacket found with the remains proved to match exactly the pajama bottoms that Dew had found at Hilldrop Crescent.

IN QUEBEC, WHILE AWAITING extradition, Crippen was lodged in a prison on the Plains of Abraham, where he seemed in good spirits and gave full play to his passion for reading. Ethel, feeling ill, was allowed initially to stay in the home of one of the Quebec inspectors, where Dew told her at last that he had found human remains in the cellar at Hilldrop Crescent. She stared at him, speechless, the expression on her face one of amazement.

Sergeant Mitchell arrived from London accompanied by two female officers to help Dew bring the captives back to England. Early one morning they smuggled Crippen and Le Neve into two closed carriages and raced over quiet, mist-shrouded country roads to a remote wharf, where all of them boarded a river steamship. No one followed. Soon afterward the little steamer intercepted the White Star liner *Megantic,* which halted and took them aboard.

Dew and Mitchell treated the captives with kindness. Dew's manner was so paternal and solicitous that Ethel teasingly called him "Father." During the voyage the inspector visited both Crippen and Le Neve in their cabins many times a day and always asked how they were faring. Crippen struck him as utterly untroubled. He ate well and slept well and conversed avidly about a broad range of subjects, though never about Belle. "He mystified me," Dew wrote. "He seemed quite happy. He gave no trouble, and never once tried the patience of Sergeant Mitchell or myself. The impression he gave was that of a man with mind completely at rest." Crippen's main preoccupation, as always, was reading. "I used to fetch his books myself from the ship's library, being careful, of course, never to get him one with a crime or murder plot," Dew wrote. "He loved novels, especially those with a strong love interest." At the Quebec prison, he had read *Barchester Towers* by Anthony Trollope, then had autographed the book and given it to a guard for a souvenir.

Dew kept Crippen and Le Neve isolated from each other. Between eight and nine o'clock each evening the *Megantic*'s captain cleared the boat deck to allow the captives exercise. Crippen walked first, Le Neve second, the timing such that they never saw each other. This arrangement was painful for Crippen, and one day he begged Dew to allow him to see Le Neve just one time—not talk to her, just look at her. "I don't know how things may go," Crippen told him. "They may go all right or they may go all wrong with me. I may never see her again, and I want to ask if you will let me see her. I won't speak to her. She has been my only comfort for the last three years."

Dew arranged it. In mid-Atlantic, at an agreed-on time, he brought Crippen to the door of his cabin. Ethel appeared at her door thirty feet away. The two looked at each other and smiled. They did not speak. "I had to be present," Dew wrote. "But somehow as I looked on I felt an interloper. Not a word was spoken. There were no hysterics on either side. Just a slight motion of the hand from one to the other. That was all."

The encounter lasted perhaps a minute. They did not see each other again for the rest of the voyage.

CRIPPEN'S TRIAL WAS HELD first and began on October 18, 1910. Four thousand people applied at the Old Bailey for tickets, so many that court authorities decided to issue passes good for only half a day, so that as many people as possible could attend. The spectators included Sir Arthur Conan Doyle and W. S. Gilbert, of Gilbert and Sullivan. During the trial a sympathetic portrait of Crippen emerged. Witnesses described him as kind and generous, Belle as volatile and controlling. Even the women of the Music Hall Ladies' Guild could not find anything bad to say about him. In the typewriters of the press the case became a darkly shaded love story—a sad, abused man finds his soul mate, who loves him back, deeply and truly.

But then came the evidence of what had been done to the victim in the cellar. On the stand Spilsbury—thirty-three years old, achingly handsome, and wearing a red carnation—testified that he had determined without doubt that the mark on the six-by-seven-inch piece of skin was indeed a scar and likely to have been caused by surgery to remove a woman's ovaries. At this point a soup plate containing the skin in question was passed among the jurors.

Misled by Spilsbury's youth and his pampered appearance, the defense attacked headlong and brought forth two physicians who swore the mark could not have been a scar. Spilsbury held fast. He spoke with such quiet confidence and aplomb that he won the jury and became the darling of the press. The episode launched him on a career without parallel in the history of forensic medicine.

The scar, the pajamas, and Crippen's purchase of hyoscine were damning, but there was broad agreement that what clinched the case for the Crown was an exchange between Crippen and the prosecuting barrister, Richard Muir, at the start of the second to last day of trial.

Muir asked, "On the early morning of the 1st of February you were left alone in your house with your wife?"

Crippen: "Yes."

"She was alive?"

"She was."

"Do you know of any person in the world who has seen her alive since?"

"I do not."

"Do you know of any person in the world who has ever had a letter from her since?"

"I do not."

"Do you know of any person in the world who can prove any fact showing that she ever left that house alive?"

"Absolutely not," he said.

The jury stayed out for twenty-seven minutes and returned with a verdict of guilty. As the judge prepared to read his sentence, he donned a black scarf.

Ethel's trial took place soon afterward, but her jury decided she had known nothing of the murder and set her free.

ON OCTOBER 25, 1910, Crippen was transferred to Pentonville Prison, in his old neighborhood. A warder took his money and jewelry, made him take off his clothes, and examined his ears and between his toes, then gave him a prison uniform. The fact of his incarceration did not stop one woman, Adele Cook, from writing to prison officials to ask if he might be allowed to write her a prescription. The reply: "The applicant should be informed that if she wishes to write letters for Crippen she may do so."

He filed an appeal but failed to reverse his conviction. In a letter to Ethel he insisted he was innocent and that someday evidence would be discovered to prove it. He acknowledged, however, that his fate was sealed. He wrote, "It is comfort to my anguished heart to know you will always keep my image in your heart, and believe, my darling, we shall meet again in another life." On November 23 he awoke to the certainty that he would never see another dawn.

Prison authorities filled out the required execution form, which they gave to the executioner, John Ellis, a village hairdresser who moonlighted as hangman. Ellis took careful note of Crippen's weight, then consulted the "Table of Drops" to determine how far Crippen's body should fall to ensure a death that was instant but not gory. Ellis was known to be an efficient hangman, though with a tendency to add a few more inches to the drop than strictly necessary.

Ellis saw that Crippen weighed 142 pounds. Next he checked the entry under "Character of prisoner's neck" and found that Crippen's neck was quite normal. Ellis saw too that his build was "proportional" and that he was only five feet, four inches tall. He set the length of drop at seven feet, nine inches.

For his last request, Crippen asked the prison governor, Major Mytton-Davies, to place a number of Ethel's letters and her photograph in his coffin. The governor agreed.

At precisely nine A.M. Ellis released the floorboard, and an instant later Crippen's neck broke, quite cleanly, at the third cervical vertebra. Happily for all present, his head remained attached.

The prison warder took note of the possessions he left behind: one overcoat, one coat, one waistcoat, one pair of trousers, two hats, four shirts, one pair of underwear, four socks, six handkerchiefs (one silk), ten collars, two bows, one pair of gloves, one gladstone bag, one toothbrush, a small amount of cash, and one pair of spectacles.

Ellis continued to moonlight as an executioner and at one point acted the role of hangman in a local play about a notorious criminal named Charles Peace. After the last performance he was allowed to take the scaffold home. When he was not hanging people or doing their hair, he demonstrated the art of execution at country fairs.

On September 20, 1932, he killed himself by slashing his own throat.

Mystery lingered around the Crippen case like tulle fog at a cemetery. An editor for the magazine *John Bull* addressed an open letter to Crippen shortly before his execution in which he asked, "Was it *your* hand which did the deed, and was it your hand *alone* which sought to destroy all traces of the tragedy?" He did not believe it possible. "Tell me," he wrote, "by what superhuman strength was the body of a heavy woman carried down below? Did you alone do that, and, in addition, dig the floor, remove the clay, cover up, rebrick and make good—you, a little, half-blind, elderly, weak and timid man? And, including the butchery, all in twenty-four hours!"

John Bull's scenario did challenge the imagination. It presupposed that Crippen killed Belle somewhere upstairs, then dragged her down to the basement. The evidence at the grave certainly suggested that at one point or another dragging occurred. There were pieces of cord and a man's handkerchief that had been tied tightly to form a loop. The handkerchief could have been secured around Belle's neck, the cord then attached to it to form a convenient handle for towing—at least, that is, until the handkerchief became torn.

But maybe Crippen never dragged Belle's *entire* corpse down to the basement. The remains of pajamas and a camisole suggest she was in nightclothes at the time of her death. Perhaps he killed her upstairs. He gave her poison, maybe in an evening brandy, and when she became disoriented, he led her to the upstairs bathroom. There, perhaps, she began to make far more noise than he had anticipated. It is possible he shot her, as the reports from neighbors suggested—though again such belated accounts must be treated with skepticism. More likely he strangled her. He tied his handkerchief tightly around her neck. She struggled but began to lose consciousness. She fell. Using all his strength, Crippen levered her body into the bathtub. He severed both her carotid arteries and waited as her body drained of blood. The tub provided an operating theater within which the gore could be contained and rinsed. He would not have risked leaving a trail of blood as he carried Belle's head, hands, feet, and bones to the Regent's Canal or the Cattle Market or some other suitable point of disposal. He brought the rest of Belle to the cellar in fragments.

There are problems with this theory as well. If Crippen had conducted his evisceration under such well-lit and controlled conditions, surely he would have recognized that he had left important pieces of evidence in the debris—the Hinde's curlers, complete with strands of hair; the portions of pajama top; the camisole; the knotted handkerchief. Their presence in the remains suggests he worked under conditions far less ideal than the bathroom afforded, and that he overlooked them because they were masked by blood and viscera and darkness. As Belle lay dead, upstairs or down, he dug the grave in the cellar, planning to rely on the earthen walls of the excavation to contain the blood. He dragged her to the grave and then began his surgery. The light was not good. Blood coated everything. He pulled from the remains Belle's head and the other portions that he hoped to get rid of, then rinsed them in the sink of the adjacent kitchen. He wrapped them carefully in oiled canvas or a raincoat and left them in the cellar, where he could be reasonably certain Ethel would not go. Over the next few nights, in installments, he carried off the head, bones, hands, and feet.

The most important question is how a man so mild and kind-hearted could resort to murder in the first place. One theory, put forth by a prominent jurist of the day, proposed that Crippen killed Belle by accident—that he deployed hyoscine merely to sedate her in order to buy himself a night of peace but miscalculated the dose. This seems improbable. Crippen knew the properties of hyoscine and knew how little of it would constitute a fatal dose. How much he actually gave her can never be known, but those familiar with the case believed he administered all five grains.

That Crippen intended to kill her cannot be doubted. What remains, then, is the likely reality that he had come to loathe Belle so completely, and to need Ethel so deeply, that when Belle lit into him for the minor infraction of failing to show Paul Martinetti to the bathroom, something in his soul fractured. Aided by gravity, he dragged Belle's corpse to the basement and in an adrenaline-powered fugue set out to remove her from the world as utterly as if she had never existed. One of the three barristers assigned to his prosecution, Travers Humphreys, later wrote, "I never looked upon Crippen as a great criminal. He made a bad mistake and paid the penalty which Society provides for those who commit the crime of which he was rightly convicted, but in another country he would I feel sure have been given the benefit of 'extenuating circumstances.' "

As to whether he had help, no one can ever know. Ethel's jury accepted without quarrel her defense that she knew nothing of the killing. And yet there were aspects of Ethel that abraded the popular image of her as an unwitting and lovestruck companion. She wrote with sophistication. She was daring and craved adventure. Richard D. Muir, who led her prosecution as well as Crippen's, seemed to have his doubts about her innocence. He wrote later, "Full justice has not yet been done."

The missing portions of Belle's body were never found, though Scotland Yard spent a good deal of time looking. Detectives probed Regent's Canal where it passed through Regent's Park. A London "sewerman" named Edward Hopper came forward and recommended that detectives examine the "intercepter" on the sewer line that drained waste from Nos. 38 and 39 Hilldrop Crescent. "We carefully examined a quantity of dirt and rubbish which was in the intercepter, but could not find any trace of flesh or bones," wrote the detective in charge, one Sergeant Cornish.

Prior experience had taught Scotland Yard that English murderers had a predilection for stuffing bodies into trunks and leaving them at train stations, so the CID asked the managers of every station in London and its suburbs to check their cloakrooms for parcels and luggage left unclaimed since early February. They found mysterious boxes and suitcases of all sizes, including a trunk with three padlocks at the Cambridge Heath Station of the Great Eastern Railway. Police opened some of the abandoned cargo, but in most cases a simple external examination sufficed. Sergeant Cornish, in charge here as well, concluded his report, "There is no bad smell attached to any of the packages, all of which we are quite satisfied contain household effects and wearing apparel."

The women of the Ladies' Guild took custody of Belle's remains from the Islington Mortuary Chapel of Ease. The Public Health Department was glad to see them go, judging them "likely to cause a serious nuisance." At precisely 3:15 on October 11, 1910, a small cortege consisting of a horse-drawn hearse and three mourning coaches set off on a slow, sad drive across the top of London to the St. Pancras Cemetery in East Finchley. Soon afterward the ladies of the guild watched as a coffin bearing their old friend was set into the earth. The police were on hand to make sure spectators did not disrupt or crowd the service, and they reported that everything "passed off quietly."

CHIEF INSPECTOR DEW saw the Crippen case as a fitting point at which to retire. His career as a detective had begun with a crime involving mutilation and murder, and now it ended with one. He felt great sympathy for Crippen and Le Neve. He wrote, "Dr. Crippen's love for the girl, for whom he had risked so much, was the biggest thing in his whole life."

He retired to a cottage called the Wee Hoose and in 1938 published a memoir, *I Caught Crippen,* in which he described the case as "the most intriguing murder mystery of the century." He dedicated the book to his son, Stanley, killed during World War I, and to his three daughters, one of whom happened to be named Ethel.

He died at his cottage on December 16, 1947.

Captain Kendall received the £250 reward from the Home Office but never cashed the check. He had it framed instead.

The great chase made Kendall a world celebrity and a star within the Canadian Pacific Railway. He rose quickly within the company and became captain of the *Empress of Ireland,* the ship aboard which he previously had been first officer and that once had taken Marconi and Beatrice to Nova Scotia. Just before two in the morning, on May 29, 1914, in almost the same location where Dew had boarded the *Montrose,* a Norwegian freighter rammed the *Empress* in the midst of a thick, sudden fog. The freighter backed away and remained afloat. The *Empress* sank in fourteen minutes, at a cost of 1,012 lives. Kendall was thrown from the bridge into the water when the ship suddenly rolled onto its side. He survived.

The toll would have been higher if not for the presence of mind of Ronald Ferguson, the *Empress*'s senior Marconi officer, who managed to send a distress call before the ship's power failed.

An inquiry absolved Kendall of blame, but the railway assigned him a desk job in Antwerp. This, however, did not long shelter him from adventure. He was there when World War I began. As the Germans raced to seize Antwerp, Kendall commandeered his old ship, the *Montrose,* and filled it and a sister ship with Belgian refugees, then used the *Montrose* to tow the latter to safety in Britain. He joined the Royal Navy and was given command of a ship, only to have it sunk by torpedo. After the war he spent another twenty years working for Canadian Pacific, again at a desk.

During the war, the Admiralty bought the *Montrose* and moored it at the entrance to Dover harbor as a guardship, but a storm tore it loose, drove it from the harbor, and destroyed it on the Goodwin Sands, where George Kemp had spent so many harrowing nights aboard the East Goodwin lightship. The *Montrose* did not go gently. One mast remained in view until June 22, 1963, when yachtsmen suddenly realized the old ship had left them at last.

Two years later the old captain left as well.

THE CASE OF DR. CRIPPEN became the subject of plays and books and drew the attention of Alfred Hitchcock, who used elements in several of his movies, including *Rope* and *Rear Window,* and in at least one episode of *Alfred Hitchcock Presents.* There is a moment in *Rear Window* when the lead character, played by Jimmy Stewart, looks out at the sinister apartment across the way and says, "That'd be a terrible job to tackle. Just how would you start to cut up a human body?"

Crippen also proved a fascination for Raymond Chandler. He wondered at the inconsistencies of the case—how anyone as clearly intelligent and methodical as Crippen could have made the mistakes he made. In a 1948 letter to a friend Chandler mused, "I cannot see why a man who would go to the enormous labor of deboning and de-sexing and deheading an entire corpse would not take the rather slight extra labor of disposing of the flesh in the same way, rather than bury it at all." Chandler did not buy the widely held notion that Crippen would have been safe if he and Ethel had stayed in London rather than fleeing after Chief Inspector Dew's initial visit. Eventually, Chandler wrote, Scotland Yard "would have come to the old digging routine." He wondered too "why a man of so much coolness under fire should have made the unconsionable error of letting it be known that Elmore had left her jewels and clothes and furs behind. She was so obviously not the person to do that."

These were mistakes that panicked men tended to make, Chandler argued. "But Crippen didn't seem to panic at all. He did many things which required a very cool head. For a man with a cool head and some ability to think he also did many things which simply did not make sense."

Chandler felt sympathy for Crippen. "You can't help liking this guy somehow," he wrote in another letter. "He was one murderer who died like a gentleman."

A play called *Captured by Wireless* made the rounds in America and for one week in April 1912 occupied the stage of the Opera House in Coldwater, Michigan, Crippen's hometown. "The play is full of comedy throughout," the *Coldwater Courier* reported. In England the case continued to engage the popular imagination for decades and sparked the creation of a second theatrical production, a musical entitled *Belle, or the Ballad of Doctor Crippen*. It debuted on May 4, 1961, at the Strand Theater in London and presented its audience with two dozen musical numbers, including "Coldwater, Michigan," "Pills, Pills, Pills," and "The Dit-Dit Song." The show lasted forty-four performances but proved a failure. England was not yet ready to mock this tragic confluence of love, murder, and invention. The *Daily Mail* headlined its review, "A Sick Joke With Music."

DURING WORLD WAR II a Luftwaffe bomb hurtling shy of its mark landed on Hilldrop Crescent, obliterating No. 39 and a good portion of the block.

The Marriage That Never Was

THE CRIPPEN SAGA DID MORE TO ACCELERATE the acceptance of wireless as a practical tool than anything the Marconi company previously had attempted—more, certainly, than any of Fleming's letters or Marconi's flashiest demonstrations. Almost every day, for months, newspapers talked about wireless, the miracle of it, the nuts and bolts of it, how ships relaying messages from one to another could conceivably send a Marconigram around the world. Anyone who had been skeptical of wireless before the great chase now ceased to be skeptical. The number of shipping companies seeking to install wireless increased sharply, as did public demand that wireless be made mandatory on all oceangoing vessels.

This effect of the Crippen case tended to be overlooked, however, because of an event a year and a half later that further sealed Marconi's success. In April 1912 the *Titanic* struck an iceberg and sank, but not before the ship's wireless operator, a Marconi employee, managed to summon help.

Marconi and Beatrice were supposed to be passengers on the *Titanic*, as guests of the White Star Line. Marconi canceled, however, and sailed a few days earlier on the *Lusitania*. Compulsive as always, he wanted to take advantage of the *Lusitania*'s public stenographer, whom he knew to be very efficient. Beatrice retained her booking. On the eve of the voyage, however, she too canceled. Their son Giulio had fallen ill with a high fever. The family was living then in a rented house called Eaglehurst, whose grounds had an eighteenth-century tower that overlooked Southampton Water. Beatrice and her daughter Degna, then three and a half, climbed to the top and watched the great ship as it left on its maiden voyage. They waved, and "dozens of handkerchiefs and scarves were waved back at us," Degna wrote. The departure saddened Beatrice. She had wanted very much to be aboard.

In remarks before the House of Commons, Lord Herbert Samuel, England's postmaster general, said, "Those who have been saved have been saved through one man, Mr. Marconi and . . . his wonderful invention." On March 8, 1913, a ship equipped with wireless set out to hunt and report the presence of icebergs—and sparked the formal inauguration in 1914 of the International Ice Patrol. Since then no ship within protected waters has been lost to a collision with an iceberg.

Marconi and Telefunken reached a truce. The companies agreed to stop challenging each other's patents; they formed a European consortium to share technology and ensure that their systems could communicate with each other. The truce did not last long. On July 29, 1914, a group of Marconi engineers paid a visit to the giant Telefunken transmitter at Nauen, said to be the most powerful in the world. Telefunken officials gave the men a tour and treated them well. As soon as the Marconi men left, the German military took control of the station and began transmitting a message commanding all German ships to proceed immediately to a friendly harbor.

As of eleven P.M., August 4, Britain and Germany were at war. Marconi's station at Poldhu sent a message to all Admiralty ships, "Commence hostilities against Germany." A team of Marconi operators began eavesdropping on all German transmissions and by the end of the war collected more than eighty million messages.

Almost at once German torpedoes began sliding through the seas off England. Marconi's annual report for 1914 stated, "Calls for assistance have been received almost daily." The wireless cabins of ships became prime targets. In 1917 a German submarine attacked the SS *Benledi* and focused its fire on the ship's wireless room, as its Marconi operator tried to reach an American warship for help. The warship arrived, the submarine fled. Afterward the *Benledi*'s captain went to the wireless cabin and found the operator still in his chair, everything in place, save for one macabre detail. His head was missing. In all, the war would kill 348 Marconi operators, most at sea.

AS MARCONI'S FAME increased and his empire expanded, his relationship with his mother, Annie, his most stalwart supporter, became more distant. She died in 1920 and was buried in Highgate Cemetery in London. Marconi did not attend her funeral. Degna wrote, "The past had been dead for him a long time."

He grew estranged as well from Beatrice. They spent more and more time apart, and soon Marconi entered an affair with another woman. For a time Beatrice and Marconi tried to preserve the illusion that their marriage was intact but eventually abandoned the effort. Marconi sold their house in Rome, and Beatrice and the children moved into the Hotel de Russie.

Marconi's affair came to an end, but Beatrice had had enough. She asked Marconi for a divorce. With reluctance, he agreed. They took temporary residence in the free city of Fiume, where in 1923 the divorce was granted. Two years later Marconi wrote to Beatrice that he was on the verge of marrying again. He was fifty-one years old; the prospective bride was seventeen. The idea that Marconi would suddenly feel driven to marry and, presumably, start another family struck Beatrice as ironic, given that he had been so consumed with work that he had barely paid attention to her and their children. She suspended her usual warmth and cordiality. "I would like to wish you every happiness but this news distresses me for I wonder after all the years we were together when your own desire expressed continually was for freedom to concentrate on your work as your family impeded and oppressed you, why you should suddenly feel this great loneliness and need of a home—this craving for fresh ties!! These ties were eventually what broke up your home and ended in our divorce. I fail to understand."

Marconi did not marry the girl. He immersed himself in work and spent more and more time aboard his yacht, *Elettra*. He again fell in love, this time with a daughter of one of Rome's most aristocratic

Catholic families, Maria Cristina Bezzi Scali. The family had ties to the so-called "black nobility," men who swore allegiance to the pope. Marconi asked her to marry him, but an obstacle immediately arose. Vatican law forbade marriage between a divorced man and a confirmed Catholic; only an annulment would allow them to proceed. Marconi investigated and found that one basis for annulment was if a man and woman married with the intention of not adhering to Catholic marital law. He might succeed, he discovered, if he could convince a church tribunal that he and Beatrice had agreed before their wedding that if the marriage proved unhappy they would seek a divorce. To make this argument, however, he would need Beatrice's help.

For the sake of the past, she agreed. As her time to testify neared, Marconi coached her on exactly what to say. His letters reprised his tendency to be oblivious to ordinary human sensitivities. In one letter he wrote, "They only want your testimony to decide, but please remember to read over my letters on the subject before you go, as it depends so much on your saying that we were agreed to divorce in the event of the marriage not being a happy one. Just simply a legal quibble on a matter of words or thoughts! Forgive very great haste, but I am still rather busy."

The church granted the annulment, and soon afterward he married Bezzi Scali.

Marconi's inventions, and advances by engineers elsewhere in the world, led quickly to the wireless transmission of voice and music. In 1920 the Marconi company invited Dame Nellie Melba to its station at Chelmsford to sing over the airwaves. At the station an engineer explained that her voice would be transmitted from the station's tower. Misunderstanding, Dame Melba said, "Young man, if you think I am going to climb up there you are greatly mistaken."

As late as 1926 wireless at sea continued to enthrall passengers. A traveler named Sir Henry Morris-Jones kept a record in his diary of a voyage on a second *Montrose*, launched by Canadian Pacific a few years earlier. "What a world we live in," he wrote. "A telegraph boy brings me a message handed in two hours before at Hull and I am 2000 miles out in the Atlantic Ocean."

Marconi realized, late, that his approach to wireless during his transatlantic quest had been a mistake. He had been obsessed with increasing the length of antennas and the power of transmitters, until he discovered through experiment that in fact very short waves could travel long distances far more readily and with far less expenditure of power. His giant stations had been unnecessary. "I admit that I am responsible for the adopting of long waves for long-distance communication," he said in 1927. "Everyone followed me in building stations hundreds of times more powerful than would have been necessary had short waves been used. Now I have realized my mistake."

Other scientists resolved the mysteries that had plagued Marconi through his early work. Oliver Heaviside, physicist and mathematician, proposed that a stratum existed in the atmosphere that caused wireless signals to bounce back to earth, and that this would account for why signals could travel very long distances over the horizon. Others confirmed its existence and dubbed it the Heaviside Layer. Scientists also confirmed that sunlight excited a region of the atmosphere known as the ionosphere, which accounted for the daytime distortion that had so plagued Marconi.

In 1933 the city of Chicago invited Marconi to attend its new world's fair, the Century of Progress Exhibition, and declared October 2 to be Marconi Day. The climax of the day occurred when Marconi tapped three dots, the letter S, into the exhibition's powerful transmitter, and stations in New York, London, Rome, Bombay, Manila, and Honolulu relayed it around the world, back to Chicago, in three minutes, twenty-five seconds.

As he aged, Marconi became aloof. At his London headquarters, Marconi House, he would only ride the elevator alone or with someone he knew, never with a stranger. He established a station to listen for signals from Mars and instructed its operators, "Listen for a regularly repeated signal." In 1923 he joined the Fascist Party and became a friend of Mussolini, though as time passed he became disenchanted with the increasing bellicosity of the Fascists and Nazis. He loathed Hitler.

On the afternoon of July 19, 1937, Marconi experienced a severe heart attack. At three the next morning he rang for his valet. "I am very sorry, but I am going to put you and my friends to considerable trouble. I fear my end is near. Will you please inform my wife?" Forty-five minutes later he was dead. The first outsider to arrive was Mussolini, who prayed at his bedside. Radio listeners around the world heard the news, which darkened an already bleak day in which the U.S. Navy announced that it had ended its search for Amelia Earhart.

That night the gloom lifted a bit, at least for those listeners who gathered around their radios and tuned in to NBC for the regularly scheduled antics of Amos n' Andy.

THROUGHOUT THE DAY of July 21, 1937, Marconi's body lay in state in the Farnesina Palace in Rome. The day was hot, the air heavy with the old-water scent of the nearby Tiber. A crowd numbering in the thousands blackened the square in front of the palace and filled the surrounding streets like spilled ink.

Beatrice came alone and uninvited. Even her children—*their* children—were not told of the funeral plans. She came incognito. She was now fifty-two and as beautiful as always. When her turn came, she moved to the bier where he lay exposed.

Once she and this man had been lovers. So much history lay between them, and now she was not even recognized, a ghost. Ten years had passed since the final humiliation of the annulment, and in that time he had seemed to abandon even the memory of her and the children.

She moved closer to the bier, and suddenly the distance that had accumulated between them shrank to nothing. She was overwhelmed and fell to her knees. Mourners passed behind her, the vanguard of a line that stretched seemingly across Rome.

At length she stood, confident that she had remained anonymous.

"I was unobserved," she wrote to Degna. "No one could have recognized me."

She exited the palace into the extraordinary heat of the afternoon and disappeared among the thousands still waiting to enter.

At six o'clock that evening, when his funeral began, wireless operators around the globe halted telegraphy for two minutes. For possibly the last time in human history, the "great hush" again prevailed.

Fleming and Lodge

In the summer of 1911 Oliver Lodge, sixty years old, began building what he called "a fighting fund" to bring a lawsuit against Marconi for infringing on his tuning patent. As of June 15, he and his allies had contributed £10,000 to the fund, more than $1 million today. Lodge wrote to William Preece, "They are clearly infringing, and we have a moral right to royalty. Accordingly I am actively bestirring myself to that end." Marconi had already approached Lodge with an offer to acquire his patents, apparently concerned that Lodge might indeed prevail in a court test, but Lodge had turned him down.

Preece, now seventy-seven, counseled caution, even though, as he put it, "I agree with every word you say and I am sure you are taking the proper attitude."

That summer Preece put aside his own antipathy toward Marconi and brokered a settlement, under which the Marconi company acquired Lodge's tuning patent for an undisclosed sum and agreed to pay him a stipend of £1,000 a year for the patent's duration. On October 24, 1911, Preece wrote to Lodge, "I am delighted to hear that matters are settled between you and the Marconi Company. I am quite sure you have done the right thing for yourself and that you will now get your deserts. But you will have to bring Marconi down a peg or two. He is soaring too much in the higher regions."

Lodge lost his youngest son, Raymond, to World War I and sought to reach the boy in the ether. He claimed success. He believed that during sittings with certain mediums he had conversed with Raymond. On one occasion, shortly before Christmas 1915, he heard his son say, "I love you. I love you intensely. Father, please speak to me." The conversation continued, and a few moments later Lodge heard, "Father, tell mother she has her son with her all day on Christmas Day. There will be thousands and thousands of us back in the homes on that day, but the horrid part is that so many of the fellows don't get welcomed. Please keep a place for me. I must go now."

Lodge published a book about his experience in 1916, called *Raymond,* in which he offered comforting advice to the bereaved: "I recommend people in general to learn and realize that their loved ones are still active and useful and interested and happy—more alive than ever in one sense—and to make up their minds to live a useful life till they rejoin them."

The book became hugely popular owing to the many parents seeking contact with sons killed in the first of the real wars.

OVER TIME THE RELATIONSHIP between Ambrose Fleming and Marconi grew more distant, but Fleming maintained his allegiance to the idea that Marconi deserved all credit for the invention of wireless. The company kept him as an adviser until 1931, when it endured one of its many periods of fiscal duress and told him his contract would not be renewed. He saw this as a new betrayal and now changed his opinion. He decided that the man who invented wireless was actually Oliver Lodge and that Lodge had first demonstrated the technology in his June 1894 lecture on Hertz at the Royal Institution.

On August 29, 1937, Fleming wrote to Lodge, "It is quite clear that in 1894 you could send and receive *alphabetical signals* in Morse Code by Electric Waves and did send them 180 feet or so. Marconi's idea that he was the first to do that is invalid."

By this point Fleming was eighty-eight years old but could not resist giving vent to a long-festered bitterness. "Marconi was always determined to claim everything for himself," he told Lodge, who was now eighty-six. "His conduct to me about the first transatlantic transmission was very ungenerous. I had planned the power plant for him and the first sending was carried out with the arrangement of circuits described in my British patent no 3481 of 1901. But he took care never to mention my name in connection with it.

"However," Fleming added, "these things get known in time and justice is done."

Coda

Voyager

Voyager

It was November 23, 1910. Southampton. A woman identified in the passenger manifest as Miss Allen walked aboard a ship, the *Majestic* of the White Star Line. She was twenty-seven years old but could easily have been mistaken for a girl in late adolescence.

For the second time in four months she felt compelled to use a false name. Though this time the circumstances were very different, the motive was the same: escape from gossip and scrutiny. It had been a whirl, London, Brussels, Antwerp, Quebec, and in that time she had felt finer, more loved, and certainly freer than ever before in her life. But now she had to leave.

On the *Majestic* she tried to bend her mind away from what had occurred that morning in London. She distracted herself with the glories of the ship and getting herself settled for the voyage. In Camden Town, she knew, a bell had rung fifteen times to mark the moment. She had heard the sound before, at Hilldrop Crescent, when the weather was right, but that was back when she felt safe and the sound of the prison bell was merely the artifact of someone else's misery, as meaningful as the distant bark of a neighbor's dog.

After arriving in New York she traveled to Toronto and adopted the name Ethel Nelson. She took a job as a typist. But Canada proved alien ground. In 1916 she braved seas traversed by German submarines and returned to London, where, as a clerk in a furniture store a few blocks from New Scotland Yard, she met a man named Stanley Smith. They married and raised two children in the peaceful middle-class community of East Croydon. In time she and Stanley became grandparents, but soon afterward he died. He never learned her true past.

A few years before her own death she received a visitor who had discovered her secret. The visitor was a novelist using the pen name Ursula Bloom, who hoped to write a novel about Dr. Crippen and the North London Cellar Murder. Ethel agreed to meet with her but declined to talk about her past.

At one point, however, Bloom asked Ethel, if Dr. Crippen came back today, knowing all she knew, would she accept marriage if he asked?

Ethel's gaze became intent—the same intensity that Chief Inspector Dew had found striking enough to include in his wanted circular.

Ethel's answer came quickly.

Notes and Sources

London 1910: *St. Paul's from Ludgate Circus,*
a photograph by Alvin Langdon Coburn.

ONE OF THE GREAT PRIVILEGES of hunting detail is the opportunity for travel to far-flung places that do not typically appear on the itineraries of tour companies. In Oxford, for example, I had the happy experience of being allowed to use the New Bodleian Library, which is only a billion years old and is not to be confused with the Old Bodleian Library. Gaining access required a bit of perseverance. Well in advance I had to fill out an application and find a "Recommender" to vouch for me. On arrival, I had to read aloud a declaration in which I swore "not to bring into the Library, or kindle therein, any fire or flame." Grudgingly, I left my blowtorch at the desk. Michael Hughes, charged with curating the library's recently acquired archive of Marconi papers, was kind enough to give me full access to the vast diary of George Kemp, even though technically the archive was closed pending completion of Hughes's work.

At the National Archives in Kew, just outside London, I entered writers' heaven. After an hour or so of acquainting myself with the archives' search and retrieval protocols and getting my "Reader's Ticket"—actually a plastic card with a bar code—I received a trove of documents accumulated by detectives of the Metropolitan Police during their hunt for Crippen and Miss Le Neve, as well as stacks of depositions from the Department of Publication Prosecutions and a small but chilling collection of records from the Prison Commission, including the "Table of Drops," which allowed me to calculate the precise distance that an Edwardian executioner would have insisted I fall in order to break my neck—four feet, eight inches. In all I collected over a thousand pages of statements, telegrams, memoranda, and reports that helped me reconstruct the hunt for Crippen and the chase that followed.

One of my greatest pleasures was simply walking London streets where the characters who appear in this book also once walked, past squares, parks, and buildings that existed in their time. There is no place quite like Hyde Park on a warm spring evening as the gold light begins to fade, no view quite so compelling as the Victoria Embankment under a bruised autumn sky. As so often occurs, I experienced strange moments of resonance where the past seemed to reach out to me as if to offer reassurance that I was on the right path. On my first research trip to London, I arrived during a week of unexpectedly hot weather. My hotel had no air conditioning. After a couple of too-still nights, I moved to a different hotel a few blocks away, the very charming and blissfully cool Academy House. I discovered the next day that the window of my room afforded me a view of Store Street, one block long, the very street where Crippen and Belle had lived after Belle's arrival in London and where for a time Ethel Le Neve kept a room.

It was easy to imagine moments when Crippen must have walked past Ambrose Fleming, whose office and laboratory at University College, London, were just a few blocks north. On some evenings, Crippen and Belle and, later, Le Neve surely must have crossed paths with Marconi himself, perhaps on the Strand or in a theater on Shaftesbury Avenue, maybe at the Cri, the Troc, "Jimmy's," or the Café Royal. One thing is certain: On an almost daily basis the waves Marconi transmitted from his early apparatus struck Crippen and Belle and the ladies of the Guild, and Chief Inspector Dew and Sir Melville, Fleming and Maskelyne, and just about everyone else who makes an appearance in this book.

I visited a number of London's many museums, and found two particularly useful in helping to conjure a vision of the past. At the Museum of London I saw one of Charles Booth's actual color-coded maps and was able to run my hand along the gleaming enameled body of a hansom cab. There I also saw Maskelyne and Cooke's most famous automaton, the whist-playing "Psycho," as well as Crippen's hearing aid and the heavy manacles that once bound his wrists. At London's Transport Museum, the little boy in me burst forth the moment I walked through the door. The place is full of vintage taximeter cabs and double-decker buses and subterranean railcars from the days when smoke and cinder filled the Tube. I had an opportunity to meet Crippen face to face, at Madame Tussaud's, in the Chamber of Horrors. He was shorter than I expected.

In the course of my research, I began studying Italian, expecting that I would need to do a lot of research in Italian archives and texts. I quickly found I was mistaken, for Marconi, "The Little Englishman," conducted his affairs, business and romantic alike, largely in England and in English. My study of Italian did give me a sense of the broader forces of heritage at work within Marconi, and quickly resolved a puzzle that confronts any monolingual American who tries to pronounce Marconi's first name. Phonetically, it's *Goo-yee-ail-mo.*

Certain published sources proved especially useful to me. On the Marconi side of the story, the most valuable was Degna Marconi's memoir, *My Father, Marconi,* one of the few works that provides a glimpse of Marconi's emotional life. Other works also proved useful, among them Richard Vyvyan's memoir, *Marconi and Wireless,* and three secondary works, Hugh Aitken's *Syntony and Spark,* Sungook Hong's *Wireless: From Marconi's Black-Box to the Audion,* and, most recent of the three, Gavin Weightman's *Signor Marconi's Magic Box.* In the special collections reading room of University College, London, I spent very pleasant days reading vitriolic back-chatter about Marconi and his claims, in letters that revealed that even the greatest intellects of the age were not above mean-spirited sniping. They just happened to be more articulate about it. Also valuable were the letters, reports, and so forth held in the archives of the Institute of Electrical Engineers, London; the Beaton Institute, Breton University, Nova Scotia; the Cape Cod National Seashore; and Archives Canada, in Ottawa.

I am grateful to Princess Elettra Marconi, Marconi's daughter from his second marriage, for allowing me to interview her at her home on the

Via Condotti in Rome, and to Marconi's grandson—Degna Marconi's son—Franceso Paresce, who spoke with me at his apartment in lovely and peaceful Munich, where he is a physicist with the European Southern Observatory. I owe thanks as well to Gabriele Falciasecca and Barbara Valotti at the Fondazione Guglielmo Marconi at Villa Griffone, where they showed me the foundation's museum and the attic in which Marconi performed his earliest experiments.

For my retelling of the Crippen story, I found three memoirs particularly useful: Chief Inspector Dew's *I Caught Crippen;* Ethel Le Neve's *Ethel Le Neve,* published as Crippen awaited execution; and Sir Melville Macnaghten's *Days of My Years.* A necessary work for anyone interested in Dr. Crippen is *The Trial of Hawley Harvey Crippen,* a more or less complete transcript of the trial published in book form as part of the Notable Trials Library. The files of the Branch County Library in Coldwater, Michigan, helped me piece together details of Crippen's childhood and family. The best popular account of the Crippen story is Tom Cullen's *Crippen: The Mild Murderer.*

Certain books gave me a good grounding in what life was like during the Edwardian era, a period typically defined as lasting from 1900 until the start of World War I, even though Edward VII died in 1910. Among the most useful: *The Edwardian Turn of Mind,* by Samuel Hynes; *The Other World,* by Janet Oppenheim, which explores Britain's early obsession with the occult; *The Edwardians,* by J. B. Priestley; and *The Edwardian Temperament, 1895–1919,* by Jonathan Rose. In the bookstore of the Museum of London I acquired a reproduction of *Baedeker's London and Its Environs 1900,* published by Old House Books, which in its more than four hundred pages provides a rich sense of London and its restaurants, hotels, subway lines, and institutions as perceived at the time. I also acquired Old House's reproduction of *Bacon's Up to Date Map of London 1902,* which gave me a good visual grasp of Edwardian London's tangle of streets and crescents. A more recent collection of maps, *London A–Z,* proved indispensable in helping me locate various obscure locales.

Though I tend to be leery of information conveyed via the Internet, I did find several websites that were credible and useful. MarconiCalling provides an easy-to-use online archive with photographs, audio recordings, early film clips, and reproductions of letters and telegrams now held by Oxford University. The website of the Edwin C. Bolles Collection on the History of London, created by Tufts University, allows visitors to search an array of books from the late Victorian era to find descriptions of particular streets and buildings, a process that otherwise would take days if not weeks. For example, one can type New Oxford Street into the site's search engine and learn all about the sordid roots of the neighborhood. The Charles Booth Online Archive of the London School of Economics presents images of the actual notebooks kept by Booth and some of his investigators, including pages that describe his walks through Hilldrop Crescent.

Regrettably, when the time came to write the final drafts of this book, I found myself forced by the demands of narrative coherence and pace to eliminate a number of compelling but useless pieces of information. Anyone obsessive enough to read the following footnotes will encounter some of these orphans, lodged here for no better reason than that I could not bear to expel them.

I do not cite a source for every fact in the book, only material that for one reason or another begs attribution, typically direct quotations and items that other authors unearthed first. And I must again remind readers: Anything between quotation marks is from a written document. All dialogue that appears in this book is taken verbatim from the sources in which it initially appeared.

NOTES

THE MYSTERIOUS PASSENGERS

1 *Captain Kendall had*: For details about Kendall's background, see Croall, *Fourteen Minutes,* 22–25.

2 *The* Montrose *was launched*: For details about the *Montrose*, see Musk, *Canadian Pacific,* 59, 74.

2 *"The Cabin accommodation"*: Canadian Pacific Railway Company, Royal Mail Steamship Lines, 1906 Summer Sailing Timetable. Canadian Pacific Steamship Line Memorabilia. In Archives Canada: MG 28 III 23.

2 *"A little better"*: Ibid.

2 *The manifest*: Kendall Statement, August 4, 1910, 1. In NA-MEPO 3/198.

3 *While on display*: Read, *Urban Democracy,* 412.

3 *"brilliant but disgusting"*: Ibid., 490.

4 *Shortly before*: Kendall Statement, August 4, 1910, 1. In NA-MEPO 3/198.

5 *"strange and unnatural"*: Ibid., 2.

5 *"I did not do anything"*: Ibid., 2.

PART I: GHOSTS AND GUNFIRE

DISTRACTION

9 *"street orderlies"*: Macqueen-Pope, *Goodbye Piccadilly,* 100.

10 *"diffusion of knowledge"*: Bolles Collection. Thomas Allen, *The History and Antiquities of London, Westminster, Southwark and Parts Adjacent* (Vol. 4). Cowie and Strange, 1827, 363.

10 *"the great head"*: Hill, *Letters,* 50.

10 *A young woman*: Haynes, *Psychical Research,* 184.

10 *Combs Rectory*: Jolly, *Lodge,* 18.

10 *"Whatever faults"*: Lodge, *Past Years,* 29. Lodge learned later in life that during one phase of his career his children were, as he put it, "somewhat afraid of me." One incident stood out. He came back late from work, tired and irritable. With his children in bed under strict instructions to keep quiet, he began marking a "thick batch" of examination papers. Suddenly,

a stream of water poured onto his windowsill from the children's room above. He became furious. "I rushed upstairs. They had just got back to bed, and said they had been watering a plant outside on their window-sill. I learned too late that it was one they had been trying to cultivate and were fond of. God forgive me, I flung it out of the window." The pot smashed on the ground. Later, he heard quiet sobbing coming from the room. He regretted the incident forever afterward (Lodge, *Past Years,* 252).

11 *"I have walked"*: Ibid., 78.

11 *"a sort of sacred place"*: Ibid., 75.

11 *"practicians"*: For one reference to the term "practician," see *The Electrician,* vol. 39, no. 7 (June 11, 1897), 1.

11 *"inappropriate and repulsive"*: Aitken, *Syntony,* 126.

11 *"As it is"*: Ibid., 112.

11 *"I became afflicted"*: Ibid., 112.

12 *"to examine"*: Haynes, *Psychical Research,* 6. At times, surely, it must have been difficult to set aside prejudice and prepossession, as when considering the feats of three sisters known widely as "The Three Miss Macdonalds." They held séances during which the table would tilt for yes and no and tap out the letters of the alphabet, a tedious process in which communicating just the word *zoo* would have required fifty-eight distinct taps. At times their tables engaged in high-velocity thumping, so much so, according to one historian of the SPR, "that one of them had to jump on it, crinoline and all, and sit there till it slowed down and stopped at last." One Macdonald sister later had a son named Rudyard, whose *Jungle Book* became one of the most beloved books of all time (Haynes, *Psychical Research,* 61).

12 *"physical forces"*: Ibid., xiv.

12 *Committee on Haunted Houses*: Ibid., 25.

12 *In Boston William James*: James's encounter with Mrs. Piper prompted him to write: "If you wish to upset the law that all crows are black you must not seek to show that no crows are, it is enough if you prove the single crow to be white. My own white crow is Mrs. Piper. In the trances of this medium I cannot resist the conviction that knowledge appears which she has never gained by the ordinary waking use of her eyes and ears and wits." (Haynes, *Psychical Research,* 83).

13 *"This," he wrote*: Ibid., 277.

13 *"thoroughly convinced"*: Ibid., 279.

13 *In his memoir*: Ibid., 184.

14 *"Well, now you can"*: Lodge, *Past Years,* 113. Lodge's biographer, W. P. Jolly, wrote of Lodge: "He was of the light cavalry of Physics, scouting ahead and reporting back, rather than the infantry of Engineering, who take and consolidate the ground for permanent useful occupation" (Jolly, *Lodge,* 113).

THE GREAT HUSH

15 *"My chief trouble"*: Marconi, *My Father,* 23. This needs a bit of qualification, for the idea of harnessing electromagnetic waves for telegraphy without wires had been proposed before, in an 1892 article in the *Fortnightly Review,* written by William Crookes, a physicist and friend of Oliver Lodge. Crookes by this time was one of Britain's most distinguished scientists, and one of its most controversial because of his interest in the paranormal. In the early 1870s he conducted a detailed investigation of Daniel Douglas Home, a medium who had held séances for Napoleon III, Tsar Alexander II, and other bright lights of the age, and was known for doing such extraordinary things as moving furniture and grabbing hot coals from fireplaces without injury (and in one case depositing said coals on the bald scalp of a séance participant, supposedly without causing harm). In the famous "Ashley House Levitation" of 1868, Home supposedly floated out one window of the séance room and back in through another. Crookes's investigations led him to conclude that Home did have psychic powers; he claimed, in fact, to have witnessed Home levitate himself several times. This did not endear him to the men of established British science, and probably accounted for why no one paid much attention to his *Fortnightly Review* article in which he discussed Heinrich Hertz's discovery of electromagnetic waves. "Here is unfolded to us a new and astonishing world—one which it is hard to conceive should contain no possibilities of transmitting and receiving intelligence," he wrote. "Rays of light will not pierce through a wall, nor, as we know only too well, through a London fog. But the electrical vibrations of a yard or more in wave-length of which I have spoken will easily pierce such mediums, which to them will be transparent. Here, then, is revealed the bewildering possibility of telegraphy without wires, posts, cables, or any of our present costly appliances." The article went virtually unread, and even Lodge appeared not to have paid it any attention (Oppenheim, *Other World,* 14–15, 35, 344, 475; Jolly, *Lodge,* 102; d'Albe, *Crookes,* 341–42).

16 "Che orecchi": Ibid., 8.

17 *"an aggregate"*: Ibid., 8.

17 *Marconi grew up*: For descriptions of Villa Griffone see Marconi, *My Father,* 6, 22, 24, 191.

17 "my *electricity"*: Ibid., 14.

17 *"One of the enduring"*: Paresce, "Personal Reflections," 3.

18 *"The expression on"*: Marconi, *My Father,* 16.

19 *"He always was"*: Maskelyne Incident, 27.

19 *"the Little Englishman"*: Ibid.

20 *Historians often*: Isted, I, 48; Jonnes, *Empires,* 19.

20 *As men developed*: Collins, *Wireless Telegraphy,* 36–37. For a good grounding in all things electrical, see Bordeau, *Volts to Hertz,* and Jonnes, *Empires.*

20 *Initially scientists*: Collins, *Wireless Telegraphy,* 36; Jonnes, *Empires,* 22.
21 *One researcher*: Jonnes, *Empires,* 29.
21 *In 1850*: Collins, *Wireless Telegraphy,* 37. The year 1850 also witnessed one of the strangest attempts at wireless communication. A Frenchman allowed two snails to get to know one another, then shipped one snail off to New York, to a fellow countryman, to test the widely held belief that physical contact between snails set up within them an invisible connection that allowed them to communicate with each other regardless of distance. They placed the snails in metal bowls marked with letters of the alphabet, and claimed that when one snail was touched against a letter, the other snail, at the far side of the ocean, likewise touched that letter. Concluding that somehow signals had been transmitted from one snail to the other, the researchers proposed the existence of an etherlike realm that they called "escargotic fluid." History is silent on the fate of the snails, though the nationality of the two researchers hints at one possible outcome (Baker, *History,* 21–22).
21 *In 1880*: Ibid., 37.
21 *He came up with*: Ibid., 8.
22 *He called it*: Massie and Underhill, *Wireless Telegraphy,* 41.
22 *Lodge's own statements*: Aitken, *Syntony,* 116, 121; Hong, *Wireless,* 46.
22 *"Whilst the issues"*: Hancock, *Wireless at Sea,* 20.
23 *"Giuseppe was punishing"*: Marconi, *My Father,* 24.
24 *The coherer "would act"*: Hong, *Wireless,* 19.
24 *"I did not lose"*: Marconi, *My Father,* 26.
24 *"he did lose his youth"*: Ibid., 26
24 *Marconi saw no limits*: Interview, Francesco Paresce, Munich, April 11, 2005.
25 *"far too erratic"*: Marconi, *Nobel,* 3.

THE SCAR

Details about Crippen's roots come mainly from Conover, *Coldwater,* 26–27, 43; Eckert, *Buildings,* 201–3; Gillespie, *A History,* 12–18, 47–49, 89–93, 127, 131; Holmes, *Illustrated,* throughout; *History,* 118, 159, 172; Massie, *Potawatomi Tears,* 270; *Portrait,* 276; Shipway, 8–13; *Michigan Business Directory,* 1863; *Trial,* 34–39, 87–130; and miscellaneous photographs, newspaper clippings, letters, and other items held in the Holbrook Heritage Room of the Branch County District Library, Coldwater, Mich.

27 *The young woman*: *Trial,* 34–35; Cullen, *Crippen,* 33–34.
28 *"I believe"*: *Trial,* 88.
28 *"I told her"*: Ibid., 35.
29 *"the coming of a colony"*: *History of Branch County, Michigan,* 118.
30 *"The pleasant drives"*: Holmes, *Illustrated.*
31 *A photograph*: This photograph appeared in the *Coldwater,* Michigan, *City Directory* of January 1920. Branch County District Library.

32 *"The devil"*: Cullen, *Crippen,* 30. Cullen, on p. 31, goes so far as to contend that the Book of Isaiah was "Hawley's favorite."
32 *"the vilest show"*: Gillespie, *History,* 89.
33 *As a homeopath*: *Trial,* 89.
33 *At Bethlehem*: Ibid., 69, 89.
34 *"I have never performed"*: Ibid., 87–88.
35 *In January 1892*: Cullen, *Crippen,* 32.
36 *"it was healed"*: *Trial,* 18–19.
36 *"There was only one"*: Cullen, *Crippen,* 35.
37 *"craved motherhood"*: Ibid., 35.
37 *"I love babies"*: Ibid., 87.
38 *Filson Young*: *Trial,* xxv–xxvi.
38 *"she was always"*: Ibid., 88.
38 *"to the outside world"*: Ibid., 126.

STRANGE DOINGS

39 *"to do the duty of two"*: Lodge, *Past Years,* 299.
40 *"Have you ever"*: Haynes, *Psychical Research,* 40. There were so many mediums roaming about the world that an American company sensed opportunity and began marketing a catalog entitled *Gambols with the Ghosts,* in which it sold various devices for use during séances, such as luminous hands and faces and a "Full, luminous female form" that would materialize slowly and then float around the room (Haynes, *Psychical Research,* 18).
40 *"Between deaths"*: Lodge, *Why I Believe,* 26.
41 *"I am not presuming"*: Lodge, *Past Years,* 297.
41 *"constantly ejaculating"*: Ibid., 295.
42 *"It was as if"*: Ibid., 297.
42 *"Every time she did this"*: Ibid., 301.
42 *"There must be some"*: Ibid., 301.
42 *"there appeared to emanate"*: Ibid., 301.
42 *"I saw this protuberance"*: Ibid., 301.
42 *"On me touche"*: Ibid., 301.
42 *"As far as the physics"*: Ibid., 302.
43 ectoplasm: Ibid., 301.
43 *"The ectoplasmic formation"*: Ibid., 302.
43 *"Let it be noted"*: Ibid., 305.
43 *"Any person"*: Ibid., 306.
43 *"Whether there is any"*: Ibid., 60.

GUNFIRE

44 *"Every time"*: Marconi, *My Father,* 27.
45 *"That was when"*: Marconi, *My Father,* 28.
45 *It was a "practician's" discovery*: Aitken, *Syntony,* 195, 286.
45 *"But," Marconi said*: Ibid., 29.

EASING THE SORE PARTS

47 *"Does your head"*: *The Weekly Courier,* Coldwater, Mich., April 6, 1895.
47 *"I will guarantee"*: *The News and Courier,* Charleston, S.C., July 6, 1898.
48 *"For Piles"*: Goodman, *Crippen,* 13.
48 *"Heed the Sign"*: Cullen, *Crippen,* 37.
48 *Later, during the Spanish-American*: See photographs at homeoint.org/photo/m2/munyonjm.htm.
48 *He called Crippen*: Cullen, *Crippen,* 38.
48 *"as docile as a kitten"*: See "Crippen Family Research Records" at familytrail.com/crippen/DrCrippen2.html
48 *"a giddy woman"*: Cullen, *Crippen,* 39.
49 *"She liked men"*: Ibid., 39.

TWO FOR LONDON

50 *"No matter how much"*: Paresce, "Personal Reflections," Part II, 1.
52 *London was still*: Details in this paragraph come mainly from Baedeker, *London,* 3, 93, 96, 97; and Massie, *Dreadnought,* xx.
52 *The new Locomotives*: Browne, *Rise,* 236.
53 *Queen Victoria herself*: Cullen, *When London Walked,* 137.
54 *"the contact of a dog's tongue"*: Ellis, *Psychology of Sex,* 21.
54 *The Duke and Duchess*: Maurois, *Edwardian Era,* 278.
54 *"probably including"*: Priestley, *Edwardians,* 66.
54 *Barbara Tuchman*: Tuchman, *Proud Tower,* 68–69.
54 *With this new awareness*: Details in this paragraph come from Tuchman, *Proud Tower,* 92–106; Browne, *Rise,* 232–35, 279; and Deghy and Waterhouse, *Café Royal,* 50–57. Malato's guidebook is excerpted on p. 57, and also in David, *Fitzrovians,* 96–97.
55 *The new location*: Cullen, *When London Walked,* 101–2.
55 *When the police moved*: Jeffers, *Bloody Business,* 93.
56 *No one called him*: Massie, *Dreadnought,* 15.
56 *Before breakfast*: Magnus, *King Edward,* 333.
56 *The prince hated*: Massie, *Dreadnought,* 15.
56 *By the late 1890s*: Rose, *Edwardian Temperament,* 165.
56 *The number of variety theaters*: Read, *Urban Democracy,* 42; Priestley, *Edwardians,* 172. Baedeker, *London,* on p. 64, puts the number of music halls in London alone at 500.
56 *She kept a cast*: Massie, *Dreadnought,* 13.

PART II: BETRAYAL

THE SECRET BOX

62 *"Maxwellians"*: For a detailed examination of Hertz's work and the respect, if not adulation, afforded him by Oliver Lodge and his camp, see O'Hara and Pricha, *Hertz.* See also Hong, *Wireless,* 44.
62 *Soon afterward*: The postal details in this paragraph come from Baedeker, *London,* 80, 122, 181.

63 *"It has gone twelve now"*: Mullis's account is quoted at length in Baker, *Preece,* 266–67.
63 *Preece and Marconi were kindred*: Hong, *Wireless,* 37–38.
64 *Preece recognized*: Aitken, *Syntony,* 210–14, 288.
64 *Later Preece would state*: Aitken, *Syntony,* 288.
64 *In a letter to his father*: Marconi to Giuseppe Marconi, April 1, 1896. "Letters," 96–97.
65 *"He promised me"*: Ibid., 97.
65 "La calma della mia": Marconi, *My Father,* 46.
66 *"an Italian had come up"*: Hong, *Wireless,* 38.
66 *"On the last day"*: Ibid., 38.
66 *"There is nothing new"*: Lodge to Preece, October 16, 1896. IEE, SC Mss. 22/213.
67 *"I think it desirable"*: Marconi to Preece, December 5, 1896. IEE, NA 13/2/02.
68 The Strand Magazine: Weightman, *Signor Marconi's,* 9; Isted, I, 53.
69 *the ambassador "even apologized"*: Marconi to Giuseppe Marconi, January 9, 1897. "Letters," 100.
70 *"the public has been educated"*: Hong, *Wireless,* 39.

ANARCHISTS AND SEMEN

72 *"I may say"*: *Trial,* 36.
72 *Just east lay Bloomsbury*: For more detail about the neighborhood and the Bloomsbury and Fitzroy Street groups, see Stansky, *December*, and David, *Fitzrovians*, respectively.
73 *"The door opened"*: Stansky, *December 1910,* 10.
73 *For years the basement*: David, *Fitzrovians,* 95.
73 *Nearby at No. 30*: David, *Fitzrovians,* 95; Deghy and Waterhouse, *Café Royal,* 54.
74 *Harrison recalled*: Cullen, *Crippen,* 42.
74 *A photograph from about this time*: Goodman, *Crippen File,* 16.
75 *"a few feeble lines"*: Cullen, *Crippen,* 42.
75 *A program from this period*: Goodman, *Crippen File,* 25.
75 *"the Brooklyn Matzos Ball"*: Cullen, *Crippen,* 44.
76 *"smoking concerts"*: *Trial,* 36.
76 *"She was always finding fault"*: Ibid., 88.
76 *"that this man visited her"*: Ibid., 36.

THE GERMAN SPY

77 *He long had resented*: André Maurois wrote of Wilhelm: "He was sensitive and ardent, and for artists might have turned out to be a desirable friend. But at the head of an Empire he was terrifying. His speeches, even his telegrams, were tirades of melodrama; his mother wished she could put a

padlock on his mouth whenever he spoke in public." (Maurois, *Edwardian Era,* 83)

77 *Kapp described him*: Kapp to Preece, March 19, 1897. IEE, SC Mss. 022. Folder 100-276.

78 *"As far as the Government"*: Marconi to Giuseppe Marconi, January 20, 1897. "Letters," 101–2.

78 *Two Americans*: Ibid., 102.

78 *In April, however*: Marconi to Preece, April 10, 1897. IEE, NA 13/2/08. Also, see Aitken, *Syntony,* 218, 222–23.

79 *On April 9, 1897*: Graham to Preece, April 9, 1897. IEE, NA 13/2/07.

79 *"I am in difficulty"*: Marconi to Preece, April 10, 1897. IEE, NA 13/2/08.

80 *Afterward he wrote*: Marconi to Giuseppe Marconi, August 8, 1897. "Letters," 102.

80 *"Marconi at the end of 1897"*: Preece's Recollections, July 26, 1937, 5. IEE, NA 13/2/24.

81 *On Friday, May 7*: Kemp Diary, May 7, 1897.

82 *"So be it"*: Kemp Diary, May 13, 1897.

82 *"I can't love him"*: Slaby to Preece, June 23, 1898. IEE, SC Mss. 22/180.

82 *"It is cold here"*: Kemp Diary, May 18, 1897.

82 *"I had not been able"*: Hancock, *Wireless,* 4.

83 *"I came as a stranger"*: Slaby to Preece, May 15, 1897. IEE, SC Mss. 22/180.

83 *In Berlin, Slaby immediately*: See Slaby to Preece, June 27, 1897. IEE, SC Mss. 22/179.

83 *He had to withdraw*: Hong, *Wireless,* 13.

83 *"The papers seem"*: Lodge to Preece, May 29, 1897. IEE, SC Mss. 22/210.

84 *"It appears that"*: Weightman, *Signor Marconi's,* 31–32.

84 *"It would be important"*: Fitzgerald to Lodge, June 21, 1897. UCL, Lodge Collection, MS Add 89/35 iii.

85 *"I have distinctly told him"*: Preece to Secretary, G.P.O., July 15, 1897. IEE, Post Office Records, English Minute No. 336170/98.

86 *"I have now constructed"*: Slaby to Preece, June 27, 1897. IEE, SC Mss. 22/179.

87 *"We are happy men"*: Slaby to Preece, June 23, 1898. IEE, SC Mss. 22/180.

BRUCE MILLER

88 *"I merely shook hands"*: *Trial,* 22.

88 *"I cannot say"*: Miller Statement, 4. NA-DPP 1/13.

89 *"When I first met her"*: Ibid., 4.

89 *"sometimes in the afternoons"*: Ibid., 5.

89 *"brown eyes"*: *Trial,* 20.

89 *Only partly true*: Deghy and Waterhouse, *Café Royal,* 22.

89 *"Anything we do"*: Miller Statement, 5. NA-DPP 1/13.

90 *"often enough to be sociable"*: *Trial,* 20.
90 *"with his Kodak"*: Miller Statement, 5. NA-DPP 1/13.
90 *"with love and kisses"*: *Trial,* 37.

ENEMIES

91 *"In fact Dr. Lodge"*: *The Electrician* 39, no. 21 (September 17, 1897), 686–87.
91 *"What we want to know"*: *The Electrician* 39, no. 25 (October 15, 1897), 832.
92 *"I hope this new attitude"*: Marconi to Preece, September 9, 1897. IEE, NA 13/2/13.
92 *"but no practical results"*: Baker, *Preece,* 275.
93 *"ignorant excitement"*: Ibid., 279.
93 *"I want to show you"*: Preece to Lodge, November 18, 1899. UCL, Lodge Collection, 89/86.
93 *"Preece's attempt"*: Lodge to Thompson, January 21, 1900. UCL, Lodge Collection, 89/104 ii.
94 *the still-pervasive skepticism:* One example: In a letter to Lodge, physicist Oliver Heaviside wrote, "I much question the usefulness of anything of Marconi's kind in practice, save in exceptional circumstances. The heliograph will carry much farther by day, + a search light by night. No doubt a special field can be found for it. But wires are the thing, in general." Heaviside to Lodge, June 23, 1897. UCL, Lodge Collection, 89/50.
94 *"Wireless is all very well"*: Marconi, *My Father,* 45.
94 *"The chief objection"*: Kelvin to Lodge, May 5, 1898. UCL, Lodge Collection, 89/107.
94 *"I think it would be"*: Kelvin to Lodge, June 11, 1898. UCL, Lodge Collection, 89/107.
95 *"In accepting"*: Kelvin to Lodge, June 12, 1898. UCL, Lodge Collection, 89/107.
95 *"Today was only the beginning"*: Muirhead to Lodge, June 4, 1898. UCL, Lodge Collection, 89/77.
95 *"This struck me"*: Jameson Davis to Lodge, July 29, 1898. UCL, Lodge Collection, 89/70.
96 ROYAL YACHT OSBORNE*:* Marconi to Lodge, August 2, 1898. UCL, Lodge Collection, 89/70.
96 *Anyone who could read:* Marconi, *My Father,* 65–67; Weightman, *Signor Marconi's,* 41–42.
97 *"go back and around"*: Marconi, *My Father,* 66; Wander, "Radio's First Home," 52.
97 *"Alas, Your Majesty"*: Ibid., 66.
97 *"H. R. H. the Prince"*: Ibid., 67.
97 *"Could you come"*: Weightman, *Signor Marconi's,* 42.

98 *"I am glad to say"*: Marconi to Lodge, August 2, 1898. UCL, Lodge Collection, 89/70.
98 *"a lady's house"*: Kemp Diary, August 24, 1898.
98 *"I wired to London"*: Ibid., August 22, 1898.
99 *"I sincerely hope"*: Marconi to Lodge, November 2, 1898. UCL, Lodge Collection, 89/70.
99 *"I much regret"*: Marconi to Lodge, October 11, 1898. UCL, Lodge Collection, 89/70.
100 *At nine* A.M.: Kemp's account of the ordeal appears in his diary entries for December 17, 1898, through January 4, 1899. Kemp Diary.
102 *In April*: Baker, *History,* 42; Faulkner, *Watchers,* 6–7.
103 *One night, during*: Marconi, *My Father,* 71–72; Vyvyan, *Marconi and Wireless,* 20.

"WERE YOU HER LOVER, SIR?"

105 *"Were you writing to her"*: This exchange appears in *Trial,* 20–21.

FLEMING

108 "Glad to send": Marconi to Fleming, March 28, 1899. UCL, Fleming Collection, 122/66.
109 *"the region of uncertain"*: Hong, *Wireless,* 56–57.
109 *"My attention has been"*: Lodge to Fleming, April 11, 1899. UCL, Fleming Collection, 122/66
109 *an "indictment against"*: Hong, *Wireless,* 57.
109 *"I made no attack"*: Fleming to Lodge, April 14, 1899. UCL, Fleming Collection, 122/66.
110 *define "my position"*: Fleming to Jameson Davis, May 2, 1899. UCL, Fleming Collection, 122/47.
110 *"I have a strong conviction"*: Ibid.

THE LADIES' GUILD COMMENCES

111 *"It was always agreed"*: *Trial,* 89.
111 *"with a free hand"*: Ibid., 97.
111 *One evening Miller*: Miller Statement, 2. NA-DPP 1/13.
112 *"I never interfered"*: *Trial,* 37.
112 *"Of course, I hoped"*: Ibid., 89.
112 *"She got an engagement"*: Ibid., 36.
112 *"She would probably"*: Ibid., 36.
112 *"There was hardly"*: Macqueen-Pope, *Goodbye Piccadilly,* 300–1.
113 *"When her hair was down"*: *Trial,* 77.
113 *"She wasn't a top-rank artist"*: Rose, *Red Plush,* 29.
113 *"To fail at even an East End"*: Machray, *Night Side,* 118.
114 *A photographer captured*: Goodman, *Crippen File,* 15.

115 *"Oh Belle does it hurt"*: Clara Martinetti Statement. Brief for the Prosecution. NA-DPP, 1/13.

"A GIGANTIC EXPERIMENT"

116 *Why bother at all*: For an excellent discussion of this, see Aitken, *Syntony*, 240–41. Aitken argues that the cable companies could have confronted any competitor with deep cuts in price. The business was lucrative, the companies profitable. They also had the capacity to handle far more business. In a price war, he argues, the cable trust would have proven a dangerous competitor.

116 *He recognized*: Interview, Francesco Paresce, Marconi's grandson, Munich, April 11, 2005. "He had no limits," Paresce told me. "I think he felt from day one that radio waves would be able to link any two points on the earth." That Marconi would propose so grand an experiment was due largely to his personality and to his appraisal of his company's prospects. "In order to win the race he could not continue as he had done before, with little steps," Paresce said. "Commercially he was realizing it wasn't working, the system really wasn't working."

Paths that might have seemed more prudent did exist, Paresce said. "You would have thought he would have pushed much harder simply to communicate with a ship farther and farther away." But to settle for this and not test his vision would have run against the grain of Marconi's character. "I think it was something that was really him," Paresce said. "He was a very stubborn man, he was a very driven man, and he was self-educated." But Marconi also understood on some deep level that what science held to be immutable law might easily with time prove to be false. "He learned very early on not to take too seriously the science of the moment," Paresce said. Marconi perhaps believed "there were enough unknowns about the problem that there was something that would come to his rescue."

However outlandish Marconi's idea might have seemed, it did have a practical dimension. "He was an extremely able media manipulator," Paresce said. "I'm almost certain that the basic reason he did it is he had to give a big impetus to his commercial operation. By making a big splash he could attract more attention to his effort, and attract the best people to help solve its problems."

117 *"I have not the slightest doubt"*: Hong, *Wireless*, 60.

117 *"When you meet Marconi"*: Quoted in Marconi, *My Father*, 76.

118 *A frightened guest*: Ibid., 76.

118 *"peculiar semi-abstract air"*: Ibid., 76.

118 *"a bit absent-minded"*: Quoted in Weightman, *Signor Marconi's*, 59.

118 *He found it*: Marconi, *My Father*, 76–77. For more details on the America's Cup episode, see pp. 77–80. See also, Weightman, *Signor Marconi's*, 60–61 and Baker, *History*, 48–49.

119 *"The shock from the sending coil"*: Quoted in Marconi, *My Father,* 80; Faulkner, *Watchers,* 7.
119 *"I noticed"*: Marconi, *My Father,* 168.
119 *"Well try using the other foot"*: Isted, I, 55.
120 *The* St. Paul *suited him*: The *St. Paul* had a twin, the *St. Louis,* which in 1907 carried a four-year-old boy named Leslie Townes Hope, later Bob Hope, from England to a new life in America. Of his emigration he later said, "I left England at the age of four when I found out I couldn't be king." Fox, *Transatlantic,* 391; Hope's quip comes from BobHope.com/bob.htm.
120 *"The Needles resembled"*: Quoted in Marconi, *My Father,* 82–83.
121 *"As all know"*: Hancock, *Wireless,* 20.
121 *Josephine Bowen Holman*: To gather details about Holman and her roots, I conducted research in the Indiana State Library, and there consulted the following sources: *Indianapolis News,* December 20, December 21, 1901; January 21, January 22, 1902; June 5, 1972; *Indianapolis Star,* December 20, 1909, May 24, 1948; August 4, 1979; *Indianapolis Star Magazine,* March 8, 1970; *Indianapolis Times,* July 23, 1937. Also, Lewis, *"Woodruff Place,"* pp. 3–8; McDonald, *Indianapolis,* 29–31; McKenzie, *Blue Book;* and *Woodruff Place Centennial,* 2, 4.
122 *"absolute certitude"*: For material on Tesla and his *Century* article, see: Cheney and Uth, *Tesla,* 87, 90, 99–100; Hong, *Wireless,* 72; the article is quoted at length in Sewall, *Wireless,* 51–52.
123 *As welcome as*: Aitken, *Syntony,* 232–35.
126 *"But greater wonders followed"*: London *Times,* October 4, 1900.
127 *"After you left"*: Marconi, *My Father,* 93.
127 *"I am thinking"*: Ibid., 93.
127 *"extreme demands on my time"*: Fleming to Flood Page, November 23, 1900. UCL, Fleming Collection, 122/47.
128 *"I am desired to say"*: Flood Page to Fleming, December 1, 1900. UCL, Fleming Collection, 122/47.
128 *"As regards any special recognition"*: Fleming to Flood Page, December 3, 1900. UCL, Fleming Collection, 122/47.
129 *In December Nevil Maskelyne*: Bartram, I, 50.
129 *And there was Lodge*: Lodge's discussions with Muirhead are cited in a letter from George Fitzgerald to Lodge, June 14, 1899. UCL, Lodge Collection, 89/35 iv.
129 *He accepted the position*: Jolly, *Lodge,* 132.

THE END OF THE WORLD

130 *"The most noticeable thing"*: Hicks, *Not Guilty,* 68.
131 *In 1898*: Massie, *Dreadnought,* 180; Clarke, *Voices,* 133.
131 *"At first there will be"*: Ibid., 134.
132 *"I wonder if"*: Weintraub, *Edward,* 387.

PART III: SECRETS

MISS LE NEVE

135 *Drouet produced*: Cullen, *Crippen*, 48.
135 *"For dolls or other girlish toys"*: Le Neve, *Ethel Le Neve*, 6.
136 *"Very soon afterwards"*: Ibid., 8.
136 *"For some reason"*: Ibid., 8.
136 *"I quickly discovered"*: Ibid., 8.
136 *On one occasion*: Ibid., 8.
137 *"With her departure"*: Ibid., 9.
137 *"Her coming was"*: Ibid., 9.
137 *An even stormier visit*: Ibid., 9.
138 *Crippen brought with him*: Cullen, *Crippen*, 61. Cullen contends Crippen also brought with him Drouet's mailing lists.
139 *"This places within"*: Goodman, *Crippen File*, 12.
139 *In time Aural Remedies*: Ibid., 13.
140 *Crippen said, "although"*: *Trial*, 37.
140 *"Give me your hand"*: Further Statement of Maud Burroughs, September 16, 1910. NA-DPP 1/13.
141 *"we had a whole day together"*: Ellis, *Black Flame*, 318.
141 *"the only person in the world"*: Le Neve, *Ethel Le Neve*, 10.
141 *"by sheer accident"*: Ibid., 12.

"THE THUNDER FACTORY"

144 *"They thought"*: Marconi, *My Father*, 100.
144 *"There was nothing"*: Thoreau, *Cape Cod*, 59.
144 *"Plenty of water"*: "Report for G. Marconi on his recent visit to America." Cape Cod National Seashore.
144 *One bit of historical resonance*: Kittredge, *Cape Cod*, 94.
145 *Cook assured Marconi*: Marconi, *My Father*, 100.
146 *"The barren aspect"*: Thoreau, *Cape Cod*, 45–46.
146 *Clouds often filled*: For weather details throughout this chapter see *Monthly Weather Review*, 49, 53, 77, 80, 85, 99, 123, 144, 182, 206, 224, 246, 272, 277, 291, 295, 318–22, 348, 380, 385, 403, 428, 433, 450, 470, 490, 493–94, 516, 536, 543, 569, 572, 596, 606, 610.
147 *"It was clear to me"*: Vyvyan, *Marconi and Wireless*, 28.
147 *"period of exceptionally severe storms"*: *Monthly Weather Review*, 99.
147 *"In view of the isolation"*: Bradfield to Executive Committee, March 30, 1906. Cape Cod National Seashore.
148 *"a vast* morgue*"*: Thoreau, *Cape Cod*, 182.
148 *"There is naked Nature"*: Ibid., 182.
148 *A photograph*: "Marconi Site." Wellfleet Historical Society.
149 *"To lose him to anyone"*: Marconi, *My Father*, 80.
149 *"I wish I had got this letter"*: Ibid., 82.
149 *He formed*: Aitken, *Syntony*, 143; Hong, *Wireless*, 46.
149 *And Marconi endured*: Aitken, *Syntony*, 246–47, fn 67 on 293.

150 *On May 21, 1901*: *Daily Graphic,* May 28, 1901, in UCL, Fleming Collection, 122/66; Faulkner, *Watchers,* 11; Hancock, *Wireless,* 29–30.
150 *Only years later*: Faulkner, *Watchers,* 11.
151 *"If you opened the door"*: *Cape Codder,* June 18, 1970, in Cape Cod National Seashore, File 4.7-2.
151 *"In August"*: Vyvyan, *Marconi and Wireless,* 28.
152 *"We used to call it"*: Crowley to Fleming, January 11, 1938. UCL, Fleming Collection, 122/3.
153 *"We had an electric phenomenon"*: Kemp Diary, August 9, 1901.
153 *"The weather is still boisterous"*: Ibid., August 14, 1901.
153 *"Caution. Very Dangerous."*: See photograph, Kemp Diary, opposite p. 154.
153 *"the thunder factory"*: Weightman, *Signor Marconi,* 170.
153 *The most important clause*: Aitken, *Syntony,* 235–36; Bartram, I, 51.

CLAUSTROPHOBIA

155 *On September 21*: *Trial,* 9.
155 *Trees lined the crescent*: The following description of Hilldrop Crescent and its surrounding neighborhood is derived primarily from the online archives of the Bolles and Booth collections, with additional detail from Baedeker, and a statement by Chief Inspector Walter Dew, in Brief for the Prosecution, 77, NA-DPP 1/13, in which he describes the layout of No. 39. I gleaned other details from two police photographs of the house and its garden, in NA-MEPO 3/198.
159 *In 1902 the prison*: Execution had been a neighborhood theme for more than a century. In the 1700s an inn stood in Camden Town by the name of Mother Red Cap, a common stop for omnibuses but also the end of the line for many condemned prisoners, who were hung in public across the street. The *Morning Post* of 1776 reported that "Orders have been given from the Secretary of State's Office that the criminals, capitally convicted at the Old Bailey, shall in future be executed at the cross road near the Mother Red Cap—the half-way house to Hampstead. . . ." One of the last things the condemned saw was the sign at the Mother Red Cap representing a woman thought to be Mother Damnable, identified in a bit of 1819 verse as a woman "so curst, a dog would not dwell with her." Bolles: Henry B. Wheatley, *London Past and Present,* John Murray, 1891.
159 *The law required*: Minutes. Executions at Pentonville. NA-HO 45/10629/200212.
160 *One immediate neighbor*: Cole to Churchill, November 11, 1910. Executions at Pentonville. NA-HO 45/10629/200212.
160 *"I do not think"*: Davies to Prison Commission, November 22, 1910. Executions at Pentonville. NA-HO 45/10629/200212.
161 *"Gee. You have got a hoo-doo"*: *Trial,* xvii.
161 *"Mrs. Crippen was strictly economical"*: Ibid., xviii.
162 *"Mrs. Crippen disliked"*: Ibid., xix.
162 *"I followed her"*: Ibid., xix.

163 *"They always appeared"*: Jane Harrison Statement. Witness, 103. NA-DPP 1/13.
163 *"Mr. and Mrs. Crippen"*: Rhoda Ray Statement. Witness, 139. NA-DPP 1/13.
163 *"somewhat hasty"*: *Trial,* 12.
163 *At one point*: *Trial,* xviii.
164 *He told his story*: Karl Reinisch Account, "Dr. Crippen on Board." Black Museum. NA-MEPO 2/10996. This document was off-limits to researchers until 2001.
165 *Another tenant, however*: *Trial,* xix.
165 *"He had to rise"*: *Trial,* xviii–xix.
166 *In June 1906*: Cora Crippen to Reinisch, June 23, 1906. Black Museum. NA-MEPO 2/10996.
166 *"He was a man"*: *Trial,* xviii.
166 *Soon after the move*: Ibid., xviii.
166 *On January 5, 1909*: Ibid., 108.
166 *"His eccentric taste"*: Ibid., xviii.
167 *"I have always hated"*: Paul Martinetti Statement. Supplemental Information, 27. (The bulk of the master document, Supplemental Information, is housed in NA-DPP 1/13, but portions, including the Martinetti Statement, appear in NA-CRIM 1/117.)
167 *"The rooms which Frankel"*: William Burch Statement. Witness, 160–61. NA-DPP 1/13.

DISASTER

168 *"At 1 p.m."*: Kemp Diary, September 17, 1901. Also, see Bussey, *Marconi's Atlantic Leap,* 34–35; Baker, *History,* 65; Fleming, "History," 39–40.
169 "PLEASE HOLD YOURSELF IN READINESS": Kemp Diary, November 4, 1901.
171 *The balloons and kites*: Hancock, *Wireless,* 32.
172 *It fell, he wrote*: Vyvyan, *Marconi and Wireless,* 28.
172 "MASTS DOWN": Flood Page to Marconi, November 29, 1901. Cape Cod National Seashore.

THE POISONS BOOK

173 *In September 1908*: Emily Jackson Statement, 44. Coroner's Depositions. NA-CRIM 1/117. Also, Emily Jackson Statement, 31. Brief for the Prosecution, NA-DPP 1/13.
173 *"I never saw the baby"*: Emily Jackson Statement, 9. Supplemental Information. NA-DPP 1/13. Also, Emily Jackson Statement, 39. Witness, NA-DPP 1/13.
174 *"I thought him quite"*: Emily Jackson Statement, 47. Coroner's Depositions. NA-CRIM 1/117.
174 *"He was the financier"*: Gilbert Rylance Statement, 81. Coroner's Depositions. NA-CRIM 1/117.

174 *Crippen continued to concoct*: William Long Statement, 84. Coroner's Depositions. NA-CRIM 1/117. William Long Statement, 17. Supplemental Information, NA-DPP 1/13.
175 *On December 15*: *Trial,* 32.
175 *"I didn't think"*: Louie Davis Statement, 101. Witness, NA-DPP 1/13; Melinda May Statement, 11. Coroner's Depositions, NA-CRIM 1/117.
176 *Belle "did not seem to know"*: Maud Burroughs Statement, 97. Witness, NA-DPP 1/13.
176 *Over the previous year*: List of Dr. Crippen's Orders. Exhibit 49, p. 44. Exhibits, NA-DPP 1/13.
176 *But Hetherington could not*: Charles Hetherington Statement. NA-DPP 1/13; *Trial,* 75–76.
177 *Hetherington relayed*: Alexander Hill Statement. NA-DPP 1/13.
177 *His company had no problem*: Ibid.
177 *Crippen "did not raise"*: *Trial,* 76–77.
177 *First the form asked*: Sale of Poisons Register Book. Exhibit 38. Exhibits, NA-DPP 1/13.

THE SECRET OF THE KITES

179 *To mask the true purpose*: Marconi, *My Father,* 104.
179 *"He reasoned"*: Vyvyan, *Marconi and Wireless,* 29.
180 "BEGIN WEDNESDAY": Fleming Notebook, December 9, 1901. UCL, Fleming Collection, 122/20.
180 *the newspaper reported*: Marconi, *My Father,* 105.
181 *In his diary*: Kemp Diary, December 11, 1901.
181 *"I should have gone"*: Ibid.
181 *"disappeared to parts unknown"*: Hancock, *Wireless,* 33.
181 *"Today's accident"*: Marconi, *My Father,* 106.
181 *"I came to the conclusion"*: Hancock, *Wireless,* 33–34.
182 *Despite his crucial role*: Hong, *Wireless,* 80.

WRETCHED LOVE

184 *She was, Jackson said, "rather strange"*: Emily Jackson Statement, 45–46. Coroner's Depositions, NA-CRIM 1/117.
185 *Her fingers twitched*: Emily Jackson Statement, 27. Brief for the Prosecution, NA-DPP 1/13.
185 *Ethel had a "horrible staring look"*: Ibid.
185 *"Go to bed"*: Emily Jackson Statement, 46. Coroner's Depositions, NA-CRIM 1/117; Further Statement of Emily Jackson, 42. Witness, NA-DPP 1/13.
185 *Ethel lay back*: Emily Jackson Statement, 28. Brief for the Prosecution, NA-DPP 1/13.
185 *"I can't let you go"*: Ibid.

185 *"For the love of God"*: Further Statement of Emily Jackson, 43. Witness, NA-DPP 1/13.
185 *"I told her"*: Emily Jackson Statement, 46. Coroner's Depositions, NA-CRIM 1/117.
185 *Mrs. Jackson assumed*: Further Statement of Emily Jackson, 43. Witness, NA-DPP 1/13.
186 *"Why worry about that now"*: Emily Jackson Statement, 46–47. Coroner's Depositions, NA-CRIM 1/117.
186 *"Don't you think"*: Emily Jackson Statement, 28. Brief for the Prosecution, NA-DPP 1/13.
186 *That night Ethel told*: Ibid.
186 taximeter: *Oxford English Dictionary,* Second Edition, 1989.
187 *"I would describe"*: Clara Martinetti Statement, 22. Coroner's Depositions, NA-CRIM 1/117.
187 *"always appeared to be very happy"*: Clara Martinetti Statement, 22. Supplemental Information, NA-DPP 1/13.
187 *"Oh make him"*: Further Statement of Mrs. Clara Martinetti, 63. Witness, NA-DPP 1/13.
188 *"I feel rather queer"*: Ibid., 64.
188 *"You call that seven o'clock"*: Ibid., 64; *Trial,* 12.
188 *Clara took off her own coat*: Clara Martinetti Statement, 18. Coroner's Depositions, NA-CRIM 1/117.
188 *Paul had two whiskeys*: Further Statement of Mrs. Clara Martinetti, 65. Witness, NA-DPP 1/13.
188 *"a funny little bull terrier"*: Ibid., 65.
188 *Crippen carved*: Ibid., 65.
188 *"the little deadlies"*: Forster, *Howards End,* 117.
188 *Belle offered cigarettes*: Further Statement of Mrs. Clara Martinetti, 66. Witness, NA-DPP 1/13.
189 *Belle told Clara*: Ibid., 66.
189 *"I had got a chill"*: Paul Martinetti Statement, 25. Coroner's Depositions, NA-CRIM 1/117.
189 *"Mr. Martinetti wanted to go upstairs"*: *Trial,* 90.
189 *"He returned looking white"*: Clara Martinetti Statement, 18. Coroner's Depositions, NA-CRIM 1/117.
189 *his hands were cold and he began to tremble*: Further Statement of Mrs. Clara Martinetti, 67. Witness, NA-DPP 1/13.
189 *"I don't think"*: Ibid., 67.
190 *"Don't come down, Belle"*: *Trial,* 13. Some accounts, likely exaggerated, have Mrs. Martinetti telling Belle, "Don't come down, you'll catch your death!"
190 *"On the night of the party"*: Clara Martinetti Statement, 18, 22. Coroner's Depositions, NA-CRIM 1/117.

190 *"Immediately after"*: *Trial,* 90.
190 *"This is the finish of it"*: Ibid., 37.
190 *"She had said this so often"*: Ibid., 37.
190 *"that I was to arrange"*: Ibid., 37.
190 *"I did not even see her"*: Ibid., 91.
191 *"his own calm self"*: Le Neve, *Ethel Le Neve,* 13.
191 *Clara asked, "How's Belle"*: Clara Martinetti Statement, 19. Coroner's Depositions, NA-CRIM 1/117; Further Statement of Mrs. Clara Martinetti, 68. Witness, NA-DPP 1/13.
191 *Belle had gone*: *Trial,* 91.
191 *The main question*: Crippen Statement, 123. Statements of Crippen and Le Neve, NA-DPP 1/13.

THE FATAL OBSTACLE

193 *"Unmistakably"*: Hancock, *Wireless,* 34.
193 *This configuration*: Kemp Diary, December 12, 1901.
193 *Marconi wrote*: Bussey, *Marconi's Atlantic Leap,* 51.
193 *During the brief periods*: Baker, *History,* 69.
194 *They began stringing*: Kemp Diary, December 19, 1901; Baker, *History,* 69.
194 *"Signals are being received"*: Bussey, *Marconi's Atlantic Leap,* 51.
194 *That night, he released*: London *Times,* December 16, 1901; Hancock, *Wireless,* 34.
195 *That Sunday*: Kemp Diary, December 15, 1901; Weightman, *Signor Marconi's,* 101.
195 *the* New York Times: Marconi, *My Father,* 104.
195 *Over the next few days*: Ibid., 104.
195 *Shares of Eastern Telegraph*: *Indianapolis News,* December 21, 1901. Indiana State Library.
195 *Ambrose Fleming*: Hong, *Wireless,* 80; *Daily Mail,* December 16, 1901. Fleming even clipped a copy and placed it, later, in his personal history. Fleming, "History," 44.
195 *Josephine Holman professed*: Weightman, *Signor Marconi's,* 113.
196 *who was headed there now*: *Indianapolis News,* December 20, 1901. Indiana State Library.

TO THE BALL

197 "Shall be in later": Le Neve, *Ethel Le Neve,* 14.
197 *"I was, of course"*: Ibid., 15–16.
198 *"He was not in a mood"*: Ibid., 16.
198 *"Has Belle Elmore really gone away"*: Ibid., 15.
198 *"I could not pretend"*: Ibid., 16.
198 *Now Crippen surprised her*: Ibid., 16–17.

198 *"a real expert in diamonds"*: Ibid., 18.
199 *He showed a clerk*: Ernest William Stuart Statement, 88. Coroner's Depositions, NA-CRIM 1/117.
199 *That night Ethel Le Neve slept*: *Trial,* 101.
199 *"Dear Miss May"*: Ibid., 23.
199 *The letter to the executive committee*: Ibid., 23–24.
200 *"He thought it would cheer us both up"*: Le Neve, *Ethel Le Neve,* 19.
200 *Lest this problem destroy*: Ibid., 19–20.
200 *The cat led her*: Ibid., 20.
201 *"Rich gowns"*: Ibid., 20–21.
201 *"I did not question"*: Ibid., 23.
201 *"There was scarcely anything"*: Ibid., 21.
201 *"From the first"*: Ibid., 21.
201 *"What is all this about"*: Clara Martinetti Statement, 23. Supplemental Information, NA-DPP 1/13.
202 *"proper engagement ring"*: Further Statement of Emily Jackson, 47. Witness, NA-DPP 1/13.
202 *"Do you know"*: Emily Jackson Statement, 15. Supplemental Information, NA-DPP 1/13.
202 *"Somebody has gone"*: Ibid., 10.
202 *Ethel began spending nights*: Emily Jackson Statement, 45. Coroner's Depositions, NA-CRIM 1/117; Emily Jackson Statement, 38. Witness, NA-DPP 1/13.
202 *She told Mrs. Jackson*: Further Statement of Emily Jackson, 47. Witness, NA-DPP 1/13; Walter Dew Statement, 31. Coroner's Depositions, NA-CRIM 1/117.
203 *Soon Ethel began giving*: Caroline Rumbold Statement, 92. Witness, NA-DPP 1/13.
203 *To her sister Nina*: Adine Prue Brock Statement, 78. Witness, NA-DPP 1/13.
203 *Yes, Ethel agreed*: Ibid., 82.
203 *1 outfit of mole*: Emily Jackson Statement, 24. Supplemental Information, NA-CRIM 1/117; Clothing Received by Mrs. Jackson from Ethel Le Neve, 71. Exhibits, NA-DPP 1/13.
204 *"Neither of us"*: Le Neve, *Ethel Le Neve,* 23–24.
204 *Built in 1873*: Baedeker, *London,* 16; Macqueen-Pope, *Goodbye Piccadilly,* 319–20.
205 *"wore it without any attempt"*: Clara Martinetti Statement, 9. Brief for the Prosecution, NA-DPP 1/13.
205 *John Nash said*: John Nash Statement, 2, in letter, Seyd to Director of Public Prosecutions, April 29, 1911, NA-DPP 1/13.
205 *Maud Burroughs saw it*: Maud Burroughs Statement, 97. Witness, NA-DPP 1/13.

205 *She recalled that Ethel*: Clara Martinetti Statement, 9. Brief for the Prosecution, NA-DPP 1/13.
205 *"I noticed that Crippen"*: John Nash Statement, 25. Witness, NA-DPP 1/13.
205 *Mrs. Louise Smythson approached*: Louise Smythson Statement, 31. Witness, NA-DPP 1/13; Louise Smythson Statement, 3. Supplemental Information, NA-DPP 1/13.
205 *"After this"*: Le Neve, *Ethel Le Neve,* 24.
206 *It made her uncomfortable*: Ibid., 24.
206 *"The pack was turning"*: Forster, *Howards End,* 246.
206 *On March 12*: Emily Jackson Statement, 38–39. Witness, NA-DPP 1/13; Emily Jackson Statement, 44. Coroner's Depositions, NA-CRIM 1/117.

"I DON'T BELIEVE IT"

207 *"I doubt this story"*: Associated Press dispatch quoted in the *Sydney Daily Post,* of Sydney, Nova Scotia, December 27, 1901. Beaton, MG 12, 214. G3. Scrapbook.
207 *"Skepticism prevailed"*: Hancock, *Wireless,* 36.
207 *"the letters S and R"*: Ibid., 36.
207 *Two days later"*: *Electrical Review,* 49, no. 1256 (December 20, 1901), 1031. Beaton, MG 12/214/A3.
208 *"It is rash"*: London *Times,* December 20, 1901.
208 *Smith watched*: For William Smith's entire account see Smith to Prof. W. J. Loudon, March 10, 1931. William Smith Papers. Archives Canada, MG 30 D18 Vol. 1; also, in the same collection, see Notes and Transcripts. Marconi Papers. Memoranda, Printed Matter, Vol. 3, File 17.
209 "Best Christmas greetings": Isted, II, 112.

NEWS FROM AMERICA

212 Dear Clara and Paul: Letter, Crippen to Martinettis, March 20, 1910. Exhibits, 21, NA-DPP 1/13.
212 Belle died yesterday: Telegram, Crippen to Martinettis. Ibid., 21.

PART IV: AN INSPECTOR CALLS

"DAMN THE SUN!"

215 *At first*: Marconi, *My Father,* 113.
215 *Black signs at three points*: Bussey, *Marconi's Atlantic Leap,* 66.
215 *"Potage Electrolytique"*: Simons, "Guglielmo Marconi," 51.
215 *Bowls of sorbet*: Weightman, *Signor Marconi's,* 110.
215 *"I am sorry"*: Isted, II, 112.
216 *In his own account of events*: Fleming, "History," no page number.
216 "ENGAGEMENT IS BROKEN": *Indianapolis News,* January 21, 1902. Indiana State Library.

216 *Later, a* News *reporter*: The following exchange appears in the *Indianapolis News,* January 22, 1902. Indiana State Library.

217 *He added a tincture*: Weightman, *Signor Marconi's,* 113.

217 *"There have been disasters"*: Ibid., 113. Early the following May, less than four months after breaking her engagement to Marconi, Josephine Holman stepped forward and announced that she had become engaged to a new man. The remarks that followed may have been made with the best intentions, but it is tempting to view them through the prism of love scorned, for Holman would have known well what other women in Marconi's life also would learn, that one salient trait of his romantic character was jealousy. "I am perfectly happy, but for one little thought," Holman wrote, "and that would vanish forever if Signor Marconi would find another love and be as happy in his choice as I feel I am in mine." *Halifax Herald,* May 8, 1902. Beaton Institute, MG 12/214.E.: Envelop/Index Cards.

217 *By the end of the day*: *Indianapolis News,* January 22, 1902. Indiana State Library.

217 *As the liner approached*: *Indianapolis News,* February 10, 1902. Indiana State Library; Bussey, *Marconi's Atlantic Leap,* 69.

217 *details of the new Canadian arrangement*: "How Marconi Came to Canada," 9–10. William Smith Papers. Notes and Transcripts. Marconi Papers. Memoranda, Printed Matter. Archives Canada, MG 30 D18 III.

218 *"Sir William Preece is"*: *Financial Times,* February 21, 1902.

219 *The* Westminster Gazette *suggested*: *Westminster Gazette,* February 26, 1902.

219 *The* Electrical Times *condemned*: *Electrical Times,* February 27, 1902.

219 *Two days later*: Details of the *Philadelphia* episode come from Bussey, *Marconi's Atlantic Leap,* 72; Marconi, *My Father,* 124–25; Weightman, *Signor Marconi's,* 124–26. Weightman quotes extensively from McClure's account, published in *McClure's Magazine.*

220 *"daylight effect"*: Marconi, *My Father,* 126; Vyvyan, *Marconi and Wireless,* 32.

220 *"Damn the sun!"*: Marconi, *My Father,* 130.

221 *That summer the* Daily Mail: Read, *Urban Democracy,* 475.

221 *So things stood when*: Baker, *History,* 95–96; Weightman, *Signor Marconi's,* 136–37.

222 *"malignant Marconiphobia"*: Weightman, *Signor Marconi's,* 137.

222 *"Marconi's whining"*: Thompson to Lodge, April 2, 1902. UCL, Lodge Collection, 89/104 ii.

222 *At Glace Bay Richard Vyvyan*: Vyvyan, *Marconi and Wireless,* 50; Marconi, *My Father,* 146.

224 *En route, during a stop*: Hong, *Wireless,* 83; Marconi, *My Father,* 131–32.

225 *Marconi blamed Fleming*: Hong, *Wireless,* 83.

225 *"It should be explained to [Fleming]"*: Quoted at length in Hong, *Wireless,* 83–84.

225 *None of this, however*: Bartram, I, 53; Hong, *Wireless,* 117.
226 *"Knowing that experiments were in progress"*: Maskelyne Incident, 2–3.

THE LADIES INVESTIGATE

227 *"a model husband"*: John Burroughs Statement, 4. Brief for the Prosecution, NA-DPP 1/13.
227 *"kind and attentive"*: Clara Martinetti Statement, 22. Coroner's Depositions, NA-CRIM 1/117.
227 *"kind-hearted humane man"*: Adeline Harrison Statement, 27. Ibid.
227 *Even before word arrived*: Michael Bernstein Statement, 90. Witness, NA-DPP 1/13.
227 *On March 30*: Louise Smythson Statement, 32–33. Witness, NA-DPP 1/13.
228 *It took him a month*: Otto Crippen to Melinda May, May 9, 1910. Copy in Melinda May Statement, 37. Witness, NA-DPP 1/13.
229 *"The smell," Jackson said*: Further Statement of Emily Jackson, 45. Witness, NA-DPP 1/13; Emily Jackson Statement, 47, 49. Coroner's Depositions, NA-CRIM 1/117.
229 *"A night or two after this"*: Further Statement of William Long, 55. Witness, NA-DPP 1/13.
229 *"I have at last"*: Le Neve to Jackson, "Sunday" (probably June 12, 1910). Letters from Le Neve to Mrs. Jackson, NA-DPP 1/13.
229 *"Have been ever so busy"*: Le Neve to Jackson, June 29, 1910. Ibid.
230 *"Still," she told Mrs. Jackson*: Ibid.
230 *"He used to come with me"*: Le Neve, *Ethel Le Neve,* 26.
230 *"So time slipped along"*: Ibid., 26.
230 *"Whilst we were talking to him"*: Clara Martinetti Statement, 25. Supplemental Information, NA-CRIM 1/117.
230 *On May 6, 1910*: Of Edward's funeral, André Maurois wrote, "The contrast of all the black with the gay spring sunshine lent a strange beauty to the streets of the capital." Maurois, *Edwardian Era,* 354.

A DUTY TO BE WICKED

232 *The first signals*: Vyvyan, *Marconi and Wireless,* 36.
233 *In the midst of it all*: Weightman, *Signor Marconi's,* 145.
233 *"It is beyond the powers"*: Details on Nevil Maskelyne and the Egyptian Hall come from the following sources: Bolles Collection. Thomas Allen, *The History and Antiquities of London, Westminster, Southwark and Parts Adjacent* (Vol. 4). Cowie and Strange, 1827, 303; Bartram, I and II, throughout; Macqueen-Pope, *Goodbye Piccadilly,* 78–81; Oppenheim, *Other World,* 25–27.
234 *In an article*: Maskelyne Incident, 2–5.
235 *"The plain question is"*: Ibid., 5.
235 *Cuthbert Hall,* Marconi's: Ibid., 7.
235 *"Clearly, Mr. Hall is between"*: Ibid., 12; Bartram, I, 54.
236 *At Glace Bay silence prevailed*: Vyvyan, *Marconi and Wireless,* 37–40.

237 "WHAT'S WRONG": *Sydney Daily Post*, Dec. 9, 1902. Beaton Institute, MG 12/214. G3.:Scrapbook.
238 *"All put cotton wool"*: Marconi, *My Father*, p. 140
238 *Times London*: Vyvyan, *Marconi and Wireless*, 38.
238 *A sudden gale*: MacLeod, *Marconi*, 78.
238 *Marconi had instructed*: Ibid., 79.
238 *Parkin crafted an account*: Weightman, *Signor Marconi's*, 147–48.
238 *"Although these three messages"*: Vyvyan, *Marconi and Wireless*, 39.
239 *The telegram as received*: Ibid., 40.
239 *Roosevelt's message*: Ibid., 40–41.
240 *Marconi's critics sensed blood*: Bartram, I, 54.
240 *"I was not concerned"*: *Westminster Gazette*, March 13, 1903. Maskelyne Incident, 17.
241 *"It was clear"*: Vyvyan, *Marconi and Wireless*, 41.
241 *In the* Morning Advertiser: Cited in *Westminster Gazette*, March 13, 1903. Maskelyne Incident, 17.
241 *One reader wrote*: *Morning Advertiser*, March 16, 1903. Maskelyne Incident, 21.
242 *"Well, we have got beyond that"*: *Westminster Gazette*, March 13, 1903. Maskelyne Incident, 17.
242 *Even as it flared*: Fleming, J.A. "A Report on Experiments," 1–7. UCL, Fleming Collection.
244 *Though somewhat wicked*: Hong, *Wireless*, 108.
244 *a silver thaw can occur*: Vyvyan, *Marconi and Wireless*, 41; see also, Baker, *History*, 82, and MacLeod, *Marconi*, 86;

BLUE SERGE

246 *For two of Belle's friends*: John Nash Statement, 2–3, in letter, Seyd to Director of Public Prosecutions, April 29, 1911. NA-DPP 1/13; John Nash Statement, 26–27. Witness, NA-DPP 1/13.
246 *Crippen told him*: The dialogue between Crippen and Nash is taken verbatim from John Nash Statement, 2–3, in letter, Seyd to Director of Public Prosecutions, April 29, 1911, NA-DPP 1/13.
247 *Two days later*: For details about Froest and Scotland Yard see Browne, *Rise*, 243–44; Jeffers, *Bloody Business*, 93; and Williams, *Hidden World*, 37.
248 *"Mr. and Mrs. Nash are not satisfied"*: Dew, *I Caught Crippen*, 8.
248 *His name was Walter Dew*: Dew, *I Caught Crippen*, throughout; Jeffers, *Bloody Business*, 116–17.
248 *"I saw a sight"*: Dew, *I Caught Crippen*, 145.
248 *"When we got back"*: The dialogue between Dew, Nash, and Froest is verbatim, from Dew, *I Caught Crippen*, 8–9.
249 *Under ordinary circumstances*: Dew, *I Caught Crippen*, 11.

249 *"What was really in the minds"*: Ibid., 9.
249 *"I think it would be just as well"*: Ibid., 9.

RATS

250 *Fleming arranged*: Details of the lecture and the intervention of Nevil Maskelyne come mainly from the Maskelyne Incident Papers, a collection of clippings and correspondence held in the archives of the Institute of Electrical Engineers, London. See in particular pp. 32–52. For overviews and additional details, see also Bartram, I, 55, and Hong, *Wireless,* 108–14.
251 *Blok was experienced*: Blok's account is quoted extensively in Hong, *Wireless,* 110.
254 *"The interference was purposely arranged"*: Maskelyne Incident, 41.
255 *"Everything went off well"*: Quoted in Hong, *Wireless,* 111.
255 *In a second letter*: Ibid., 111.
255 *On June 11, 1903*: Maskelyne Incident, 32.
256 *"Sir," Maskelyne wrote*: Ibid., 33.
257 *As the* Morning Leader *of June 15*: Ibid., 43.
257 *In an interview in*: Ibid., 35.

AH

259 *"curiously enough"*: Dew, *I Caught Crippen,* 10.
259 *"a great favorite"*: Chief Inspector Dew. Report to Criminal Investigation Division, July 6, 1910, NA-MEPO 3/198.
259 *Maud Burroughs described*: Ibid., 7.
259 *"The story told by"*: Ibid., 1.
259 *"most extraordinary"*: Ibid., 15.
259 *"without adopting the suggestion"*: Ibid., 15.
260 *"Is Dr. Crippen at home"*: Ibid., 11.
260 *"She was not pretty"*: Ibid., 11.
260 *"Who are you"*: "Further Report of Chief Inspector Dew," 1, NA-DPP 1/13. Dew himself offers several different accounts, in different reports and statements, of how this initial contact unfolded.
260 *"Unfortunate the doctor is out"*: Dew, *I Caught Crippen,* 12.
261 *Ethel's recollection*: Le Neve, *Ethel Le Neve,* 28–33. The dialogue between Dew and Ethel appears here as Ethel retold it in her memoir.
263 *At Albion House*: Le Neve, *Ethel Le Neve,* 33–34.
263 *"insignificant little man"*: Dew, *I Caught Crippen,* 12. A slightly different account appears in Further Report of Chief Inspector Dew, 1, NA-DPP 1/13.
263 *"I am Chief-Inspector Dew"*: Ibid., 13.

THE GIRL ON THE DOCK

265 *On Nova Scotia he faced a choice*: MacLeod, *Marconi,* 90.
265 *They agreed also*: Baker, *History,* 96.
266 *The previous December*: Ibid., 98.

266 *"Notwithstanding the great mass"*: Sewall, *Wireless,* 89.
266 *"At thirty"*: Marconi, *My Father,* 151.
266 *The fact that he was Italian*: Ibid., 168.
266 *In the summer of 1904*: For more detail on Beatrice and her background, see Marconi, *My Father,* 155–62, and Weightman, *Signor Marconi,* 182–85.
267 *To her, it seemed lovely*: Marconi, *My Father,* 161–62.
267 *"the dress she had on was* awful*"*: Ibid., 155.
267 *He fled for the Balkans*: Ibid., 164.
268 *Stricken with the grief*: Ibid., 164.
268 *Without telling Beatrice*: Ibid., 164.
268 *"It's so serious"*: Ibid., 165.
269 *Troubling news drifted back*: Ibid., 166–67.
270 *"What can you be thinking"*: Ibid., 167.
270 *"She was a born flirt"*: Ibid., 168.
270 *"I have not mentioned this"*: Baker, *History,* 107.
270 *In 1904, while seeking*: Jolly, *Lodge,* 153.

HOOK

271 *The fastest ocean liners*: Fox, *Transatlantic,* 308.
271 *The government began talks*: Clarke, *Voices,* 133.
271 *"Is it not becoming patent"*: Childers, *Riddle,* 308.
271 *Ever since the turn of the century*: Hynes, *Edwardian,* 22.
271 *A royal commission found*: Browne, *Rise,* 279–83.
272 *The government investigated*: Hynes, *Edwardian,* 22–23.
272 *A month later the government launched*: Ibid., 32–33.
272 *In London on the night*: Dunbar, *J. M. Barrie,* 170.
272 *The* Daily Telegraph *would call*: Ibid., 170.
272 *"Do you believe in fairies"*: Barrie, *Peter Pan,* 115.
273 *"How still the night is"*: Ibid., 117.

PART V: THE FINEST TIME

THE TRUTH ABOUT BELLE

277 *"From his manner"*: Dew, *I Caught Crippen,* 13–14.
277 *"Meanwhile," she wrote*: Le Neve, *Ethel Le Neve,* 35.
277 *Crippen ordered a steak*: Dew, *I Caught Crippen,* 15.
277 *"I realized that she had gone"*: Crippen Statement, 123. Statements of Crippen and Le Neve, NA-DPP 1/13.
278 *"I was impressed"*: Dew, *I Caught Crippen,* 14.
278 *"The girl showed"*: Ibid., 19–20.
278 *"He told you a lie"*: Le Neve, *Ethel Le Neve,* 35.
279 *"I was stunned"*: Ibid., 36.
279 *"There was not enough"*: Dew, *I Caught Crippen,* 19.
279 *"I seemed to be living"*: Le Neve, *Ethel Le Neve,* 36.

279 *"I certainly had no suspicion"*: Dew, *I Caught Crippen,* 21.
279 *"In the bedroom"*: Ibid., 21.
280 *"What were these men doing"*: Le Neve, *Ethel Le Neve,* 36.
280 *"The place was completely dark"*: Dew, *I Caught Crippen,* 21.
280 *"Of course I shall have to find"*: Ibid., 22.
281 *"I did not absolutely think"*: Chief Inspector Dew Statement, 81. Brief for the Prosecution, NA-DPP 1/13. In his memoir, on p. 21, Dew says essentially the same thing: "I certainly had no suspicion of murder. You don't jump to the conclusion that murder has been committed merely because a wife has disappeared and a husband has told lies about it."
281 *He told at least one observer*: *Trial,* xxx.
281 *Ethel felt great relief*: Le Neve, *Ethel Le Neve,* 36–38.
281 *"For mercy's sake"*: Ibid., 38.
281 *"My dear," he said*: Ibid., 39.

THE PRISONER OF GLACE BAY

282 *Beatrice and Marconi married*: Weightman, *Signor Marconi's,* 191–92.
282 *Marconi gave Beatrice*: Marconi, *My Father,* 169.
282 *They fought*: Marconi, *My Father,* 172.
283 *They moved to something far grander*: Ibid., 175; Baedeker, *London,* 9; Weightman, *Signor Marconi's,* 194.
283 *His ship-to-shore business*: Baker, *History,* 105.
284 *One clause of the agreement*: Hong, *Wireless,* 148; "Memorandum of Agreement." May 26, 1905. UCL, Fleming Collection, 122/47.
284 *The new station at Whittle Rocks*: Cash Book, Vol. 33, March 1905. Marconi Wireless Telegraph Co. of Canada Ltd. Archives Canada, MG 28 III 72.
284 *In April 1905*: Ibid., April 1905.
284 *Each got one*: Ibid., April 1905.
284 *In August 1905*: Ibid., August 1905 and August 1904.
284 *In 1904 Glace Bay*: Balance Sheets, Vol. 22, Glace Bay. See balance sheets for years ended January 31, 1904, and 1908. Marconi Wireless Telegraph Co. of Canada Ltd. Archives Canada, MG 28 III 72.
285 *For the first time*: MacLeod, *Marconi,* 95; Marconi, *My Father,* 176.
285 *As she put it*: Marconi, *My Father,* 168.
285 *He gave her many*: Ibid., 176.
286 *The crew of 415*: Fox, *Transatlantic,* 318–20.
286 *"When her husband did emerge"*: Marconi, *My Father,* 176.
286 *One day Beatrice entered*: Ibid., 177.
287 *The new station*: MacLeod, *Marconi,* 93; Vyvyan, *Marconi and Wireless,* 43.
287 *The news made Marconi furious*: Marconi, *My Father,* 178–79.
288 *"The stillness of winter"*: Vyvyan, *Marconi and Wireless,* 48–49.
288 *Beatrice did not agree*: Marconi, *My Father,* 178.

288 *He volunteered his own fortune*: MacLeod, *Marconi,* 96; Marconi, *My Father,* 180.
288 *At Poldhu he inaugurated*: MacLeod, *Marconi,* 96–98.
289 *At last, at nine o'clock*: Vyvyan, *Marconi and Wireless,* 44.
289 *He began to see a pattern*: Baker, *History,* 112–13; Vyvyan, *Marconi and Wireless,* 44.
289 *He realized now*: Vyvyan, *Marconi and Wireless,* 44–45.

LIBERATION

291 *Ethel looked "rather troubled"*: This and subsequent dialogue come from Adine True Brock Statement, 83. Witness, NA-DPP 1/13.
292 *Dew composed a circular*: Dew, *I Caught Crippen,* 23.
293 *"You will look a perfect boy"*: Le Neve, *Ethel Le Neve,* 43.
293 *"Dear Sid"*: Walter William Neave Statement, 88. Witness, NA-DPP 1/13.
293 *"It was not a good fit"*: This and subsequent dialogue come from Le Neve, *Ethel Le Neve,* 43–44.
294 *Crippen reassured her*: Ibid., 48.
294 *To enhance her costume*: Ibid., 44.
294 *"I was terribly self-conscious"*: Ibid., 44–45.
295 *"Strange as it may seem"*: Ibid., 45.
295 *"Oh, the pretty English boy"*: Ibid., 48.
295 *Crippen identified himself*: Police Reports as to Enquiries at Antwerp. NA-DPP 1/13.
295 *The innkeepers noticed*: Ibid.
296 *Later that Sunday*: Dew, *I Caught Crippen,* 24.

A LOSS IN MAYFAIR

297 *He envisioned*: MacLeod, *Marconi,* 80–83; Vyvyan, *Marconi and Wireless,* 45.
297 *"I was almost too young"*: Marconi, *My Father,* 180.
298 *"Our darling little baby"*: Weightman, *Signor Marconi,* 195.
298 *Now he endured*: Marconi, *My Father,* 180–81.
298 *Eventually he found one*: Weightman, *Signor Marconi,* 196.
298 *Beatrice's sister*: Marconi, *My Father,* 181.
298 *On April 3, 1906*: H. Kershaw to Fleming, April 3, 1906. UCL, Fleming Collection, 122/48.
299 *During this time*: Marconi, *My Father,* 181.
299 *Standing on his head*: Ibid., 181–82.
299 *He recruited Marconi's opponents*: Hong, *Wireless,* 166.
299 *"We find that the administration"*: Hozier to Lodge, May 11, 1906. UCL, Lodge Collection, 89/77.
300 *Muirhead arranged*: Muirhead to Lodge, June 10, 1906. Ibid.

300 *Impressed anew*: Lodge, *Past Years,* 283–84.
300 *In 1906, in response*: Massie, *Dreadnought,* 481–82; Read, *Urban Democracy,* 475.
300 *That year*: Clarke, *Voices,* 144–52.
300 *One witness*: Ibid., 145.
300 *The publisher of the German-language edition*: Ibid., 148.
300 *On September 11, 1908*: Marconi, *My Father,* 186.

AN INSPECTOR RETURNS

301 *"Will you do me"*: *Trial,* 30.
301 *Long chose not to mention*: Further Report of Chief Inspector Dew, 7. NA-DPP 1/13.
301 *a five-chambered revolver*: Ibid., 4.
302 *He composed detailed descriptions*: Ibid., 4.
302 *The hotel's owner noticed*: Police Reports as to Enquiries at Antwerp. NA-DPP 1/13.
302 *Ethel loved touring*: Le Neve, *Ethel Le Neve,* 49.
302 *On Tuesday Dew ordered*: "Further Report of Chief Inspector Dew," 4. NA-DPP 1/13.
303 *"Even in bed"*: Dew, *I Caught Crippen,* 27.
303 *At last Long disclosed*: Further Report of Chief Inspector Dew, 7. NA-DPP 1/13.
303 *He and Mitchell worked*: Dew, *I Caught Crippen,* 27.
303 *One of the bricks*: Further Report of Chief Inspector Dew, 5. NA-DPP 1/13.
303 *Mitchell went to the garden*: Dew, *I Caught Crippen,* 28.

THE MERMAID

304 *In future years*: Interview, Princess Elettra Marconi; also, Marconi, *Marconi My Beloved,* 159.
304 *Suddenly Beatrice appeared*: Marconi, *My Father,* 188–89.

THE MYSTERY DEEPENS

306 *"The stench was unbearable"*: Dew, *I Caught Crippen,* 28.
306 *He and Froest set out*: Macnaghten, *Days of My Years,* 195.
307 *His task would be*: Thomas Marshall Statement, 50. Brief for the Prosecution, NA-DPP 1/13.
307 *The men concentrated*: Arthur Mitchell Statement, 4. Further Information, NA-DPP 1/13.
307 *As Dew would note*: Particulars of Human Remains, 23–24. Witness, NA-DPP 1/13.
307 *There was nothing*: Thomas Marshall Statement, 43. Supplemental Information, NA-DPP 1/13. Marshall said, "We found not one single bone, no head, no arms or leg."

307 *"Someone had simply carved"*: Walter Dew Statement, July 18, 1910, 38. NA-DPP 1/13.

307 *The scope of the challenge*: Dew, *I Caught Crippen,* 30.

308 *"From the doctor's chair"*: Macnaghten, *Days of My Years,* 195.

309 *He titled the circular*: Dew, *I Caught Crippen,* 29; for a photograph of the circular, see Goodman, *Crippen File,* 10.

309 *The detectives returned*: Alfred Henry Sargent Statement, 156, and Francis Barclay Statement, 158–59. Witness, NA-DPP 1/13.

310 *"I asked him several times"*: Le Neve, *Ethel Le Neve,* 50.

THE DYNAMITE PRIZE

311 *"The Telefunken Wall"*: Baker, *History,* 131.

311 *A company memorandum*: "Traffic Between Clifden and Glace Bay from October 10, 1907 to June 27, 1908." August 4, 1908. William Smith Papers, Vol. 1. Archives Canada, MG 30 D18.

311 *Another company report*: "Analysis of Clifden Traffic from 4th January 1908 to 15th August 1908." William Smith Papers, Vol. 2. Archives Canada, MG 30 D18.

312 *"I might mention"*: Marconi, *Nobel,* 1, 2.

312 *He acknowledged*: Ibid., 40.

312 *Nor had he found*: Ibid., 27–28.

312 *"It would almost appear"*: Ibid., 41.

313 *"Whatever may be its present shortcomings"*: Ibid., 44.

FIVE JARS

314 *Dew also retrieved*: Walter Dew Statement, July 18, 1910, 39. NA-DPP 1/13.

315 *"There was one large mass"*: Augustus Joseph Pepper Statement, 40. Brief for the Prosecution, NA-DPP 1/13.

315 *Amid the discarded skin*: Ibid., 46.

315 *"False hair"*: Ibid., 41.

315 *As Pepper probed*: *Trial,* 48; W. H. Willcox, Report, 4. NA-DPP 1/13.

316 *Ethel grew weary*: This and subsequent dialogue and detail: Le Neve, *Ethel Le Neve,* 49–53.

317 *His search also turned up*: Walter Dew Statement, July 18, 1910, 39. NA-DPP 1/13.

317 *"It was the one big topic"*: Dew, *I Caught Crippen,* 33.

318 *"Not a day passed"*: Ibid., 36–37.

318 *On this score*: Browne, *Rise,* 250, 258.

318 *The most important lesson*: Macnaghten, *Days of My Years,* 98.

318 *"I did what I could"*: Dew, *I Caught Crippen,* 36.

318 *On Friday, July 15*: Emily Jackson Statement, 38–40. Witness, NA-DPP 1/13.

319 *They revisited Clara Martinetti*: Clara Martinetti Statement, 63–68. Ibid.

319 *They interviewed Marion*: Marion Louisa Curnow Statement, 72–73. Ibid.
319 *"They were on exceedingly good"*: Emily Cowderoy Statement, 104. Ibid.
320 *that Sir Melville Macnaghten believed*: Macnaghten, *Days of My Years,* 196–97.
320 *Mrs. Ginnette and the police*: Cullen, *Crippen,* 69; Fox, *Transatlantic,* 405.
321 *"Up till today"*: Cullen, *Crippen,* 70.
321 *"The reverend gentleman"*: Ibid., 74.
321 *"Many a man"*: Ibid., 72.
321 *Afterward, in the hall*: Dew, *I Caught Crippen,* 31–32.

TESTAMENT

323 *"Marconi-Atlantic"*: Marconi, *My Father,* 192.
324 *"It was without"*: Le Neve, *Ethel Le Neve,* 52.

PART VI: PURSUIT BY THUNDER

THE ROBINSONS

327 *"The whole ship"*: Le Neve, *Ethel Le Neve,* 52–53.
327 *"I felt so sure"*: Ibid., 53.
327 *The captain also produced*: *Trial,* 187.
328 *The ship's open-sea velocity*: Ibid., 188.
328 *"So with a rug"*: Le Neve, *Ethel Le Neve,* 54.
328 *He found their hats*: *Trial,* 187; also, Cullen, *Crippen,* 126.
328 *"I warned him"*: *Trial,* 187.

SUICIDE

329 *In Chicago police arrested*: *New York Times,* July 23, 1910.
330 *In Marseille a shipping agent*: Ibid.
330 *In Halifax, Nova Scotia*: Ibid., July 24, 1910.
330 *From Brussels came*: Ibid., July 23, 1910.
330 *"Many meek looking men"*: Ibid., July 24, 1910.
331 *On the night of Wednesday*: Ibid., July 22, 1910.

A MESSAGE FROM THE SEA

331 *In talking with Robinson*: Henry George Kendall Statement, 2. NA-MEPO 3/198; also, Croall, *Fourteen Minutes,* 25, and Jeffers, *Bloody Business,* 125.
331 *Once, a gust of wind*: *Trial,* 188.
331 *Kendall invited the Robinsons*: *Trial,* 188.
331 *"This ruse was successful"*: Ibid., 188.
332 *"I was then fully convinced"*: Henry George Kendall Statement, 2. NA-MEPO 3/198.
332 *One afternoon Kendall spotted*: *Trial,* 188.
332 Have strong suspicions: Henry George Kendall Statement, 3. NA-MEPO 3/198; Jeffers, *Bloody Business,* 126; see photograph of Marconigram in Goodman, *Crippen Files,* 28.

"MR. DEWHURST"

334 *"It was eight o'clock"*: Dew, *I Caught Crippen,* 37.
335 *"Read it to me"*: Jeffers, *Bloody Business,* 126.
335 *"What do you think"*: This and subsequent dialogue is from Dew, *I Caught Crippen,* 39.
335 *"It was a serious step"*: Macnaghten, *Days of My Years,* 199.
335 *Moreover, the Murder Squad*: Ibid., 229–31.
335 *Macnaghten worried*: Ibid., 199.
335 *"Here is your authority, Dew"*: Dew, *I Caught Crippen,* 39.
336 *"That night could not fail"*: Macnaghten, *Days of My Years,* 199.
336 *An officer with the Liverpool police*: Telegram. Head Constable Leonard Dunne to Macnaghten, July 22, 1910. NA-MEPO 3/198.
336 *Only the ship's captain*: Dew, *I Caught Crippen,* 40.
336 *He was met*: Telegram. Head Constable Leonard Dunne to Macnaghten, July 22, 1910. NA-MEPO 3/198.
336 *"It was hopeless"*: Dew, *I Caught Crippen,* 40.
336 *"I assumed an air"*: Macnaghten, *Days of My Years,* 199–200.
337 *Macnaghten's anxiety increased*: Ibid., 200.
337 *New York police*: Inspector John H. Russell, Police Department of the City of New York, to Macnaghten, July 22, 1910. NA-MEPO 3/198.
337 *A French rail guard*: *New York Times,* July 20, 1910.
337 *A traveler on an English train*: Charles Jones to Head Constable, Cardiff City Police, July 15, 1910. NA-MEPO 3/198.
337 *In Brussels a Scotland Yard detective*: Central Officer's Special Report: Re John Robinson and John Robinson Junior. July 24, 1910. NA-MEPO 3/198.
337 *The innkeeper's wife*: *Daily Mail,* July 27, 1910. Reproduced in Goodman, *Crippen File,* 35.

AN INTERCEPTED SIGNAL

338 *"The doctor was as calm"*: Le Neve, *Ethel Le Neve,* 53.
338 *She imagined the letter*: Ibid., 54.
339 *"What is Inspector Dew doing"*: MarconiCalling. Search Crippen. See, "Kendall's Message Reaches Scotland Yard."

CAGE OF GLASS

340 *"It is believed that"*: *Daily Telegraph,* July 25, 1910. Reproduced in Goodman, *Crippen File,* 28.
340 *The story consumed*: *Trial,* xxxi.
340 *"At noon to-day"*: *Daily Mail,* July 26, 1910. Reproduced in Goodman, *Crippen File,* 31.
341 *One article speculated*: *Daily Mail,* July 25, 1910. Reproduced in Goodman, *Crippen File,* 29.

341 *"Mysterious voices"*: *Daily Mirror*, July 27, 1910. Reproduced in Goodman, *Crippen File*, 33.
341 *A French newspaper*: Quoted in Goodman, *Crippen File*, 37.
341 *"The people, who have a sure instinct"*: Priestley, *Edwardians*, 200.
342 *"The King's Poisoner"*: Willcox, *Detective-Physician*, 324.
342 *He took the first steps*: *Trial*, 68.
342 *"The remains"*: Ibid., 66.
342 *He succeeded in locating*: William Henry Willcox Statement, 58. Brief for the Prosecution, NA-DPP 1/13.
342 *At No. 46 Brecknock Road*: Lena Lyons Statement, 133–35. Witness, NA-DPP 1/13.
343 *Another neighbor, Franziska*: Franziska Hachenberger Statement, 135A. Ibid.
343 *The most detailed report*: Frederick Evans Statement, 136–38. Ibid.
344 *Crutchett tracked down*: William Curtis Statement, 162–63. Ibid.
345 *On Wednesday, July 27*: Cullen, *Crippen*, 135; Jeffers, *Bloody Business*, 126–27.
345 *"What the devil"*: Jeffers, *Bloody Business*, 127.
346 *"Speaking for myself"*: London *Times*, July 29, 1910.

QUIVERING ETHER

347 *"Kindly wireless"*: Cullen, *Crippen*, 135.
347 *"too good a thing to lose"*: *Trial*, 187.
348 *"My dear," he told her*: This and subsequent dialogue come from Le Neve, *Ethel Le Neve*, 55–56.
348 *On Friday, July 29*: *Trial*, 187–88.
350 *"There was something"*: London *Times*, August 1, 1910.
350 *"The suspect fugitive"*: Reproduced in Goodman, *Crippen File*, 37.
350 *"What a wonderful invention"*: *Trial*, 188.

THE ST. MARY'S CAT

351 *At St. Mary's Hospital*: *Trial*, 71; Willcox, *Detective-Physician*, 28; William Henry Willcox Statement, 58–65. Brief for the Prosecution, NA-DPP 1/13.
351 *"It is necessary"*: *Trial*, 70.
351 *He found, for example*: Willcox, *Detective-Physician*, 27.
352 *He knew of only*: William Henry Willcox Statement, 60. Brief for the Prosecution, NA-DPP 1/13.
352 *When exposed to*: *Trial*, 71.
353 *Adopted by a medical student*: Willcox, *Detective-Physician*, 31.

WHISPERS

354 *Jones proved himself*: *New York Times*, July 30, 1910.

THE INSPECTOR ARRIVES

355 *He was appalled*: Walter Dew Report, August 2, 1910. NA-MEPO 3/198.

355 *On shore Dew*: Ibid.; Dew, *I Caught Crippen,* 41. The Marconi station at Father Point offered an example of the costs and problems that accumulated as Marconi expanded his ship-to-shore empire. The Father Point station began operation on December 22, 1906, and almost immediately things began going awry, as recorded in the station's log. Pipes froze. Engines failed. One entry reads, "Pump pipe thawed out by removing suction chamber and thrusting a red hot iron pipe down the other."

The record for 1907 is full of similar interruptions. Engines broke down. Signals grew weak and spontaneous disruptions denoted by the letter X became commonplace. "Xs fierce," the operator wrote one night. And again, "Xs bad all pm." Weather bedeviled the station. The cruelest month was April 1908, a model of meteorological perversity.

The entry for Saturday, April 4, reads: "Hurricane from West . . ."

For Thursday, April 9: "Hurricane from East . . ."

See Log Book of Father Point, Quebec, 1906–1914. Archives Canada, MG 28 III 72 Vol. 81.

356 *"The lighthouse foghorn"*: Dew, *I Caught Crippen,* 41.

356 *"Now I don't pretend"*: Ibid., 42.

356 *He called all the reporters*: Ibid., 42; Walter Dew Report, August 2, 1910. NA-MEPO 3/198.

357 *Even the home secretary*: Central Officer's Special Report, July 30, 1910. NA-MEPO 3/198.

357 *They learned, for example*: John William Stonehouse Statement, 143–44. Witness, NA-DPP 1/13.

358 *Later the clerk called*: Central Officer's Special Report, August 1, 1910. NA-MEPO 3/198.

A BOAT IN THE MIST

359 *"The last night was dreary"*: Priestley, *Edwardians,* 199.

359 *"I don't think I will"*: Le Neve, *Ethel Le Neve,* 56.

360 *Inside the lining*: Walter Dew Statement, 75. Brief for the Prosecution, NA-DPP 1/13.

360 *The ship's surgeon*: *New York Times,* August 1, 1910.

360 *As a precaution*: Priestley, *Edwardians,* 199.

TREACHEROUS WATERS

361 *Dew realized*: Dew, *I Caught Crippen,* 42–43.

362 *Kendall led the party*: Dew, *I Caught Crippen,* 44; Priestley, *Edwardians,* 199.

362 *"During my long career"*: Dew, *I Caught Crippen,* 43.

362 *Crippen, he wrote, "had been caught"*: Ibid., 44.
363 *"I am Chief Inspector Dew"*: Ibid., 44.

EPILOGUE: INTO THE ETHER

THE TABLE OF DROPS

367 *"If the fatal dose"*: *Trial,* 69.
367 *Investigators made another*: Dew, *I Caught Crippen,* 62.
368 *Dew's manner was so paternal*: Le Neve, *Ethel Le Neve,* 60.
368 *"He mystified me"*: Dew, *I Caught Crippen,* 56–57.
368 *At the Quebec prison*: C. L. Gauvrea to Superintendent, Scotland Yard, December 9, 1959. Black Museum, NA-MEPO 3/3154.
369 *Dew kept Crippen*: Dew, *I Caught Crippen,* 57.
369 *"I don't know how things may go"*: Ibid., 54.
369 *"I had to be present"*: Ibid., 55.
369 *Four thousand people*: Willcox, *Detective-Physician,* 28.
369 *The spectators included*: Jeffers, *Bloody Business,* 129.
369 *On the stand Spilsbury*: Browne and Tullett, *Scotland Yard,* 53–54.
370 *At this point a soup plate*: *Trial,* xxxii; Jeffers, *Bloody Business,* 128.
370 *Muir asked*: *Trial,* 94.
371 *A warder took his money*: Memorandum, W. Middleton to Governor of Pentonville Prison, October 25, 1910. NA-PCOM 8/30.
371 *The fact of his incarceration*: Memorandum HM Prison Brixton, September 19, 1910. NA-PCOM 8/30.
371 *"It is comfort"*: Ellis, 316. During Crippen's incarceration, an old man applied to be hanged in his place, arguing that his own life was not worth as much as that of a doctor. The offer was declined. Browne, *Travers Humphreys,* 78.
371 *Ellis was known to be*: Memorandum to Commissioners, March 11, 1914. Execution Record, Execution of Josiah Davies, March 10, 1914. NA-PCOM 8/213. One notorious series of executions conducted by a hangman named Berry had demonstrated the worth of attending carefully to the physics of the process. He tried three times to hang a convicted killer named John Lee, and three times failed, prompting a judge to commute Lee's sentence to life. Chastened, Berry resolved to correct his mistake by adding a little extra distance to the drop for future executions. His next subject was a killer named Robert Goodale. The noose tore Goodale's head off. A year later, while trying to hang a murderer named David Roberts, he allowed too little distance. Roberts struggled in midair until prison authorities killed him by other means. See Browne, *Rise,* 180.
371 *"Character of prisoner's neck"*: Execution Record, Execution of Hawley Harvey Crippen, November 23, 1910. NA-PCOM 8/30.
372 *The prison warder*: Inventory, Crippen's Clothing, August 21, 1911. NA-PCOM 8/30.

372 *Ellis continued to moonlight*: Rochdale Folk, at manchesterhistory.net/rochdale/ellis.html

372 An editor for: "To H. H. Crippen, Condemned Cell, Pentonville Gaol." November 19, 1910. Newspaper Extracts, NA-HO /44/1719/ 195492.

373 *One theory*: Hicks, *Not Guilty,* 83.

374 *"I never looked upon"*: Humphreys, *Criminal Days,* 113.

374 *"Full justice has not yet been done"*: Browne and Tullett, *Scotland Yard,* 58.

374 *"We carefully examined"*: Central Officer's Special Report: Murder of Cora Crippen. Information, September 1, 1910. NA-MEPO 3/198.

375 *"There is no bad smell*: Central Officer's Special Report: Special Enquiry at Railway Stations re Crippen, September 16, 1910. NA-MEPO 3/198.

375 *The Public Health Department*: Alfred Edwin Harris, Medical Officer of Health, to Sir Melville Macnaghten, October 7, 1910. NA-MEPO 3/198.

375 *At precisely 3:15*: Memorandum: "I beg to report the funeral cortege. . . ." NA-MEPO, 3/198.

375 *"Dr. Crippen's love"*: Dew, *I Caught Crippen,* 47.

375 *the Wee Hoose*: Cullen, *Crippen,* 197.

375 *"the most intriguing murder mystery"*: Dew, *I Caught Crippen,* 7.

376 *Just before two in the morning*: Canada's national archives contain a great trove of material on the *Empress* disaster. See in particular: Commission of Enquiry into Wreckage of Empress of Ireland, June 16, 1914. Archives Canada, RG 42 Vol. 351.

376 *As the Germans raced*: Musk, *Canadian Pacific,* 74; Croall, *Fourteen Minutes,* 229–30.

376 *He joined the Royal Navy*: Croall, *Fourteen Minutes*, 230.

376 *One mast remained*: Musk, *Canadian Pacific,* 74.

377 *Alfred Hitchcock*: Hitchcock, "Juicy Murders," 23; Massie, *Potawatomi Tears,* 277; "Hitchcock's Favorite Crime," members.aol.com/vistavsion/doctorcrippen.html.

What Hitchcock found particularly appealing about the Crippen saga was its subtlety. "The Crippen case was fraught with understatement, restraint, and characteristic British relish for drama," he wrote. He called understatement "an occupational tradition of English police. With the most atrocious criminals, they never bluster up and say, 'O.K.—we gotcha!' They say: 'I beg your pardon, but it seems that someone has been boiled in oil. We wondered if you'd mind answering a few questions about it. . . .' " Hitchcock, "Juicy Murders," 23.

377 *Crippen also proved a fascination*: Gardiner and Walker, *Raymond Chandler,* 197–98.

377 *A play called* Captured by Wireless: *Coldwater Courier,* April 12, 1912. Holbrook Heritage Room, Branch County District Library, Coldwater, Michigan.

378 *"A Sick Joke With Music"*: Cullen, *Crippen,* 202–3.

378 *During World War II*: A bomb also struck New Scotland Yard and destroyed several floors, including the police commissioner's office. Happily, he was not in at the time. Browne, *Rise,* 360.

THE MARRIAGE THAT NEVER WAS

380 *They waved*: Marconi, *My Father,* 197–99.

380 *Since then no ship*: "What is the economic value of the International Ice Patrol?" U.S. Coast Guard at www.uscg.mil/lantarea/iip/FAQ/Org6.shtml

380 *The companies agreed*: Baker, *History,* 135; Aitken, *Syntony,* 284.

380 *As soon as the Marconi men left*: Baker, *History,* 158–59. As war loomed, Sir Edward Grey, Foreign Secretary, looked out the window of his office and watched as the lamps in St. James's Park were lit. As tears filled his eyes, he spoke one of the saddest sentences of history: "The lamps are going out all over Europe; we shall not see them lit again in our lifetime." Tuchman, *Guns,* 122.

380 *Marconi's station at Poldhu*: Baker, *History,* 159.

380 *"Calls for assistance"*: Marconi Co. of Canada Annual Report, Year-Ended January 31, 1915. File 191, 2–48. Annual Reports. Archives Canada, MG 28 III 72 Vol. 6.

380 *In 1917 a German submarine*: Hancock, *Wireless,* 91. Marconi himself may have been a target of the Imperial German Navy. In April, 1915, Marconi booked passage on the *Lusitania* and sailed to New York to testify in a patent lawsuit his company had brought against a competitor. While he was there, German officials warned that if the *Lusitania* reentered English waters, it would be torpedoed. A rumor circulated that the Germans planned to capture Marconi. On May 7, a German submarine did indeed sink the *Lusitania.*

Later that month, Marconi learned that Italy had entered the war. He was still in New York. He excused himself from the patent fight and sailed back to London aboard his old favorite, the *St. Paul.* He took the German threat against him seriously, however, and traveled incognito. From England he crossed to Italy, where he was put in charge of the Italian army's wireless operations. Baker, *History,* 171.

381 *"The past had been dead"*: Marconi, *My Father,* 232.

381 *Marconi sold their house*: According to Degna Marconi, the sale was devastating. Degna recalled one day walking past the house just as the move was under way. "The door of our old home stood open and I went into the front hall. Most of the furniture had been taken away, and in its place were crates covered with dust. In a corner a few books we had once loved had been dumped like so much trash. Lamps with broken shades and letters in my father's handwriting littered the floor. Mother, too sad to attend to the home that was being destroyed, had left the packing to the servants." Degna

wrote this in 1962. "I still feel grief for myself, a child standing alone in that derelict house." Marconi, *My Father,* 234.

381 *"I would like to wish"*: Marconi, *My Father,* 252.

382 *"They only want"*: Ibid., 269.

382 *"Young man"*: Baker, *History,* 185.

383 *"What a world we live in"*: Sir Henry Morris-Jones. "Diary of a visit to Canada and the U.S.A., 1926." Archives Canada, MG40 M22 Microfilm Reel A-1610.

383 *"I admit that I am responsible"*: Aitken, *Syntony,* 272.

383 *The climax of the day*: Marconi, *My Father,* 294.

383 *As he aged, Marconi became aloof*: Baker, *History,* 295.

383 *"Listen for a regularly repeated signal"*: Isted, I, 54.

384 *"I am very sorry"*: *Indianapolis Times,* July 20, 1937. Indiana State Library.

384 *Amelia Earhart*: Ibid.

384 *That night the gloom*: Ibid.

385 *"I was unobserved"*: Marconi, *My Father,* 311

FLEMING AND LODGE

386 *"a fighting fund"*: Lodge to Preece, June 15, 1911. Baker Collection: Further Papers of Sir William Henry Preece. IEE-NAEST 021.

386 *"They are clearly infringing"*: Ibid.

386 *"I agree with every word"*: Baker, *Preece,* 304–5.

386 *"I am delighted to hear"*: Preece to Lodge, October 24, 1911. UCL, Lodge Collection, 89/86.

386 *"I love you"*: Lodge, *Raymond,* 205.

387 *"Father, tell mother"*: Ibid., 207.

387 *"I recommend people"*: Ibid., 342.

387 *The book became hugely popular*: Lodge's biographer, W. P. Jolly, put it nicely: "Seldom can a work of research and philosophy have been more opportunely published, when almost everyone in England was mourning the loss of some friend or relative." Jolly, *Lodge,* 205.

387 *"It is quite clear"*: Fleming to Lodge, August 29, 1937. UCL, Lodge Collection, 89/36.

387 *"Marconi was always determined"*: Ibid.

CODA

VOYAGER

391 *For the second time*: Cullen, *Crippen,* 191.

391 *After arriving in New York*: Ibid., 199–201.

Archives and Bibliography

ARCHIVES

Archives Canada. Ottawa.

Beaton Institute. Breton University, Nova Scotia.

Bolles, Edwin C. The Edwin C. Bolles Collection on the History of London. Tufts University. Online Archive. http://www.perseus.tufts.edu/cache/perscoll_Bolles.html

Booth, Charles. Charles Booth Online Archive. London School of Economics. http://www.booth.lse.ac.uk/

Cape Cod National Seashore. Eastham, Mass.

Institute of Electrical Engineers (IEE). London.

Kemp Diary. Marconi Archive. Bodleian Library. Oxford.

MarconiCalling. Marconi Online Archive. http://www.MarconiCalling.com

National Archives (NA), Britain. Kew, England:

> Central Criminal Court/Old Bailey (CRIM)
> Department of Public Prosecution (DPP)
> Treasury (T)
> Home Office (HO)
> Metropolitan Police (MEPO)
> Prison Commission (PCOM)

University College of London (UCL), Special Collections:

> Fleming Collection
> Lodge Collection

Wellfleet Historical Society, Wellfleet, Mass.

BIBLIOGRAPHY

A-Z London (edition 5A). Geographer's A-Z Map Co., 2002.

Aitken, Hugh G. J. *Syntony and Spark—The Origins of Radio.* John Wiley, 1976.

Alverstone, Viscount. *Recollections of Bar and Bench.* Longmans, Green, 1915.
Bacon's Up to Date Map of London 1902. Old House Books, undated.
Baedeker, Karl. *London and its Environs, 1900.* Old House Books, 2002 (1900).
Baker, E. C. *Sir William Preece.* Hutchinson, 1976.
Baker, W. J. *A History of the Marconi Company.* Methuen, 1970.
Barrie, J. M. *Peter Pan and Other Plays.* Charles Scribner's Sons, 1930.
Bartram, Graeme. "Wireless and the Art of Magic, Part I." *Bulletin of the British Vintage Wireless Society* 28, no. 4 (Winter 2003), pp. 48–55.
———. "Wireless and the Art of Magic, Part II." *Bulletin of the British Vintage Wireless Society* 29, no. 1 (Spring, 2004), pp. 42–45.
Bayless, Raymond. *Voices from Beyond.* University Books, 1976.
Biel, Steven. *Down with the Old Canoe.* W. W. Norton, 1996.
Birkenhead, Lord (F. E. Smith). *Famous Trials of History.* George H. Doran, 1926.
Blunt, Wilfrid Scawen. *My Diaries.* Alfred A. Knopf, 1922.
Bodanis, David. *Electric Universe.* Crown, 2005.
Booth, Martin. *The Doctor, the Detective and Arthur Conan Doyle.* Hodder & Stoughton, 1997.
Bordeau, Sanford. *Volts to Hertz.* Burgess, 1982.
Braga, Gioia Marconi. *A Biography of Guglielmo Marconi.* Marconi Foundation. Web: MarconiFoundation.Org
Browne, Douglas G. *The Rise of Scotland Yard.* George G. Harrap, 1956.
———. *Sir Travers Humphreys.* George G. Harrap, 1960.
Browne, Douglas G., and E. V. Tullett. *The Scalpel of Scotland Yard.* E. P. Dutton, 1952.
Brownlow, Arthur H., ed. *Cape Cod Environmental Atlas.* Boston University, 1979.
Bryant, John H. *Heinrich Hertz: The Beginning of Microwaves.* Institute of Electrical and Electronic Engineers, 1988.
Bussey, Gordon. *Marconi's Atlantic Leap.* Marconi Communications, 2000.
Cheney, Margaret, and Robert Uth. *Tesla, Master of Lightning.* MetroBooks, 1999.
Chesterton, G. K. *The Autobiography of G. K. Chesterton.* Sheed & Ward, 1936.
Childers, Erskine. *The Riddle of the Sands.* Modern Library, 2002 (1903).
Clarke, I. F. *Voices Prophesying War 1763–1984.* Oxford, 1966.
Coldwater, Michigan, City Directory. 1949.
Collier, Price. *England and the English from an American Point of View.* Charles Scribner's Sons, 1909.
Collins, A. Frederick. *Wireless Telegraphy: Its History, Theory and Practice.* McGraw, 1905.
———. *Manual of Wireless Telegraphy.* John Wiley, 1906.
Conover, J. S. *Coldwater Illustrated.* Conover, 1889.

Constable, Tony. "Oliver Lodge and the Hertz Memorial Lecture of 1894." *Bulletin of the British Vintage Wireless Society* 19, no. 1 (n.d.), pp. 21–22.

Corazza, Gian Carlo. "Marconi's History." *Proceedings of the IEEE* 86, no. 7 (July 1998), pp. 1307–11.

Croall, James. *Fourteen Minutes.* Stein & Day, 1979.

Cullen, Tom. *Crippen: The Mild Murderer.* Penguin, 1988.

———. *When London Walked in Terror.* Houghton Mifflin, 1965.

Daily Climatological Data, Sydney, Nova Scotia 1870–1971. Environment Canada, 1973.

d'Albe, E. E. Fournier. *The Life of Sir William Crookes.* D. Appleton, 1924.

David, Hugh. *The Fitzrovians.* Michael Joseph, 1988.

de Falco, Alessandro. *Santa Maria degli Angeli e dei Martiri: Incontro di Storie.* BetaGamma, 2005.

Deghy, Guy, and Keith Waterhouse. *Café Royal.* Hutchinson, 1956.

Dew, Walter. *I Caught Crippen.* Blackie & Son, 1938.

Douglas, John, and Mark Olshaker. *The Cases That Haunt Us.* Scribner, 2000.

Doyle, Arthur Conan. *Memories & Adventures.* Greenhill Books, 1988 (1924).

Dunbar, Janet. *J. M. Barrie: The Man Behind the Image.* Houghton Mifflin, 1970.

Dunlap, Orrin E. *Marconi: The Man and His Wireless.* Macmillan, 1937.

Eastwood, Sir Eric, ed. *Wireless Telegraphy.* John Wiley & Sons, 1974.

Eckert, Kathryn Bishop. *Buildings of Michigan.* Oxford, 1993.

Ellis, Havelock. *Studies in the Psychology of Sex: Sexual Selection in Man* (vol. 4). F. A. Davis, 1921.

Ellis, J. C. *Black Flame.* Hutchinson & Co., undated.

Falciasecca, Gabriele, and Barbara Valotti. *Guglielmo Marconi: Genio, Storia e Modernità.* Giorgio Mondadori, 2003.

Faulkner, Brian. *Watchers of the Waves.* Radio Bygones, undated.

Felstead, Sidney Theodore. *Famous Criminals and Their Trials.* George H. Doran, 1926.

Fleming, J. A. *A Report on Experiments Carried out at Poldhu on Wednesday, March 18, 1903.* Marconi Wireless Telegraph Co., March 23, 1903.

———. "The History of Transatlantic Wireless Telegraphy by Marconi Methods." Vol. I. Unpublished ms. UCL, Fleming Collection, 122/64.

Forster, E. M. *A Room with a View.* Penguin Books, 2000 (1908).

———. *Howards End.* Penguin Books, 2000 (1910).

Fox, Stephen. *Transatlantic.* HarperCollins, 2003.

Fried, Albert, and Richard M. Elman. *Charles Booth's London.* Pantheon, 1968.

Furbank, P. N. *E. M. Forster: A Life* (vol. 1). Secker & Warburg, 1977.

Gardiner, Dorothy, and Kathrine Sorley Walker, eds. *Raymond Chandler Speaking.* Allison & Busby, 1984 (1962).

Gay, Peter, ed. *The Freud Reader.* Norton, 1995 (1989).

Gillespie, Carolyn L. *A History of the Tibbits Opera House, 1882–1904.* Doctoral thesis, Kent State, June 1925.

Goldenberg, Susan. *Canadian Pacific: A Portrait of Power*. Methuen, 1983.
Goodman, Jonathan, ed. *The Crippen File*. Allison & Busby, 1985.
Goodwin, W. D. *One Hundred Years of Maritime Radio*. Brown, Son & Ferguson, 1995.
Gore, John. *Edwardian Scrapbook*. Evans Brothers, 1951.
Hancock, H. E. *Wireless at Sea*. Marconi International Marine Communication Co., 1950.
Haynes, Renée. *The Society for Psychical Research, 1882–1982*. Macdonald & Co., 1982.
Hicks, Seymour. *Not Guilty, M'Lord*. Cassell, 1939.
Hill, J. Arthur. *Letters from Sir Oliver Lodge*. Cassell, 1932.
History of Branch County, Michigan. Everts & Abbott, 1879.
Hitchcock, Alfred. "Juicy Murders, and Why We Love 'Em So." *Everybody's Magazine,* June 8, 1957.
Holmes, Nathaniel. *An Illustrated City Directory of Coldwater, Michigan*. Conover Engineering and Printing, 1894
Hong, Sungook. *Wireless: From Marconi's Black-Box to the Audion*. MIT Press, 2001.
Humphreys, Travers. *Criminal Days*. Hodder & Stoughton, 1946.
Hynes, Samuel. *The Edwardian Turn of Mind*. Princeton, 1968.
Isted, G. A. "Guglielmo Marconi and the History of Radio, Part I." *GEC Review* 7, no. 1 (1991).
———. "Guglielmo Marconi and the History of Radio, Part II." *GEC Review* 7, no. 2 (1991).
James, Frank A. J. L. "The Royal Institution of Great Britain: 200 Years of Scientific Discovery and Communication." *Interdisciplinary Science Reviews* 24, no. 3 (1999).
———. *"The Common Purposes of Life": Science and Society at the Royal Institution of Great Britain*. Ashgate, 2002.
Jeffers, H. Paul. *Bloody Business*. Pharos Books, 1992.
Jolly, W. P. *Sir Oliver Lodge*. Fairleigh Dickinson, 1975.
Jonnes, Jill. *Empires of Light*. Random House, 2003.
Kestner, Joseph A. *The Edwardian Detective, 1901–1915*. Ashgate, 2000.
King, William Harvey, ed. *History of Homœopathy*. Lewis Publishing, 1905.
Kittredge, Henry C. *Cape Cod*. Houghton Mifflin, 1968.
Lacouture, John. "Siasconset Wireless Stations." *Bulletin of the British Vintage Wireless Society* 17, no. 3 (June 1992), pp. 29–32.
Lane, Margaret. *Edgar Wallace*. Hamish Hamilton, 1964 (1938).
Le Neve, Ethel. *Ethel Le Neve: Her Life Story With the True Account of Their Flight and Her Friendship for Dr. Crippen, Told by Herself*. Daisy Bank Printing, 1910.
"Letters from Guglielmo Marconi to His Father 1896–1898." GEC *Review,* 12 no. 2 (1997).

Lewis, David V. "A Survey of Woodruff Place History, Nov. 28, 1972." Indiana State Library.
Lodge, Sir Oliver. *Raymond, or Life and Death*. George H. Doran, 1916.
———. *Ether & Reality*. Hodder & Stoughton, 1926.
———. *Why I Believe in Personal Immortality*. Doubleday/Doran, 1929.
———. *Past Years*. Hodder & Stoughton, 1931.
Lord, Walter. *A Night to Remember*. Bantam, 1997 (1955).
Lyons, Sir Henry. *The Royal Society, 1660–1940*. Cambridge, 1944.
Machray, Robert. *The Night Side of London*. Bibliophile Books, 1984 (1902).
MacLeod, Mary K. *Marconi: Whisper in the Air*. Lancelot Press, 1995.
Macnaghten, Sir Melville. *Days of My Years*. Edward Arnold, 1914.
Macqueen-Pope, W. *Goodbye Piccadilly*. Drake Publishers, 1972.
Magnus, Philip. *King Edward the Seventh*. Penguin, 1979 (1964).
Marconi, Degna. *My Father, Marconi*. McGraw-Hill, 1962.
Marconi, Guglielmo. *Wireless Telegraphy: Lecture Given Before the Liverpool Chamber of Commerce, Monday, 24 February, 1908*. Marconi Wireless Telegraph Co., 1908.
———. *Transatlantic Wireless Telegraphy: Lecture Given Before the Royal Institution, Friday Night, 13th March, 1908*. Marconi Wireless Telegraph Co., 1908.
———. *Nobel Lecture, Dec. 11, 1909*. Fondazione Guglielmo Marconi, undated.
Marconi, Maria Cristina. *Marconi My Beloved*. Dante University of America Press, 1999.
Maskelyne (Nevil) Incident Papers. IEE (S.C. Mss. 17).
Maskelyne, Nevil, and David Devant. *Our Magic*. George Routledge, undated.
Massie, Larry B. *Potawatomi Tears & Petticoat Pioneers*. Priscilla Press, c. 1992.
Massie, Robert K. *Dreadnought*. Random House, 1991.
Massie, Walter W., and Charles R. Underhill. *Wireless Telegraphy and Telephony Popularly Explained*. D. Van Nostrand, 1908.
Masterman, C.F.G. *The Condition of England*. Methuen, 1910 (1909).
Maurois, André. *The Edwardian Era*. D. Appleton-Century, 1933.
McDonald, John P. *Lost Indianapolis*. Accadia, 2002. Indiana State Library.
McKenzie, Anna, ed. *The Greater Indianapolis Blue Book, 1898–1899*. Brown-Merrill. Indiana State Library.
Michigan Business Directory. 1863.
Mitchell, W. H. *Canadian Pacific and Southampton*. World Ship Society, 1991.
Moffett, Cleveland. "Signor Marconi and Wireless Telegraphy." *Windsor Magazine* (1899).
Monthly Weather Review. U.S. Weather Bureau 29 (1902).
Musk, George. *Canadian Pacific Afloat: 1883–1968*. Canadian Pacific, 1968.
Norman-Butler, Belinda. *Victorian Aspirations*. George Allen, 1972.

"Notes." *Electrician* 39, no. 7 (June 11, 1897), pp. 207–8.
Notes on Naval Progress. Office of Naval Intelligence, General Printing Office, July 1901.
O'Hara, J. G., and W. Pricha. *Hertz and the Maxwellians.* Peter Peregrinus/Science Museum, London, 1987.
O'Neill, John J. *Prodigal Genius: The Life of Nikola Tesla.* Ives Washburn, 1944.
Oppenheim, Janet. *The Other World.* Cambridge, 1985.
Paresce, Francesco. "Personal Reflections on an 'Italian Adventurer.'" http://www.marconifoundation.org.
Paterson, John. *Edwardians.* Ivan R. Dee, 1996.
Petrie, Sir Charles. *The Edwardians.* W. W. Norton, 1965.
Place-Names and Places of Nova Scotia. Public Archives of Nova Scotia, 1967.
Pocock, R. E., and G.R.M. Garratt. *The Origins of Maritime Radio.* Her Majesty's Stationery Office, 1972.
Portrait and Biographical Album of Branch County, Michigan. Chapman, 1888.
Preece, W. H. "Signalling Through Space Without Wires." *Electrician* 39, no. 7 (June 11, 1897), pp. 216–18.
———. "'Wireless' Telegraphy.'" *Page's Magazine,* August 1902, pp. 131–36.
Priestley, J. B. *The Edwardians.* Harper & Row, 1970.
Read, Donald. *The Age of Urban Democracy.* Longman, 1994 (1979).
Rose, Clarkson. *Red Plush and Greasepaint.* Museum Press, 1965.
Rose, Jonathan. *The Edwardian Temperament 1895–1919.* Ohio University, 1986.
Rosenthal, Michael. *The Character Factory.* Pantheon, 1984.
Schneer, Jonathan. *London 1900.* Yale, 1999.
Seitz, Frederick. *The Cosmic Inventor: Reginald Aubrey Fessenden.* American Philosophical Society, 1999.
Sewall, Charles Henry. *Wireless Telegraphy.* D. Van Nostrand, 1904.
Sexton, Michael. *Marconi: The Irish Connection.* Four Courts Press, 2005.
Shipway, William H. and Lena. *History of the First Methodist Church (Coldwater, Mich.), 1832–1958.* Branch County District Library.
Simons, R. W. "Guglielmo Marconi and Early Systems of Wireless Communication." *GEC Review* 11, no. 1 (1996), pp. 37–55.
Stansky, Peter. *On or About December 1910.* Harvard, 1996.
Tedeschi, Enrico. *Guglielmo Marconi in London.* Hove Books, 1998.
Thoreau, Henry David. *Cape Cod.* W.W. Norton, 1951 (1865).
Traxel, David. *1898.* Alfred A. Knopf, 1998.
The Trial of Hawley Harvey Crippen. Notable Trials Library, 1991 (1920).
Tuchman, Barbara W. *The Guns of August.* Ballantine, 1994 (1962).
———. *The Proud Tower.* Ballantine, 1996 (1962).
Vyvyan, R. N. *Marconi and Wireless.* EP Publishing, 1974 (1933).
Walkowitz, Judith R. *City of Dreadful Delight.* Virago Press, 1992.
———. *People.* Doubleday, 1929.

Wallace, Edgar. *The Four Just Men.* Dover Publications, 1984 (1905).
Wander, Tim. *Marconi on the Isle of Wight.* TRW Design & Print, 2000.
———. "Radio's First Home." *Bulletin of the British Vintage Wireless Society* 18, no. 4 (n.d.), 51–52.
Weightman, Gavin. *Signor Marconi's Magic Box.* HarperCollins, 2003.
Weintraub, Stanley. *Edward the Caresser.* Free Press, 2001.
Wells, H. G. *The Time Machine and The War of the Worlds.* Random House, 1983 (1895, 1898).
———. *Tono-Bungay.* Modern Library, 2003 (1909).
West, Nigel. *GCHQ: The Secret Wireless War, 1900–86.* Weidenfeld & Nicolson, 1986.
Westman, Eric. "Marconi's 'Forgotten' Transmission." *Bulletin of the British Vintage Wireless Society* 11, no. 3 (January 1987), p. 37.
———. "Marconi's Tests in 1897." *Bulletin of the British Vintage Wireless Society* 11, no. 4 (March 1987), p. 52.
Whatley, Michael E. *Marconi: Wireless on Cape Cod.* 1987.
———. *Common Trailside Plants of Cape Cod National Seashore.* Eastern National, 2003.
Willcox, Philip H. A. *The Detective-Physician.* William Heinemann Medical Books, 1970.
Williams, Guy R. *The Hidden World of Scotland Yard.* Hutchinson, 1972.
Woodruff Place Centennial, 1872–1972. Official Centennial History Booklet, Jan. 1, 1972. Indiana State Library.
Ybarra, T. R. *Caruso.* Harcourt, Brace, 1953.

Acknowledgments

I am one fortunate author. Not only is this my fourth book with the same publisher, Crown, but it's my fourth with my beloved editor, Betty Prashker, and my agent, friend and consigliere, David Black. Once again all have proven to be steadfast allies, not flinching—at least not much—even when the manuscript arrived six months late. Betty has a remarkable ability to ease an author's anxiety. She has edited so many books by so many fine writers that when she says, "Don't worry, it will all be fine," you know that indeed there is cause for calm.

At Crown my books have always received maximum support, thanks to the enthusiasm of Jenny Frost, Steve Ross, and Tina Constable, and their secret weapon, the legion of ardent book reps—evangelists, really—who escort Crown's books into the world. Whitney Cookman made the book jacket beautiful; Janet Biehl, copy editor and savior, made it coherent. Penny Simon, supreme publicist, took on the all-important task of placing this book in the minds of readers. Special thanks go to Lindsey Moore, assistant editor, for cheerily serving as intermediary and finder.

I owe my greatest debt to my wife, Christine Gleason, and my daughters and dog for keeping me sane and relatively stable. It is hard to take yourself too seriously when you have three daughters all in or near their teens, especially if two of them are learning to drive. My wife once again demonstrated her innate talent for editing. She knows that when she receives my manuscript, she suddenly possesses a great deal of power, but she uses that power wisely—though those periodic trains of zzzzz's in the margin did now and then wrench my soul from its moorings. She was right, though. As always.

I thank my friends Carrie Dolan and Robin Marantz Henig, both excellent writers, for also reading critical portions of the manuscript and advising me on how to adjust the narrative to enhance clarity and pace.

I am grateful also to my Italian teacher, Robert—*Roberto*—Strait, whose gift for acquiring language is exceeded only by his knack for conveying its secrets to his students. Italian is a gorgeous, dynamic language. Even the simplest phrase, if delivered with gusto, can sound magnificent.

My travels for this book occurred at a time when public opinion of America could not have been lower, but I was always treated with kindness and generosity. In Cape Breton, Nova Scotia, everyone is your friend. Immediately. In Italy, everyone wants to feed you. In Britain, every question I asked was met with warmth and humor. And the tea, as always, was marvelous.

Illustration Credits

Page 7:
Guglielmo Marconi portrait reproduced by
courtesy of Essex Record Office.

Page 133:
Portrait of kite launch courtesy of
The Bodleian Library, University of Oxford, MSS Marconi.

Page 275:
Portrait of Beatrice O'Brien reproduced
by courtesy of Essex Record Office.

Page 393:
Alvin Langdon Coburn, *St. Paul's from Ludgate Circus*,
from the book *London*, 1910; photogravure;
Collection of the Prentice and Paul Sack Photographic Trust,
courtesy of the San Francisco Museum of Modern Art.

Front endpaper:
Bacon's 1902 Map of London © Old House Books
(www.OldHouseBooks.co.uk).

Rear endpaper:
Map of the North Atlantic by Mapping Specialists, Ltd.

INDEX

About the Author

Erik Larson is the author of *The Devil in the White City,* which won an Edgar Award for nonfiction, was a finalist for a National Book Award, and remained on the bestseller lists of the *New York Times* for well over two years. He also wrote the best-selling *Isaac's Storm,* about a hurricane that destroyed Galveston, Texas, in 1900. He has written for a variety of national magazines and is a former staff writer for the *Wall Street Journal* and *Time* magazine. He lives in Seattle with his wife and three daughters, and a golden retriever named Molly.

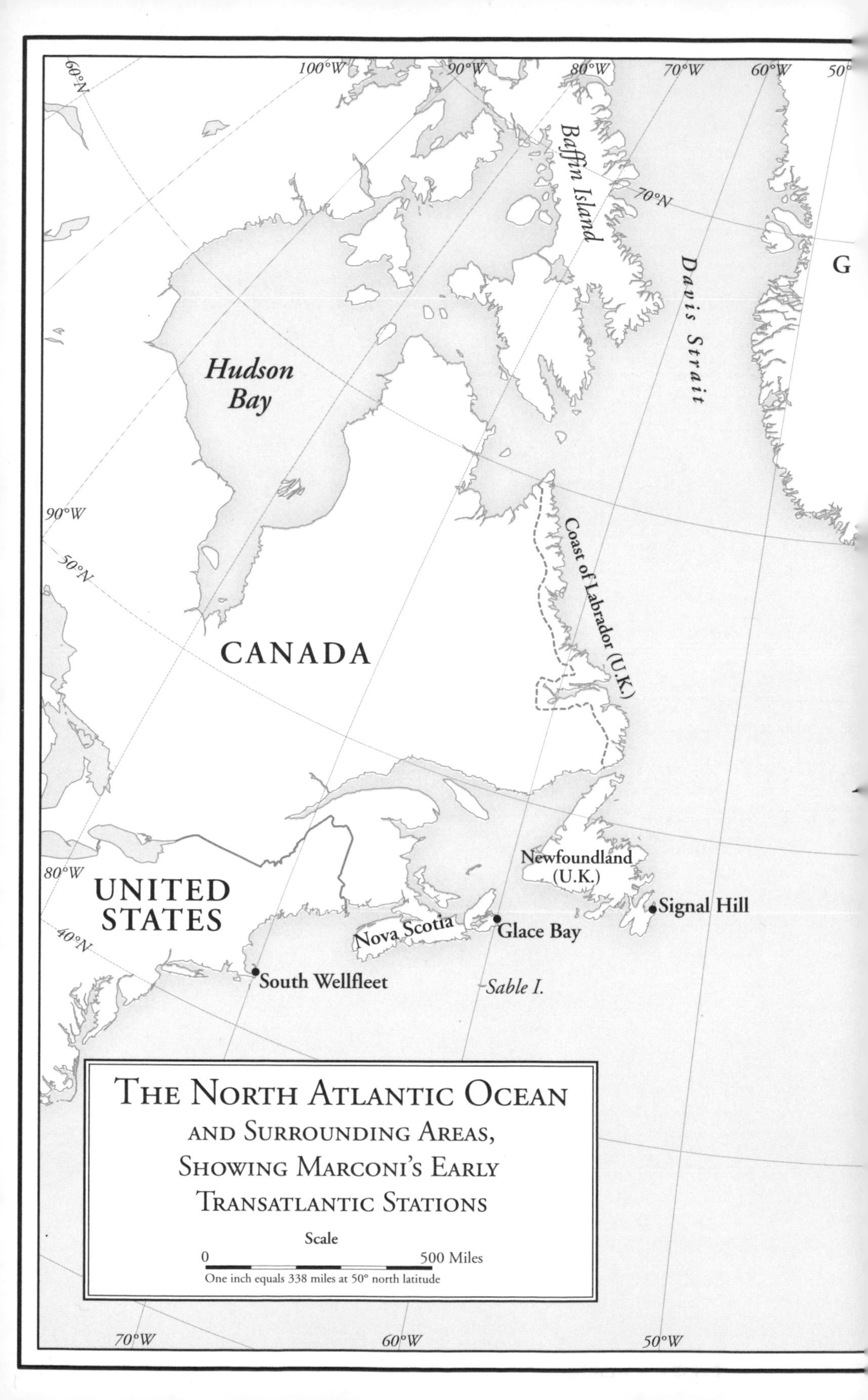
100°W
90°W
80°W
70°W
60°W
60°N
70°N
Baffin Island
Davis Strait
G
Hudson Bay
90°W
50°N
CANADA
Coast of Labrador (U.K.)
Newfoundland (U.K.)
Signal Hill
80°W
UNITED STATES
Nova Scotia
Glace Bay
40°N
South Wellfleet
Sable I.
THE NORTH ATLANTIC OCEAN AND SURROUNDING AREAS, SHOWING MARCONI'S EARLY TRANSATLANTIC STATIONS
Scale
0
500 Miles
One inch equals 338 miles at 50° north latitude
70°W
60°W
50°W